Cancer Sensitizing Agents for Chemotherapy

pH-INTERFERING AGENTS AS CHEMOSENSITIZERS IN CANCER THERAPY

VOLUME 10

Cancer Sensitizing Agents for Chemotherapy Series

Series Editor: Benjamin Bonavida, PhD

Volume 1: Nitric Oxide (Donor/Induced) in Chemosensitization
Edited by Benjamin Bonavida, PhD

Volume 2: Role of Nutraceuticals in Cancer Chemosensitization
Edited by Alok Chandra Bharti and Bharat Bhushan Aggarwal

Volume 3: Targeting Cell Survival Pathways to Enhance Response to Chemotherapy
Edited by Daniel E. Johnson, PhD

Volume 4: Protein Kinase Inhibitors as Sensitizing Agents for Chemotherapy
Edited by Zhe-Sheng Chen and Dong-Hua Yang

Volume 5: Breaking Tolerance to Pancreatic Cancer Unresponsiveness to Chemotherapy
Edited by Ganji Purnachandra Nagaraju, PhD, DSc

Volume 6: Improving the Therapeutic Ratio in Head and Neck Cancer
Edited by Randall J. Kimple, MD, PhD

Volume 7: Drug Efflux Pumps in Cancer Resistance Pathways: From Molecular Recognition and Characterization to Possible Inhibition Strategies in Chemotherapy
Edited by Alejandro Sosnik, PhD and Reina Bendayan, PharmD

Volume 8: Drug Resistance in Colorectal Cancer: Molecular Mechanisms and Therapeutic Strategies
Edited by Chi Hin Cho and Tao Hu, PhD

Volume 9: Novel Therapies in Head and Neck Cancer: Beyond the Horizon
Edited by Maie A. St. John, MD, PhD, FACS and No-Hee Park, DMD, PhD

Volume 10: pH-Interfering Agents as Chemosensitizers in Cancer Therapy
Edited by Claudiu T. Supuran, PhD and Simone Carradori, PhD

Volume 11: Overcoming Ovarian Cancer Chemoresistance
Edited by Goli Samimi, PhD, MPH and Christina Annunziata, MD, PhD

Volume 12: Biological Mechanisms and the Advancing Approaches to Overcoming Cancer Drug Resistance
Edited by Andrew Freywald, PhD and Franco J. Vizeacoumar, PhD

Volume 13: Therapeutic Strategies to Overcome ALK Resistance in Cancer
Edited by Luc Friboulet, PhD

Upcoming Volumes:

New Targeting in the Reversal of Resistant Glioblastomas
Edited by Ali S. Arbab

Terpenoids as Chemosensitizers Against Drug-resistant Tumors
Edited by Thomas Efferth

Reversal of Glioblastoma Resistance to Chemotherapy
Edited by Ramasamy Paulmurugan and Tarik F. Massoud

Epigenetic Regulation in Cancer Chemotherapy
Edited by Chunfu Wu and Lihui Wang

Chemotherapy Sensitizing Agents for Prostate Cancer
Edited by K. Ketola

Chemosensitization by TRK Inhibitors
Edited by Dirk Hempel

Chemosensitization by BTK Inhibitors
Edited by Varinder Kaur

Chemosensitization by Anti-angiogenic Agents
Edited by Lucia Morbidelli

Chemosensitization by IDH1 Inhibitors
Edited by Robert Musiol

Chemosensitization by EGFR Inhibitors
Edited by M. Pancione

Chemosensitization by IDH2 Inhibitors
Edited by Sanjee Singh

Chemosensitization by Bcl2 Inhibitors
Edited by Yu Yu

Chemosensitization by BTK Inhibitors (II)
Edited by Jean Koff

Cancer Sensitizing Agents for Chemotherapy

pH-INTERFERING AGENTS AS CHEMOSENSITIZERS IN CANCER THERAPY

VOLUME 10

Edited by

CLAUDIU T. SUPURAN, PhD

NEUROFARBA Department, Section of Pharmaceutical and Nutraceutical Sciences, University of Florence, Florence, Italy

SIMONE CARRADORI, PhD

Department of Pharmacy, "G. d'Annunzio" University of Chieti-Pescara, Chieti, Italy

ELSEVIER

ACADEMIC PRESS

An imprint of Elsevier

Academic Press is an imprint of Elsevier
125 London Wall, London EC2Y 5AS, United Kingdom
525 B Street, Suite 1650, San Diego, CA 92101, United States
50 Hampshire Street, 5th Floor, Cambridge, MA 02139, United States
The Boulevard, Langford Lane, Kidlington, Oxford OX5 1GB, United Kingdom

Library of Congress Cataloging-in-Publication Data
A catalog record for this book is available from the Library of Congress

British Library Cataloguing-in-Publication Data
A catalogue record for this book is available from the British Library

ISSN: 2468-3183

ISBN: 978-0-12-820701-7

For information on all Academic Press publications
visit our website at https://www.elsevier.com/books-and-journals

Publisher: Stacy Masucci
Senior Acquisitions Editor: Rafael Teixeira
Editorial Project Manager: Samantha Allard
Production Project Manager: Punithavathy Govindaradjane
Cover Designer: Greg Harris

Typeset by SPi Global, India

Cover Image Insert

Artistic representation of the extracellular and intracellular pH regulation in cancer and normal cells. This phenomenon is critical to their survival and growth. In metabolically active cancer cells, upregulation of membrane transporters and enzymes to extrude acids is utilized to tightly control pHi (light blue color) at the expense of acidifying pHe (light red color). Defects in blood supply and perfusion could further exacerbate extracellular acidosis and influence the metabolic activity, therapy response, and attitude to migration/invasion/metastasis.

Aims and scope for series "Cancer sensitizing agents for chemotherapy"

Current cancer management strategies fail to adequately treat malignancies with chemotherapy, with multivariable dose-restrictive factors such as systemic toxicity and multidrug resistance; hence, limiting therapeutic benefits, quality of life, and complete long-term remission rates. The resistance of cancer cells to anticancer drugs is one of the major reasons for the failure of traditional cancer treatments. Cellular components and dysregulation of signaling pathways contribute to drug resistance. If modulated, such perturbations may restore the drug response and its efficacy. The recent understanding of the molecular mechanisms and targets that are implicated for cancer chemoresistance has paved the way to develop a large battery of small molecules (sensitizing agents) that can target resistance and, thus, allow their combination with chemotherapeutic drugs to be effective and to reverse chemoresistance. A large variety of chemotherapy-sensitizing agents have been developed and several have been shown to be effective in experimental models and in cancer patients.

The main objective of the proposed series "Cancer Sensitizing Agent for Chemotherapy" is to publish individualized and focused volumes whereby each volume is edited by an invited expert Editor(s). Each volume will dwell on specific chemosensitizing agents with similar targeting activities. The combination treatment of the sensitizing agent with chemotherapy may result in a synergistic/additive activity and the reversal of tumor cells resistance to drugs.

The Editor(s) will compile nonoverlapping review chapters on reported findings in various cancers, both experimentally and clinically, with particular emphasis on underlying biochemical, genetic, and molecular mechanisms of the sensitizing agent and the combination treatment.

The scope of the series is to provide scientists and clinicians with updated and clinical information that will be valuable in their quest to investigate, develop, and apply novel combination therapies to reverse drug resistance and, thereby, prolong survival and even cure in cancer patients.

Dr. Benjamin Bonavida, PhD (Series Editor)

About the Series Editor

Dr. Benjamin Bonavida, PhD (Series Editor), is currently distinguished research professor at the University of California, Los Angeles (UCLA). He is affiliated with the Department of Microbiology, Immunology and Molecular Genetics, UCLA David Geffen School of Medicine. His research career, thus far, has focused on investigations in the fields of basic immunochemistry and cancer immunobiology. His research investigations have ranged from the biochemical, molecular, and genetic mechanisms of cell-mediated killing and tumor cell resistance to chemo-immuno cytotoxic drugs. The reversal of tumor cell resistance was investigated by the use of various selected sensitizing agents based on molecular mechanisms of resistance. In these investigations, there was the newly characterized dysregulated NF-κB/Snail/YY1/RKIP/PTEN loop in many cancers that was reported to regulate cell survival, proliferation, invasion, metastasis, and resistance. Emphasis was focused on the roles of the tumor suppressor Raf kinase inhibitor protein (RKIP) and the tumor promoter Yin Yang 1 (YY1) and the role of nitric oxide as a chemo-immuno-sensitizing factor. Many of the aforementioned studies are centered on the clinical challenging features of cancer patients' failure to respond to both conventional and targeted therapies.

The editor has been active in the organization of regular sequential international miniconferences that are highly focused on the roles of YY1, RKIP, and nitric oxide in cancer and their potential therapeutic applications. Several books edited or coedited by the editor have been published. In addition, the editor has been the series editor of books (over 23) published by Springer on *Resistance to Anti-Cancer Targeted Therapeutics*. In addition, the editor is presently the series editor of three series published by Elsevier/Academic Press on *Cancer Sensitizing Agents for Chemotherapy, Sensitizing Agents for Cancer Resist to Cell Mediated Immunotherapy*, and *Breaking Tolerance to Anti-Cancer Immunotherapy*. Lastly the editor is the editor in chief of the Journal *Critical Reviews in Oncogenesis*. The editor has published over 500 research publications and reviews in various scientific journals of high impact.

Acknowledgments: The Editor wishes to acknowledge the excellent editorial assistance of Ms. Inesa Navasardyan who has worked diligently in the completion of this volume, namely, in both the editing and formatting of the various contributions of this volume.

The Series Editor acknowledges the Department of Microbiology, Immunology and Molecular Genetics and the UCLA David Geffen School of Medicine for their continuous support. The Series Editor also acknowledges the assistance of Mr. Rafael Teixeira, Acquisitions Editor for Elsevier/Academic Press, and the excellent assistance and support of Ms. Samantha Allard, Editorial Project Manager for Elsevier/Academic Press, for their continuous cooperation throughout the development of this book.

Aims and Scope of the Volume

This book project provides a detailed overview of the whole world of "chemosensitizers for the treatment of cancer" spanning from biochemical and structural features to pharmacology and drug design, finally including technological applications of such tools. It is structured with innovative book setting outlines and a clear distinction between experimental and clinical results, in order to aid the reader to gradually enter into the different issues of such a wide and intricate topic.

The continuous discovery and assessment of the role played by old/new synthetic drugs, natural compounds, and technological applications thereof led to the urgent need of classification in terms of biological activity, mechanism of action, clinical outcomes, cancer cell lines sensible to the treatment and potentialities to better orient research in this field. Moreover, all the aspects relevant for Medicinal Chemistry (drug design, structure-activity relationships, permeability data, cytotoxicity, appropriate statistical procedures, molecular modeling studies) are strictly considered. It is more than ever necessary to properly gather all issues over chemosensitizers to guide researchers to always more concrete applications for cancer treatment.

About the Volume Editors

Claudiu T. Supuran is professor of medicinal and pharmaceutical chemistry at the University of Florence, Italy, since 1995. He was a visiting scholar at University of Florida, Gainesville, USA; at Griffith University, Brisbane, Australia; and visiting professor at University of La Plata, Argentina; at Federal University of Rio de Janeiro, Brazil; and at UNSW, Sydney, Australia (several times). His research is in the medicinal chemistry/biochemistry of metalloenzymes, mainly carbonic anhydrases, field in which he has made contributions to the design of many novel classes of enzyme inhibitors and activators, deciphering their mechanism of action at molecular level, discovery of new isoforms and their role in disease (cancer, obesity, epilepsy, neuropathic pain and cognition), discovery and characterization of carbonic anhydrases from various organisms (bacteria, fungi, protozoans, sponges, corals, plants, diatoms, vertebrates, etc.). Other research interests include X-ray crystallography of metalloenzymes, heterocyclic chemistry, chemistry of sulfonamides, sulfamates, and sulfamides, biologically active organoelement derivatives, metal-based drugs, cyclooxygenases, serine proteases, matrix metalloproteinases, bacterial proteases, and computational chemistry. He has published more than 1700 papers in these fields and his Hirsch index is 142, with >85,000 citations. One of the compounds discovered in his laboratory (SLC-0111) is in Phase II clinical trials for the treatment of advanced metastatic solid tumors in Canada/USA.

Simone Carradori's research interests are devoted to the design, synthesis, and biological evaluation of natural and synthetic compounds endowed with different mechanisms of action in the field of oncology. These contributions led to the discovery of new biologically active agents characterized by in vitro, in vivo, and in silico studies as well as to the drug repositioning in cancer therapy. He has published more than 170 papers in these fields (Hirsch index 35, citations ~4000) spanning from carbonic anhydrase inhibitors, epigenetic modulators, kinesin inhibitors, and antioxidant/anti-inflammatory compounds.

Preface

Despite the continuous efforts made in targeted therapy, precision medicine, and other recent developments, the tumor cell resistance to most clinically used drugs easily leads to clinical failure, which is one of the major obstacles in successful cancer therapy in many patients. This phenomenon could be intrinsic to a specific cell line (primary resistance) or can be the consequence of initially efficacious therapy, which may later develop drug resistance after repeated or eventually not optimally scheduled treatment cycles. Such a fatal outcome has requested combination therapy protocols to overcome a faster resistance development rate. The use of such drug cocktails, acting on several targets, led to improved survival rates, made therapeutic options efficacious for longer periods, and limited the side effects associated with higher single drug dosages. During the past decades, the tumor microenvironment was observed to display multidrug resistance phenomena, despite combinations of structurally and functionally different drug classes. These are mainly driven by drug efflux pumps or diverse families of transporters (such as the P-glycoprotein, ABC transporters, etc.).

The most innovative strategy designed to fight multifactorial chemoresistance in combination therapies is represented by the use of chemosensitizing agents. These agents can limit or reverse specific mechanisms of drug resistance toward standard anticancer drugs on the basis of a deeper biological knowledge of the tumors. Moving from this evidence, scientists discovered alternative chemotypes based on naturally occurring and synthetic scaffolds. Medicinal chemists were the most involved researchers and their efforts opened new scenarios in this rapidly changing field. These approaches can be divided into several mechanisms of action, namely, "drug-target site mechanisms" (DNA repair inhibitors, inhibitors of specific target proteins, redox modulators, chemosensitizers of endocrine resistance, chemosensitizers of resistance toward tyrosine kinase inhibitors, epigenetic inhibitors, aptamers, therapeutic nucleic acids), "resistance mechanisms operating upstream of the actual drug target" (inhibitors of P-glycoprotein and other ABC transporters, inhibitors of microenvironment-mediated drug resistance, and chemosensitizers for drug-resistant cancer stem cells), and "resistance mechanisms operating downstream of the actual drug target" (inhibitors of negative regulators in apoptosis, autophagy, and other modes of programmed cell death).

Moreover, beyond the rational considerations to enhance pharmacodynamics, special attention has been devoted to modulate the pharmacokinetics of well-known drugs, especially regarding the drug delivery systems. Novel nanotechnologies have been explored applying a systematic approach along with the high-throughput screening of vast libraries. This allowed a broader network of expertise to overcome a common problem and permitted the disposal of a wider arsenal for the clinicians.

This book project is new in this field and of significant interest and importance for clinicians, scientists, students, pharmaceutical companies, and health providers. This complex hot topic has been analyzed from three different points of view. An update of the biological aspects inherent to cancer resistance mechanisms and pH modulators in the tumoral microenvironment is reported in the first part (Chapters 1–3). The second part is devoted to the description of the role of the most important pH-interfering agents along with an in-depth analysis of the recent literature following a Medicinal Chemistry approach. We suggested new drug candidates or assessed the clinical impact of well-recognized drugs as Proton Pump Inhibitors (Chapter 4), carbonic anhydrase IX and XII inhibitors (Chapters 5 and 6), modulators of efflux pumps and transporters (Chapter 7), and alternative lysosomal pH modulators (Chapter 8). To better comprehend the state-of-the-art of this topic, we also included a patent survey on chemosensitizers (Chapter 9) and the emerging role of natural compounds (Chapter 10). Each chapter reviewed on chemosensitizing agents with common targeting activities and focused on experimental results, which can be translated into clinical ones. The third part discussed both the in silico approaches to explore new chemical space or to validate the *in vitro*/in vivo bioactivity (Chapter 11) and the proposed delivery systems to finely modulate the pharmacodynamics and targeting of these chemosensitizers (Chapter 12).

Prof. Claudiu T. Supuran, Florence (Italy)
Dr. Simone Carradori, Chieti (Italy)
May 15, 2020

Contents

Contributors xix

I

General overview of the topic: An update

1. Tumors and their microenvironment

Paola Chiarugi and Luigi Ippolito

Introduction 4
The role of tumor microenvironment 5
References 11

2. pH regulators of the tumoral microenvironment: A general overview

Shawn C. Chafe, Paul C. McDonald, and Shoukat Dedhar

Introduction 14
MCT1 and MCT4 15
Carbonic anhydrases 16
NHE1 18
Sodium bicarbonate cotransporters and anion exchangers 19
pH metabolons 20
Effect of inhibiting pH regulators on migration, invasion, and metastasis 21
Effect of inhibiting pH regulation on response to chemotherapy 22
pH and mitochondria 23
Conclusions 24
References 24

3. Use of pH-interfering agents as chemosensitizers: Clinical studies survey

Joonas Haapasalo, Kristiina Nordfors, and Seppo Parkkila

Carbonic anhydrases 36
Proton pump inhibitors 37
Monocarboxylate transporters 38
Ionophores 38
Other pH-interfering agents 39
Conclusion 40
Acknowledgments 40
References 40

II

pH-Interfering agents as chemosensitizers

4. Proton pump inhibitors and other pH-buffering agents

Alessio Nocentini and Stefano Fais

Introduction 48
Tumor acidity and chemoresistance 48
V-ATPases and proton pump inhibitors 50
Preclinical data supporting the use of proton pump inhibitors in cancer treatment 52
Clinical outcomes of proton pump inhibitors in association to anticancer therapies 54
Monocarboxylate transporter (MCT) inhibitors 55
Inhibitors of Na^+/H^+ exchanger (NHE), Na^+/HCO_3^- cotransporter (NBC), and anion exchanger (AE) 56
Conclusions 57
References 57

5. Design, synthesis, and biological evaluation of selective hCA IX inhibitors

Fabrizio Carta, Daniela Vullo, and Andrea Angeli

Introduction 64
Antibodies and antibody fragments which target CA IX 65

Small molecules which target CA IX 69
Conclusion 73
References 74

6. Small molecules and monoclonal antibodies as selective hCA XII inhibitors: An update

Simone Carradori and Claudiu T. Supuran

Introduction 80
Small molecules and monoclonal antibodies as selective hCA XII inhibitors 81
Conclusion 92
References 92

7. Efflux pumps, NHE1, monocarboxylate transporters, and ABC transporter subfamily inhibitors

Utpal K. Mondal and Marc A. Ilies

Introduction 96
Classes of transporters and efflux pumps 97
Conclusions and perspectives 112
Acknowledgments 112
References 112

8. Innovative pH-interfering agents in association with anticancer agents

Daniela Secci and Paolo Guglielmi

Introduction 122
pH-interfering agents with uncommon mechanism in association with anticancer drugs 123
Conclusion 127
References 127

9. Patent survey on chemosensitizers (2015–2019)

Ahmed M. Shabana and Suleyman Akocak

Introduction 130
Chemosensitizers inhibiting the activity of drug efflux transporters 131
Chemosensitizers inhibiting DNA repair mechanisms 137
Chemosensitizers as modulators of apoptotic pathways 139
Conclusions and perspectives 142
References 142

10. Natural compounds as chemosensitizers: A lesson from plants

Liora Colobatiu, Laura Gavrilas, and Andrei Mocan

Introduction 148
Chemoresistance: A therapeutic challenge 149
Combined approach in cancer treatment: Natural compounds as chemosensitizers 150
Curcumin 152
Resveratrol 155
Epigallocatechin-3-gallate 157
Quercetin 157
Conclusion 158
References 158

III

Chemosensitizing agents: Computational tools and technological approaches

11. Molecular docking studies and virtual drug screening of chemosensitizers

Agostino Bruno and Gabriele Costantino

Introduction 170
Carbonic anhydrase (CA) 172
Hypoxia-inducible factor (HIF) 176
Glucose transporter 1 (GLUT1) 178
Conclusions 179
References 179

12. Nanostructures and innovative delivery systems for overcoming cancer resistance

Jean-Yves Winum

Introduction 186
pH-sensitive nano-objects targeting tumor cells 188
Nanostructure targeting CA IX/CA XII 190
Conclusion 197
References 197

Index 203

Contributors

Suleyman Akocak Department of Pharmaceutical Chemistry, Faculty of Pharmacy, Adiyaman University, Adiyaman, Turkey

Andrea Angeli NEUROFARBA Department, Section of Pharmaceutical and Nutraceutical Sciences, University of Florence, Florence, Italy; Centre of Advanced Research in Bionanoconjugates and Biopolymers Department, "Petru Poni" Institute of Macromolecular Chemistry, Iasi, Romania

Agostino Bruno Department of Food and Drug, University of Parma, Parma, Italy

Simone Carradori Department of Pharmacy, "G. d'Annunzio" University of Chieti-Pescara, Chieti, Italy

Fabrizio Carta NEUROFARBA Department, Section of Pharmaceutical and Nutraceutical Sciences, University of Florence, Florence, Italy

Shawn C. Chafe BC Cancer Research Institute, Vancouver, BC, Canada

Paola Chiarugi Laboratory of Tumor Microenvironment, Department of Experimental and Clinical Biomedical Sciences, University of Florence, Florence, Italy

Liora Colobatiu Iuliu Hatieganu University of Medicine and Pharmacy, Cluj-Napoca, Romania

Gabriele Costantino Department of Food and Drug, University of Parma, Parma, Italy

Shoukat Dedhar BC Cancer Research Institute; Department of Biochemistry and Molecular Biology, University of British Columbia, Vancouver, BC, Canada

Stefano Fais Department of Oncology and Molecular Medicine, Italian National Institute of Health, Rome, Italy

Laura Gavrilas Iuliu Hatieganu University of Medicine and Pharmacy, Cluj-Napoca, Romania

Paolo Guglielmi Department of Drug Chemistry and Technology, Sapienza University of Rome, Rome, Italy

Joonas Haapasalo The Hospital for Sick Children, Arthur and Sonia Labatt Brain Tumor Research Center, Toronto, ON, Canada; Tampere University Hospital, Unit of Neurosurgery, Tampere, Finland

Marc A. Ilies Department of Pharmaceutical Sciences and Moulder Center for Drug Discovery Research, Temple University School of Pharmacy, Philadelphia, PA, United States

Luigi Ippolito Laboratory of Tumor Microenvironment, Department of Experimental and Clinical Biomedical Sciences, University of Florence, Florence, Italy

Paul C. McDonald BC Cancer Research Institute, Vancouver, BC, Canada

Andrei Mocan Iuliu Hatieganu University of Medicine and Pharmacy, Cluj-Napoca, Romania

Utpal K. Mondal Department of Pharmaceutical Sciences and Moulder Center for Drug Discovery Research, Temple University School of Pharmacy, Philadelphia, PA, United States

Alessio Nocentini NEUROFARBA Department, Section of Pharmaceutical and Nutraceutical Sciences, University of Florence, Florence, Italy

Kristiina Nordfors The Hospital for Sick Children, Arthur and Sonia Labatt Brain Tumor Research Center; Division of Hematology and Oncology, The Hospital for Sick Children, Toronto, ON, Canada

Seppo Parkkila Faculty of Medicine and Health Technology, Tampere University; Fimlab Ltd, Tampere University Hospital, Tampere, Finland

Daniela Secci Department of Drug Chemistry and Technology, Sapienza University of Rome, Rome, Italy

Ahmed M. Shabana Institute for NanoBio-Technology, Johns Hopkins University, Baltimore, MD, United States; Department of Pharmaceutics and Industrial Pharmacy, Faculty of Pharmacy, Cairo University, Cairo, Egypt

Claudiu T. Supuran NEUROFARBA Department, Section of Pharmaceutical and Nutraceutical Sciences, University of Florence, Florence, Italy

Daniela Vullo Department of Chemistry "Ugo Schiff", University of Florence, Florence, Italy

Jean-Yves Winum IBMM, Univ Montpellier, CNRS, ENSCM, Montpellier, France

PART I

General overview of the topic: An update

CHAPTER

1

Tumors and their microenvironment

Paola Chiarugi and Luigi Ippolito

Laboratory of Tumor Microenvironment, Department of Experimental and Clinical Biomedical Sciences, University of Florence, Florence, Italy

Abstract

It is well acknowledged that tumors display extensive cellular and molecular heterogeneity. Despite of cell-intrinsic (genetic and epigenetic) changes, tumor cells evolve into high-grade malignancies, along with the tumor microenvironment representing a landscape of cellular and structural components. Malignant cells experience a dynamic and complex scenario of communications with nonmalignant stromal cells, recruited and/or corrupted *in loco* by tumor-derived soluble factors, cytokines, metabolites, and matrix-remodeling agents. Together with acidity and hypoxia, microenvironmental signals favor tumor progression and exacerbate malignant phenotype, from the tumor initiation to the metastatic spreading.

Abbreviations

CA carbonic anhydrase
CAF cancer-associated fibroblastic cell
CREB cAMP response element-binding protein
ECM extracellular matrix
EMT epithelial-mesenchymal transition
HIF hypoxia-induced factor
IFN interferon
IL interleukin
MAPK mitogen-activated protein kinase
MMP matrix metalloproteinase
NF-κB nuclear factor kappa-light-chain-enhancer of activated B cells
pHe extracellular pH
PI3K phosphoinositide 3-kinase
pRB retinoblastoma protein
Tc cytotoxic T cells
TGF transforming growth factor
Th helper T cells
TME tumor microenvironment
TNF tumor necrosis factor
Treg regulatory T cells
VEGF vascular endothelial growth factor

pH-Interfering Agents as Chemosensitizers in Cancer Therapy
https://doi.org/10.1016/B978-0-12-820701-7.00010-5

Conflict of interest

No potential conflicts of interest were disclosed.

Introduction

Hanahan and Weinberg have successfully summarized the main traits of cancer aggressiveness in their two seminal papers. They simply describe that the complexity of cancer biology can be condensed in a small number of fundamental principles that are biological capabilities shared by cancer cells during the multistep process of tumor formation and progression [1].

The *hallmarks* of cancer aggressiveness are:

- *Self-sufficiency in growth signals.* Cancer cells have the ability to grow without hormonal signals, likely by producing these signals by themselves (autocrine signaling), by constitutively activating signaling pathways associated to hormones, or by silencing termination signals.
- *Evading growth suppressors.* Cancer cells resist to inhibitory signals that might otherwise stop their growth (an event occurring in normal proliferative cells), eliminating tight control of cell division through tumor suppressor genes. This allows cell division also if DNA is damaged or during contact with other neighboring cells.
- *Evading apoptosis.* Cancer cells lose the ability to activate a programmed cell death, a process known as apoptosis. Nevertheless, cells may become completely abnormal, they do not undergo apoptosis. Cancer cells are able to do so by altering the mechanisms to detect damages and circumventing cell cycle, or nutrient and adhesion checkpoints.
- *Limitless replicative potential.* Cancer cells indefinitely proliferate. Cancer cells escape natural senescence and undergo immortalization, being apparently capable of indefinite growth and division. Cancer cells bypass the senescence barrier by manipulating telomerases to increase the length of telomeres or by disabling their pRB and p53 tumor suppressor proteins, with the consequent emergence of a genome alterations, strongly correlated with DNA damage to multiple genes controlling cell division (oncogenes) and tumor suppressors.
- *Sustained angiogenesis.* Cancer cells stimulate the growth of blood vessels to supply nutrients for themselves. As a growing tumor requires new blood vessels to receive adequate oxygen and nutrients, cancer cells acquire the ability to orchestrate the machinery needed for a new vasculature by activating the so-called angiogenic switch, unbalancing the net production of proangiogenic factors over the antiangiogenic ones.
- *Tissue invasion and metastasis.* Cancer cells are able to invade local tissue and spread to distant sites. Malignant tumors undergo a multistep process allowing (i) local invasion of these cells from the primary site into the surrounding tissues, then (ii) the intravasation and the consequent facing the harsh environment of the circulatory system, (iii) the extravasation and engraftment into a distant organ. Cancer cell motility is enhanced through activation of epithelial-mesenchymal transition (EMT) or mesenchymal-amoeboid transition, two epigenetic pathways allowing adaptations of motility styles to different tissue microenvironment.

- *Metabolic deregulation*. Most cancer cells exhibit abnormal metabolic demands, fluxes, and bioenergetics, as revealed in the early twentieth century in the Warburg hypothesis (i.e., massive upload of glucose, which is diverted into glycolytic collateral pathways as well as in lactate fermentation, even under aerobic conditions). Recent advances showed a peculiar metabolic plasticity of cancer cells, beyond the glycolytic pathway. Such metabolic deregulation showed by cancer cells spans from different nutrients exploitation in a restricted environment, to the nutrient competition, until the epigenetic regulation exerted by metabolites such as succinate, fumarate, and 2-hydroxyglutarate (namely, oncometabolites).
- *Evading the immune system*. Cancer cells are able to avoid immune-surveillance, the natural mechanism of cancer detection by the immune system. Indeed, cancer cells elicit a complex response driving immune effector cells inhibition, while sustaining Treg activation, with the final outcome to be hidden to the immune system.
- *Sustained inflammation*. Cancers have been described as "wounds that do not heal," mainly due to sustained local chronic inflammation, caused by recruiting several accessory populations within the neoplasm. Chronic inflammation has been correlated with sustained angiogenesis and inhibition of functional immune response, thereby explaining the role of inflammation in sustaining cancer aggressiveness. ECM (extracellular matrix), hypoxia end acidity are all environmental factors enhancing tumor inflammation, thereby concurring to increase cancer aggressiveness (see below).

The role of tumor microenvironment

It has been largely clarified that most of the hallmarks of cancer are allowed and sustained to varying degrees through contributions from different stromal cell types (Fig. 1).

Cellular and structural elements organize the so-called tumor microenvironment (TME) [2]. Whereas cancer had previously been viewed as a heterogeneous disease involving aberrant mutations in tumor cells, it is now evident that tumors are also heterogeneous because of nongenetic features, provided by the microenvironmental cellular and structural composition. Indeed, in response to evolving environmental conditions and oncogenic signals from growing tumors, TME dynamically changes over the course of cancer progression, underscoring the need to consider his influence on events, often spatially and temporally distant such as metastasis, and to understand how tumor cells drive the construction of their own niche. Indeed, the current conceptual model (Fig. 2) proposes that, in the course of tumor progression, cancer cells exert increasingly stronger influence over their microenvironment.

As the tumor grows, it recruits or reprograms associated cells that dampen immune response, supply growth factors, nutrients and oxygen, transport waste products, and remodel the ECM to further tumor growth and spread. Specifically, incipient neoplasia initiates a crosstalk by recruiting and activating stromal cell types that assemble into an initial preneoplastic stroma, which, in turn, responds reciprocally by enhancing the neoplastic phenotypes of the nearby cancer cells. The cancer cells again feed signals back to the stroma, continuing the reprogramming of normal stromal cells to serve the budding neoplasm, contributing to tumor hypoxia, acidosis, and increased interstitial fluid pressures. Ultimately,

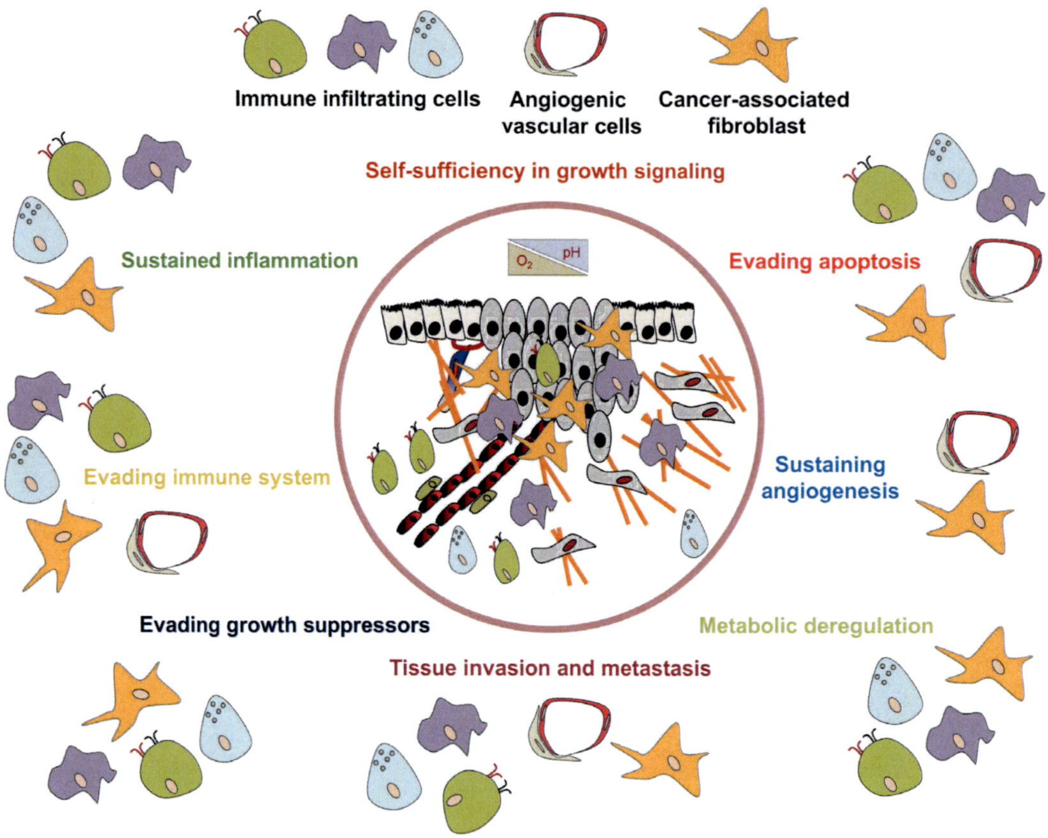

FIG. 1 Contributions of activated/recruited stromal cells to the hallmarks of cancer. Most of hallmarks of cancer involve contributions by stromal cells of the tumor microenvironment. The stromal cells can be divided into three general classes, depicted here by their involvement in specific hallmarks.

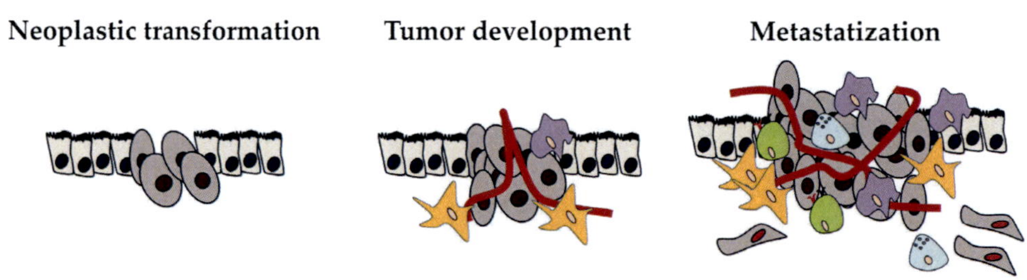

FIG. 2 Schematic view of tumor progression accompanied with stromal cells. Normal tissues are composed of the basement membrane and a layer of luminal epithelial cells. During *neoplastic transformation*, the epithelial cells are epigenetically and phenotypically altered and their number decreases, potentially due to degradation of the basement membrane, and transformed cells start to unlimitedly grow. At the same time, the number of stromal cells - including CAFs, immune, and endothelial cells—increases during the *tumor development*. The massive growth of tumor cells, together with their intensive relationship with resident and recruited stromal cells, leads to the loss of basement membrane resulting in invasive carcinoma, in which tumor cells can invade surrounding tissues and can migrate to distant organs, eventually leading to *metastatization*.

signals originating in the tumor stroma enable cancer cells to invade normal adjacent tissues and disseminate.

The complexity of tumor-stroma interactions and the role of the TME has been actively researched and reviewed over the last decades. On a cellular level, the TME comprises a broad spectrum of subpopulations. For simplification, it has been grouped into three compartments including (i) Cancer-Associated Fibroblastic cells (CAFs), (ii) Angiogenic Vascular Cells and (iii) Infiltrating Immune Cells [2].

Therefore, tumor properties depend on complex interactions between cancer epithelial cells and the surrounding stromal compartment within the TME. The stromal compartment comprises multiple nonmalignant cells such as fibroblasts, myofibroblasts, endothelial and immune cells, as well as different biophysical features such as hypoxia, acidity, extracellular matrices, all of them acting together in order to promote cancer progression and metastasis. Tumor stroma composition is similar to that observed during the wound healing process and it is commonly referred to as "reactive stroma," which is associated with an increased number of fibroblasts, enhanced capillary density, and collagen and fibrin deposition. Also, unlike cancer cells that are transformed cells due to several genetic mutations, stromal cells are mostly genetically intact. Furthermore, the tumor microenvironment is dramatically active in shaping the tumor therapeutic responses and resistance to different therapies, thereby explaining the recent interest to target the different populations of cells within tumor microenvironment.

Cancer-associated fibroblasts (CAFs)

In cancer tissues fibroblasts are the predominant cell type and are the primary producers of the noncellular scaffolds—the extracellular matrix. Fibroblasts are responsible for the deposition of the fibrillar ECM—type I, type III, and type V collagen, and fibronectin—and contribute to the formation of the basement membrane by secreting type IV collagen and laminin. Fibroblasts are also an important source of ECM-degrading proteases such as matrix metalloproteinases (MMPs), continually and dynamically remodeling the connective tissue and the EMC around them. During wound repair, fibroblasts are responsible for orchestrating healing, and for this purpose they become "activated" concerning an increased proliferation and alterations in both phenotype and secretory capacity. Expression of alpha-smooth muscle actin allows these cells to migrate into areas of damage and contract for tissue restitution. Fibroblasts serve as a scaffold and secrete increased levels of ECM proteins, growth factors, and chemotactic factors, thereby coordinating the influx of inflammatory cells and vascular progenitor cells as well as supplying the scaffold structure for cell growth and proliferation. Fibroblasts within tumors have an activated phenotype, and closely resembling fibroblasts in wound healing. CAFs are identified within tumor stroma by their spindled shape and the expression of α–smooth actin, vimentin, desmin, and fibroblast activation protein. CAFs are associated with cancer cells in all stages of cancer progression by promoting tumor growth, angiogenesis, and metastatic process. There are several theories about the origin of CAFs, including activation of tissue-resident fibroblasts, the migration and activation of marrow-derived fibroblastoid precursors and mesenchymal stem cells, as well as local cancer cells or epithelial cells that transform into fibroblasts by undergoing epithelial-to-mesenchymal transition.

The role of CAFs in promoting cancer has been demonstrated in established tumors, carcinoma in situ, during cancer onset and in orchestrating metastasis. Cancer cells and CAFs establish a close crosstalk based on mutual interplay: carcinoma cells elicit a reactive response in the stroma, which, in turn, enhances cancer cell motility, resistance to apoptosis, resistance to therapy, thus influencing tumor progression. Intriguingly, either at primary sites or in vasculature vessels, cancer cells can release microvesicles or soluble factors to adapt initial metastatic sites into "premetastatic niches." Indeed, systemic factors attract bone marrow-derived macrophages and hematopoietic progenitor cells that are accompanied by CAFs and endothelial cells to remodel tissue and eventually cause lung metastasis.

Alongside their role in the initiation of neoplasia, CAFs play a mandatory role in progression of tumors toward malignancy by regulating motility and invasion of cancer cells, thus positively affecting their metastatic spread toward distant organs. The contribution of CAFs to the invasive process of cancer cells is mainly due to their ability to induce EMT and to remodel the ECM. Indeed, CAFs are sources of ECM degrading proteases such as MMPs, which allow cancer cells to escape the primary tumor site by degrading ECM proteins. In addition, CAF-induced EMT has been correlated with the activation of a cancer stem-like phenotype both in prostate and in breast cancer. This feature confers to cancer cells the self-renewal capability that is crucial for clonal expansion during metastatic dissemination. Finally, CAFs escort cancer cells while disseminating to other organs, as they are associated to circulating cancer cells and sustain their survival suppressing their apoptosis due to lack of cell contact to ECM proteins (a process referred as *anoikis* resistance).

In the context of their prometastatic role, increasing evidence showed that CAFs orchestrate tumor angiogenesis in a variety of tumor types. First, CAFs in different TMEs can produce a number of proangiogenic signaling proteins and promote dynamics of ECM, the framework in which angiogenic growth factors are sequestered. Secondly, CAFs are able to stimulate the recruitment of endothelial precursor cells from bone marrow, as well as produce chemoattractants for proangiogenic macrophages, neutrophils, and other myeloid cells, thereby indirectly orchestrating tumor angiogenesis. Finally, CAFs have been reported as key actors in improvement of tumor hemodynamics, allowing a more favorable biodistribution of drugs, mainly causing a reduction of interstitial fluid pressure in the neoplasm.

Angiogenic vascular cells

It is well established that vascularization of tumors has a central role in attenuating cell death that would otherwise result from hypoxia. Indeed, several studies report an increase of proliferation of cancer cells following activation of the angiogenic switch, as well as an increase of apoptosis. The role of angiogenesis in limiting apoptosis is illustrated by the effects of vascular agents that destroy the tumor vasculature, causing hypoxia and massive cell death inside treated tumors.

Several tumors express high levels of the vascular endothelial growth factor (VEGF), a proangiogenic factor, signaling through VEGF-Receptor 2, rendering the vasculature permeable, allowing blood diffusion, and at the same time lowering barriers for intravasation of cancer cells into blood circulation. VEGF effects on tumor vasculature include reductions of pericytes coverage, causing a looser association with endothelium and facilitating metastatic dissemination of cancer cells. Hypoxia inside and around tumor vessels also contributes to metastatic dissemination of cancer cells through the actions of genes regulated by hypoxia-induced factors (HIFs), including VEGF and inducible nitric oxide synthase. In addition, the aberrant morphology of the de novo angiogenic tumor vasculature facilitates transit of cells across the vascular wall in both directions, allowing intravasation of cancer cells, but also extravasation in different organs.

Immune infiltrating cells

The role of immune system is to screen tissue homeostasis, to protect against infectious pathogens, and to eradicate damaged cells, including emerging tumor cells. Seminal studies revealed the involvement of adaptive and innate immunity in cancer immune editing. The immunologic system can defeat nascent tumors, a task accomplished by different inflammatory cells. To escape the immune surveillance, cancer cells evolve several phenotypic changes and ambient adaptations. Also, the immune microenvironment helps cancer cells to select the dominant/survival cells that repopulate neoplasms. Several studies have indeed shown that the depletion of macrophages during tumor induction inhibited tumor growth. Several stromal cells, mainly CAFs, can affect the role of immune cells, mainly acting through secretion of cytokines and growth factors.

The immune system is divided into adaptive immunity and innate immunity. Adaptive immune cells include T and B cells, while innate immune cells consist of dendritic cells, natural killer cells, macrophages, granulocytes, and mast cells. T cells, divided into $CD4^+$ T (helper T cells, Th) and $CD8^+$ T (cytotoxic T cells, Tc) cells, secrete antitumor factors as IFN-γ, TNF-α, and IL-17. Natural killer cells are devoted to recognition of infected and malignant target cells. Dendritic cells express costimulatory molecules and innate inflammatory cytokines, leading to enhancement of Th and Tc lymphocyte responses. Conversely, the main immunosuppressive cells that actively infiltrate tumors are regulatory T (Treg) cells, myeloid-derived suppressor cells, and M2 polarized protumorigenic macrophages.

ECM within the microenvironment

ECM is a dynamic environment composed of structural proteins, including collagens, laminins, fibronectins, proteoglycans, and hyaluronans, as well as cytokines and growth factors secreted by both stromal and tumor cells. Depending on the different organization and locations, ECM has complementary effects on the development and metastasis of tumors in diverse ways [3]. First, ECM may be modified by stromal and cancer cells through extracellular secretion, thereby affecting cancer cells survival, due to altered recognition of proper ECM proteins. Indeed, adherent cancer cells are conditioned to integrin-mediated adhesion to ECM proteins to survive, as these adhesions are able to inhibit *anoikis*, an apoptotic death due to lack of proper ECM binding. During tumor progression, several metastatic cancer cells develop the ability to self-survival to *anoikis*, a feature called *anoikis* resistance, very useful in order to survive in later stages of tumor dissemination, during circulation in the vasculature or infiltration into secondary organs endowed with a different ECM composition with respect to the tissue of origin. In addition, the complete maturation of collagen and associated proteins is mandatory to develop the proper metastatic niche. In this line, inhibition of lysyl oxidases or of prolyl hydroxylases, enzymes devoted to maturation of ECM proteins, is able to block metastatic dissemination of several cancers, indicating that cancer cells need a mature ECM to move. Both CAFs and macrophages infiltrating growing tumors produce several cytokines, as TGF-β1 and IL-6, thereby helping in recruiting endothelial cells and favoring de novo angiogenesis, granting oxygen and nutrient supply. Third, ECM and its soluble factors can help neoplasms to avoid immune surveillance, as both TGF-β1 and IL-6 are acknowledged to favor the differentiation of Th cells into Tregs. Fourth, the ECM is greatly affected by hypoxia and acidity, two additional extracellular structural factors selecting aggressive tumor cells with survival advantages.

Hypoxia and acidity

Hypoxia, a phenomenon very common in a vast majority of malignant tumors, is a nonphysiological level of oxygen tension, leading to advanced but dysfunctional vascularization and acquisition of EMT phenotype, finally leading to enhanced cell mobility and metastasis. Hypoxia is able to reprogram cancer cell metabolism toward glycolysis and oxygen independence and contributes to therapy resistance by inducing sustained cell survival signaling [4]. Hypoxic conditions activate a complex signaling network in sensitive cells, mainly through HIFs-mediated transcriptional pathways, active on survival, metabolic, proliferative, and motile routes, as well as on induction of de novo angiogenesis, mainly through activation of VEGF. Besides VEGF, PI3K, MAPK, NF-κB, metalloproteases are examples of HIF target genes in the regulation of tumor blood vessel formation, metastasis, and development of the resistance to therapy.

Acidosis is again a central feature of the TME. It directly regulates tumor cell invasion by affecting immune cell function, clonal cell evolution, and drug resistance [5]. Because of increased glucose consumption due to their Warburg metabolism, H^+ production is commonly increased in cancers. This effect is joined with reduced tissue perfusion and culminates in an acidic extracellular pH (pHe) in malignant tumors (pH 6.5–6.9). Acidic pHe has been correlated with release and activation of cysteinyl- or aspartyl- proteases, enhancing EMT of cancer cells, local invasion, and tissue remodeling. The resulting acidic environment, which is toxic to normal cells but not to cancer cells, promotes in cancer tissue the release of VEGF and ECM-associated growth factors, leading to increase angiogenesis and inhibition of immune response. Cancer adaptations to acidic microenvironment include upregulation of the sodium-hydrogen exchanger-1 or carbonic anhydrases (CAs). How acidosis induces such phenotypes is only partially understood, but protonation of critical pH-sensitive residues of cytoskeleton and membrane bound proteins is largely acknowledged. Moreover, acidosis may also contribute to aggressive cancer phenotypes through modulation of transcriptional pathways, as it regulates histone acetylation, H^+-dependent activation of HIF-1 and CREB transcription factors, as well as with H^+-mediated regulation of alternative splicing of key supervisors of metastasis.

References

[1] Hanahan D, Weinberg RA. Hallmarks of cancer: the next generation. Cell 2011;144:646–74.

[2] Hanahan D, Coussens LM. Accessories to the crime: functions of cells recruited to the tumor microenvironment. Cancer Cell 2012;21:309–22.

[3] Leight JL, Drain AP, Weaver VM. Extracellular matrix remodeling and stiffening modulate tumor phenotype and treatment response. Annu Rev Cancer Biol 2017;1:313–34.

[4] Petrova V, Annicchiarico-Petruzzelli M, Melino G, Amelio I. The hypoxic tumour microenvironment. Oncogenesis 2018;7:10.

[5] Corbet C, Feron O. Tumour acidosis: from the passenger to the driver's seat. Nat Rev Cancer 2017;17:577–93.

CHAPTER

2

pH regulators of the tumoral microenvironment: A general overview

Shawn C. Chafe[a], *Paul C. McDonald*[a], *and Shoukat Dedhar*[a,b]

[a]BC Cancer Research Institute, Vancouver, BC, Canada [b]Department of Biochemistry and Molecular Biology, University of British Columbia, Vancouver, BC, Canada

Abstract

Cancer cells need to adapt to an ever-evolving environment to meet energy demands in an ecosystem where nutrient and oxygen levels are in constant flux. Rapid growth, enhanced metabolic rates, and the preference of glycolysis for energy production lead to increased production of metabolic byproducts in the form of lactate, protons, and carbonic acids. In order to survive, cancer cells adapt by upregulating expression of membrane transporters and enzymes that facilitate extrusion of lactate and protons and hydrate carbon dioxide to maintain an alkaline intracellular pH. Increased expression of MCT1/4, CAII/IX/XII, NHE1, and sodium bicarbonate transporters facilitate buffering of intracellular pH while acidifying the extracellular milieu. Defects in blood supply and perfusion exacerbate extracellular acidosis, which has implications for metabolic activity, therapy response, and migration/invasion/metastasis. Here we will review the current state of pH regulation in the tumor microenvironment, advantages to interfering with the function of pH regulatory machinery, the impact of pH homeostasis on response to therapy and its impact on metabolism.

Abbreviations

AE anion exchanger
CA carbonic anhydrase
COX cytochrome *c* oxidase
DMOG dimethyloxalylglycine
ECM extracellular matrix
EMT epithelial to mesenchymal transition
ETC electron transport chain
FDG fluorodeoxyglucose
GLUT1 glucose transporter 1
HIF hypoxia-inducible factor

https://doi.org/10.1016/B978-0-12-820701-7.00016-6

LDHA	lactate dehydrogenase A
MCT	monocarboxylate transporter
mTOR	mammalian target of rapamycin
NAF-1	nutrient autophagy factor 1
NBC	sodium bicarbonate cotransporter
NHE1	sodium hydrogen exchanger 1
OXPHOS	oxidative phosphorylation
PD-1	programmed cell death protein 1
PDK1	pyruvate dehydrogenase kinase 1
PET	positron emission tomography
PHD	prolyl hydroxylase domain
ROS	reactive oxygen species
SLC	solute carrier
SNP	single nucleotide polymorphism
TCA	tricarboxylic acid
TME	tumor microenvironment
UTR	untranslated region

Conflict of interest

PCM and SD are coinventors of SLC-0111 and declare no other conflicts of interest. SCC declares no conflict of interest.

Introduction

The microenvironment of solid tumors plays a role in shaping its behavior [1]. As a result of gain-of-function mutations in oncogenes or loss-of-function mutations in tumor suppressors, trigger sustained cellular growth and metabolic changes leading to a preference for glycolysis for energy production [2]. Cellular growth beyond 150 μm from the nearest capillary exceeds the oxygen diffusion limit [1]. Cells beyond this distance are consequently exposed to less oxygen (hypoxia) [3]. Hypoxia triggers a global transcriptional response driven by the hypoxia-inducible factors (HIFs) through stabilization of the oxygen-dependent alpha subunits of HIF1 and HIF2. As part of the adaptation, angiogenesis is stimulated to try and circumvent the limited supply of oxygen, however in contrast to normal tissues tumors produce abnormal, leaky blood vessels [3]. The abnormal vasculature caused by tumor angiogenesis leads to defects in perfusion and increased interstitial pressure. In addition, metabolic phenotype is influenced due to increased transcription of the HIF targets GLUT1, PDK1, and LDHA. Cancer cells become more reliant upon glycolysis for energy production, decreasing flux of pyruvate through the TCA cycle and increasing its conversion to lactate. Increased accumulation of lactate needs to be extruded from the cell in order to maintain an alkaline intracellular pH (pHi), which is required for maintenance of cell proliferation, survival, and proper metabolic function [1]. In contrast to normal cells ($pH_i \sim 7.2$), cancer cells maintain a $pH_i > 7.4$. In order to cope with the stress of increasing intracellular accumulation of acidic metabolic byproducts and maintain this alkaline pH_i, cancer cells upregulate expression of cell surface transporters that aid in lactate (MCT1/4) and proton extrusion (proton pumps, NHE's), bicarbonate generation (carbonic anhydrases) and transport (NBCs, AEs) to maintain intracellular acid base homeostasis. On the other hand, the continued extrusion of metabolic acids acidifies the extracellular environment, which accumulate due to the poor perfusion resulting from the aberrant vasculature. Normal cells face an extracellular pH (pH_e) of 7.4, whereas the pH_e of cancer cells falls between 6.5 and 7.0 and sometimes even lower [4]. The ability to adapt to the changing pH conditions gives cancer cells a survival advantage over normal cells. The acidic microenvironment triggers secretion and activation of proteases leading to increased degradation of ECM, decreased cell-cell contacts, Epithelial-Mesenchymal Transition (EMT), increased migration and invasion, acquisition of stem cell phenotypes, and impacts response to chemotherapy. The breadth of response to acidosis is perhaps not surprising given that all cellular proteins are sensitive to changes in proton concentrations [5–7]. Indeed acidosis triggers a transcriptional rearrangement due to epigenetic changes [8] eliciting a response that appears distinct from that influenced by hypoxia [9–11]. This metabolic rearrangement may lead to the identification of new therapeutic vulnerabilities that can be exploited for cancer therapy through the development of rational targeted strategies.

In this chapter we will review the current landscape surrounding pH regulation in the tumor microenvironment and its impact on tumor growth and progression. We will summarize the impact on cellular phenotype in response to targeted disruption/inhibition of those critical pH regulators mentioned here, and discuss the impact of pH regulation on response to chemo- and immunotherapy and potential for combinatorial approaches, especially with inhibitors to the cell surface carbonic anhydrases IX and XII (CAIX/XII), to overcome therapeutic resistance. Lastly, we will focus on the impact of pH regulation on mitochondrial function, the most important organelle in the cell for energy production and cell death decisions.

MCT1 and MCT4

Cancer cells relying on glycolysis to support survival and rapid proliferation in hypoxia produce large amounts of acidic byproducts, particularly lactate [12–15]. Cancer cells enhance the efflux of lactate through the upregulation of members of the monocarboxylate (MCT) family of lactate/H^+ symporters [16]. There are 14 members of the MCT family; two members of this family, MCT1 and MCT4, are upregulated in several cancers, including breast, colorectal, lung, kidney, and glioblastoma [16–18], and function to primarily regulate lactate transport across the plasma membrane [14, 19]. These transporters can also transport other monocarboxylates, including pyruvate, acetoacetate, and β-hydroxybutyrate together with H^+ [18]. The MCTs rely on the chaperone CD147/Basigin for proper folding, insertion into the membrane, and transport function. A recent study also suggested that the transmembrane serine protease TMPRSS11B was required for optimal lactate efflux from lung cancer cells by inducing the solubilization of CD147 [20]. However, while this could have implications for the transport metabolon (see below) further characterization of this solubilization is required. Similar to their distribution pattern in muscle, MCT1 and MCT4 tend to be differentially located in cancer cells. MCT4 is upregulated in hypoxia as a direct target of HIF-1α and functions to export lactate from hypoxic and glycolytic tumor cells [15, 16]. MCT1, in contrast, is expressed on oxidative tumor cells (and muscle) and functions to both import and export lactate [18]. MCT1 and MCT4 have been shown to be involved in migration, invasion, and metastasis furthering the importance of targeting their expression in cancer [18, 21, 22].

Long thought to be a metabolic waste product, the finding that lactate can be taken up by cancer cells and utilized to fuel respiration has changed that dogma [23]. Initially described as metabolic symbiosis between glycolytic and oxidative cancer cells, this phenomenon has now been shown to occur between cancer cells and CAF's [24], endothelial [25] and immune cells and referred to as the "reverse Warburg effect." Furthermore, this phenomenon is not restricted to preclinical models, as metabolomic profiling of lung cancer patients has provided evidence that lactate is utilized as a fuel [26]. As the MCTs are a family of passive transporters, excess accumulation of lactate in the TME would eventually inhibit lactate efflux if not removed. Recent findings also implicate the gap junction protein Connexin-43 in mediating lactate flux between cells in areas where efflux may have been blocked due to excess extracellular accumulation [27]. Nevertheless, the relationship that exists between MCT1 and MCT4 in regulating lactate levels in hypoxia has suggested it would be prudent to target both of these transporters for cancer therapy. Indeed, efforts are underway to target metabolic symbiosis in resistance to antiangiogenic therapies. Studies so far have focused on the genetic manipulation of MCT4 or have interfered with glycolysis or mTOR signaling [28, 29]. Further studies are needed with direct pharmacologic intervention of MCT1 and MCT4. A recent in vitro study demonstrated that simultaneous inhibition of both MCT1 and MCT4 was lethal when combined with the complex I inhibitor metformin [30]. While a specific inhibitor to MCT4 may be in preclinical development [31], the MCT1 specific inhibitor AZD3965 has progressed through preclinical evaluation and is set to be evaluated clinically (NCT01791595) [32]. Given the expression of MCTs in normal tissues, including the brain, it will be interesting to see how clinical development of MCT inhibitors progresses.

Carbonic anhydrases

There have been 16 mammalian carbonic anhydrases (CAs) characterized to date, 15 in humans, all displaying a broad distribution in tissue expression and subcellular localization [33, 34]. Three of the CAs characterized to date have been found to lack catalytic function and have been referred to as carbonic anhydrase-related proteins (CARPs). The CAs are a family of zinc metalloenzymes that catalyze the reversible hydration of CO_2 to HCO_3^- and H^+ in a very efficient manner [35, 36]. This catalytic function plays an important role in tumorigenicity, pH homeostasis, electrolyte secretion, bone resorption, calcification and in metabolic pathways such as gluconeogenesis, lipogenesis, and ureagenesis [33]. In the context of cancer, many of the CAs have not received very much attention likely due to their ubiquitous expression throughout the body [36]. The cytosolic isoform CAII has been studied in cancer and associated with good prognosis in some cancers and worse prognosis in others. However, given its widespread expression in normal tissues targeting CAII may be difficult. In contrast, two membrane-bound CAs, CAIX and CAXII, are highly overexpressed in cancer and play a role in pH regulation, migration, invasion, metastasis, tumor growth, cancer stem cell function, and response to therapeutic interventions [32].

CAIX is organized with extracellular facing proteoglycan domain (PG) and catalytic domains, a single pass transmembrane domain and an intracellular cytoplasmic tail [37]. Analysis of the crystal structure of CAIX revealed that it functions as a homodimer held together by a disulfide bond [38]. CAXII contains a similar domain structure to CAIX with the exception of the PG domain [39, 40]. Studies have identified that the PG domain, which is rich in acidic amino acid residues, acts as a proton buffer permitting CAIX function in acidic environments and to also play a role in cell spreading [35, 38, 41], although recent work has suggested that the difference in amino acid composition in the catalytic domains of CAIX and CAXII may explain why CAIX is better suited to function optimally in the pH range found within the TME of hypoxic solid tumors [42]. Both CAIX and CAXII have been shown to be induced following exposure to hypoxia in an HIF1α [43]. Indeed, CAIX is a major effector of the HIF1-mediated transcriptional response to tumor hypoxia and its critical role in tumor progression is well recognized [44–46]. It is highly expressed in the hypoxic regions of many types of solid tumors, has limited expression in normal tissues, and is a well-established marker of poor prognosis across many solid cancers [44, 46]. Through its enzymatic function CAIX controls an intracellular and extracellular acid-base balance that regulates both survival and invasive properties [46–49]. The HCO_3^- produced by CAIX re-enters the cell through bicarbonate transporters and anion exchangers, thereby buffering intracellular acidosis and facilitating tumor cell survival and growth. The H^+ participates in the generation of an increasingly acidic extracellular environment (Fig. 1). Recent investigation with hyperpolarized ^{13}C-magnetic resonance spectroscopy (MRS) supports a role for CAIX in the acidification of pH_e in colorectal cancer [50]. Further study with ^1H-MRS identified that CAIX acts as a pH-stat maintaining an acidic extracellular pH and establishing an upper pH_e limit of 6.93 in these tumors [51]. Consistent with its role in regulating pH, several studies have demonstrated that perturbing CAIX function in hypoxia elicits biological consequences that impede cancer progression and demonstrate its utility as a therapeutic target (discussed below).

While CAXII catalyzes the same reaction as CAIX, studies directly demonstrating its role in the control of pH_e of solid tumors are lacking. However, studies have suggested that CAXII can compensate for loss of CAIX [52, 53] and combined loss of CAIX and NHE1 expression [54]. Indeed, genetic disruption of both *CA9* and *NHE1* increased tumor growth relative to disruption of either gene individually and these tumors contained increased CAXII expression [54]. To date, this has only been shown in select models and further investigation is required to generalize this response. Nevertheless, these studies provide evidence of the importance of CAXII in pH regulation for tumor survival and suggest that therapeutic interventions targeting both CA isoforms simultaneously could be advantageous. CAXII has been shown to play a role in migration and invasion, play a role in chemoresistance, and be involved in a reciprocal relationship with the hedgehog pathway, which have implications for disease progression [55–59]. The prognostic value of CAXII expression in solid tumors is murkier than it is for CAIX [36]. CAXII expression has been associated with favorable prognosis in breast [60] and lung cancer [61], while also being associated with worse prognosis in gliomas [62] and both oral [63] and esophageal squamous tumors [64]. Continued investigation into the role of CAXII in tumor biology should yield insight into cancers best suited for its intervention.

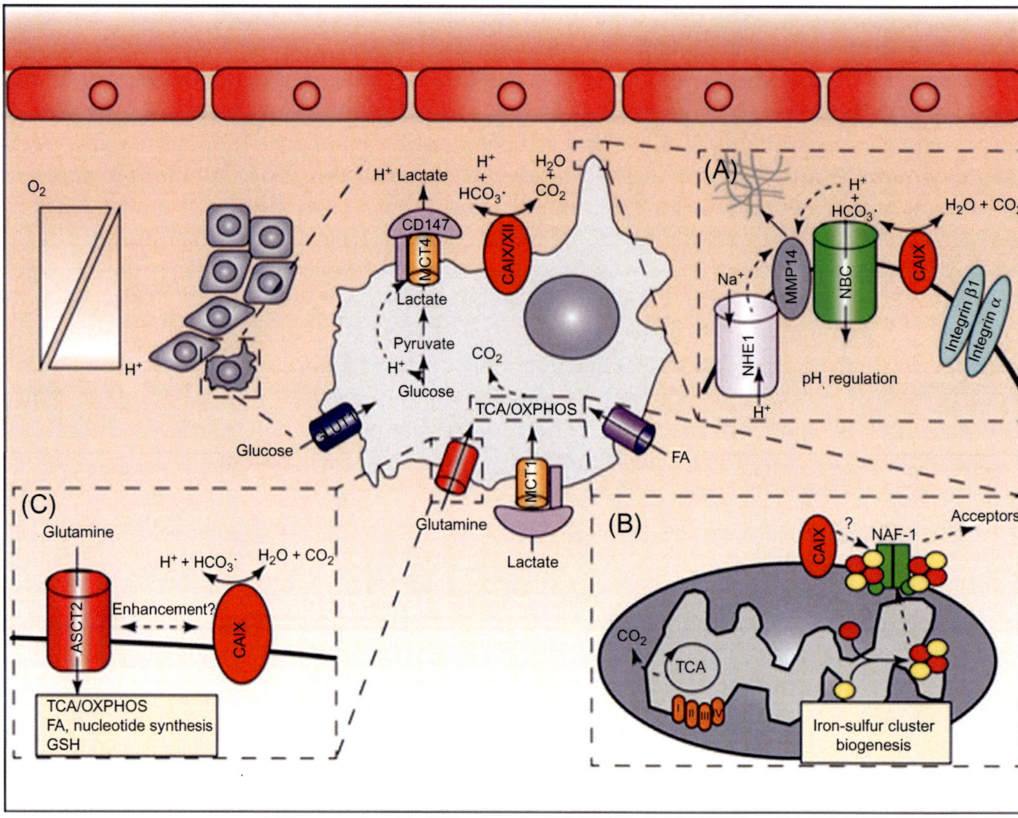

FIG. 1 pH regulation in the tumor microenvironment. (A) Acidosis-mediated stimulation of ECM breakdown and the pH regulatory components that localize to the lamellipodia. Proteomic analysis identified CAIX interacts with integrins, the sodium bicarbonate transporter NBCn1, and the matrix metalloprotease MMP14. NHE1 plays a well-established role maintaining pHi while contributing to pericellular acidification. (B) Mitochondrial metabolism plays a critical role in acidosis-mediated metabolic adaptations. Glutamine, lactate, and fatty acids are used to drive TCA cycle flux to meet energy demands. The mitochondria are also sites of iron-sulfur cluster formation. *Red balls*—iron atoms; *Yellow balls*—sulfur atoms. NAF-1 was identified in the CAIX interactome. (C) Hypoxia and acidosis trigger increased uptake and utilization of glutamine through the membrane transporter ASCT2. ASCT2 was found to be part of the CAIX interactome. In addition to driving the TCA cycle, glutamine is also used to drive reductive carboxylation to produce intermediates for lipid biosynthesis, is used in nucleotide biosynthesis and is also a key component in the synthesis of glutathione and is thus critical for maintaining redox homeostasis.

NHE1

The sodium hydrogen exchangers are a family of 9 isoforms. NHE1 is the most important isoform for pH_i regulation and the most heavily studied in cancer since being implicated in malignant transformation [65]. NHE1 functions to maintain intracellular pH homeostasis by pumping protons out of the cell in exchange for sodium ions and is very important in maintaining the reverse pH gradient common in cancer cells. NHE1 senses changes in intracellular pH through an internal allosteric proton binding regulatory site [66]. If the pH_i drops below a threshold it triggers activation of the protein and increases transport of protons out of the cell [67]. This effectively determines the pH set point of the cells where NHE1 becomes activated. In normal cells this is ~ pH7.2, but in cancer cells where NHE1 has an increased affinity for protons, this occurs at a more alkaline pH (>7.4) thereby helping to maintain an alkaline pH_i. NHE1 activity is stimulated in response to growth factors and has a heavily phosphorylated c-terminal tail suggesting additional layers of regulation. Recent findings identified phosphorylation of T779 is critical for transport activity and that calcineurin is the phosphatase regulating this site [68]. NHE1 is hypoxia induced in cardiomyocytes, but evidence of this response in cancer cells is lacking [47]. Depletion of NHE1 or treatment with the inhibitor amiloride or its derivatives has been shown to block tumor growth [69, 70]. However, targeting NHE1 in cancer could be challenging and be accompanied by toxicity issues given that NHE1 is ubiquitously expressed and plays important physiological roles throughout the body and knock-out mice die early after birth [71, 72].

Sodium bicarbonate cotransporters and anion exchangers

The bicarbonate transporters are encoded by two protein families: SLC4 and SLC26. The SLC4 family of transporters consists of 10 genes (SLC4A1-5,7-11), whereas the SLC26 family consists of 7 genes (SLC26A2-4,7-9,11). The transporters can be divided into two categories based on their dependence on sodium: the sodium-independent anion exchangers SLC4A1-3 (AE1-3) and SLC26A2-4,7-9,11 and the sodium-dependent transporters, which includes sodium dependent cotransporters (NBC) SLC4A4 (NBCe1), SLC4A5 (NBCe2), SLC4A7 (NBCn1) and SLC4A10 (NBCn2) and the sodium-dependent anion exchangers SLC4A8 (Cl^- HCO_3^- exchange) and SLC4A9 (AE4). The NBCs transport Na^+ and HCO_3^- across the membrane in 1:1 (NBCn1 - electroneutral) or 1:2/1:3 (NBCe1, NBCe2 – electrogenic) stoichiometry [73, 74]. Altered expression of several bicarbonate transporters has been detected across a spectrum of solid tumors [75]. NBCn1 has been the most studied of the group. Its expression has been shown to be elevated in breast cancer as well as contain an SNP in its 3′ UTR that has been associated with increased risk of developing breast cancer [76–80]. Consistent with this, genetic loss of NBCn1 delayed tumor growth in carcinogen-induced and oncogene-driven mouse models of breast cancer [81, 82]. Bicarbonate transport was reported to be unaffected by hypoxia or exposure to the PHD inhibitor DMOG [83]. However, expression of bicarbonate transporters was increased in a panel of cell lines investigated in response to exposure to severe hypoxia (0.1% O_2) [84]. Consistent with previous findings on NBCe1, NBCe1 and AE4 showed the most robust response to hypoxic exposure [85]. Genetic depletion of AE4 was found to significantly delay growth of breast and glioblastoma tumors suggesting an important role for this transporter in tumor physiology. The fact that bicarbonate flux was unhindered in hypoxia suggests that this important pH regulatory machinery continues to provide critical pH_i-buffering capacity in the acidic microenvironment present in hypoxic regions presenting a therapeutic opportunity [83]. Interestingly, transporter expression varied across the models interrogated and even within models of the same cancer type, suggesting that careful consideration be given to model selection for biological characterization of the bicarbonate transporters in tumor physiology, but also suggests that the family needs to be considered as a whole before drawing conclusions on regulation of

HCO_3^- transport in a particular cancer type. Targeting bicarbonate transport has faced difficulty when it comes to selectivity and delivery of the inhibitors produced to date, with promising drug candidates failing to reproduce efficacy observed in cells grown in 2D in 3D models [86, 87]. However, model selection may be important as a recent study suggests efficacy with S0859 in 3D spheroid models [84]. Nevertheless, further development is needed in this area especially with in vivo study of these compounds, considering the importance of bicarbonate transport in normal physiology [88].

pH metabolons

Consistent with the discussions above on the importance of pH_i regulatory machinery in maintaining acid-base homeostasis, hypotheses have been developed coupling production and transport of H^+, lactate, and HCO_3^- into and out of the cell in a highly coordinated manner in transport "metabolons." Metabolons have been defined as complexes that mediate sequential reactions on a substrate that transfer metabolites directly between binding partners preventing the metabolites from establishing equilibrium with bulk cellular fluid [89]. The carbonic anhydrases have been found to interact with bicarbonate transporters, anion exchangers and monocarboxylate transporters in these functional transport complexes. The carbonic anhydrase transport metabolons have been shown to play a role in metabolism, pH regulation, and cell migration (reviewed in [90]). The first description of a CA in a transport metabolon was CAII with AE1 [91]. CAII binds to an acidic motif in the c-terminal tail of AE1 and facilitates transport of bicarbonate. This interaction was found to be dependent on the enzymatic activity of CAII as treatment with the broad spectrum carbonic anhydrase inhibitor acetazolamide or expression of catalytically inactive CAII V143Y variant reduced AE1 Cl^- HCO_3^- exchange [91]. CAII has also been shown to bind sodium bicarbonate transporters, NBCe1 and NBCn1, and sodium hydrogen exchangers, NHE1 and NHE3 (reviewed in [90]).

Carbonic anhydrase IX has also been identified in transport metabolons. CAIX was found to interact with AE1/2/3 in HEK293 cells via its catalytic domain. Expression of CAIX increased bicarbonate transport of all 3 transporters, with the greatest enhancement observed for AE2 [92]. The interaction of CAIX with AE2 and NBCe1 was also observed in the lamellipodia of cervical cancer cells [93]. Additional proteomic analysis of the CAIX interactome in breast cancer has identified an interaction with NBCn1 [94, 95], although further investigation is needed to characterize the biological significance of these interactions. The interaction of CAIX with bicarbonate transporters, whether through direct physical association or through proximity as part of a larger complex [93, 94], would facilitate the transfer of bicarbonate generated back into the cell to titrate intracellular pH especially in lamellipodia and invadopodia to facilitate cellular migration and invasion [93, 94].

CAIX has also been shown to participate in transport metabolons independent of its enzymatic activity. CAIX was found to interact with MCT1 and MCT4 via the Ig1 domain of CD147/Basigin [96]. This association has been demonstrated to occur via the proteoglycan domain of CAIX and be independent of enzymatic activity [97]. The proteoglycan domain of CAIX contains 18 glutamate and 8 aspartate residues that can aid in proton shuttling from the enzyme to nearby protonatable substrates [35, 97]. Indeed, CAIX has been suggested to function as a proton antenna to facilitate continued proton and lactate extrusion by the transporters [97]. Similarly, CAII was found to associate with both MCT1 and MCT4 on the cytoplasmic face of the plasma membrane, where the association is independent of CAII enzymatic activity, but requires Glu69 and Asp72 to facilitate proton transfer to the transporters to enhance proton and lactate extrusion efficiency [98]. Interestingly, the interactions between the CAs with the MCTs is influenced by a single histidine residue: His64 on CAII and His200 on CAIX. Both histidine residues function as proton shuttles within the catalytic domains shuttling H^+ away from the active site, but may still facilitate the electrostatic interaction between CAII and the acidic c-terminal tails of MCT1/4 and CAIX with the acidic Glu31 and Glu73 of CD147 [96, 98].

In light of the number of studies demonstrating the existence of the transport metabolon between the CAs and bicarbonate exchangers and MCTs, opposition to this theory exists questioning the direct interaction of these proteins [99]. Further investigation is warranted, although, regardless of whether a direct physical association exists, interfering with pH regulation has clear therapeutic potential.

Effect of inhibiting pH regulators on migration, invasion, and metastasis

Enhanced metabolism coupled with reduced perfusion leads to an accumulation of acidic metabolites in the interstitium of solid tumors. The compensatory upregulation of membrane transporters that function to maintain an alkaline pH_i such as NHE1, MCT4, and CAIX, further contributes to the acidification the pH_e. Decreased pH_e leads to enhanced secretion and activation of proteases that remodel the ECM to facilitate tumor cell migration and invasion [4]. Tumor cells with an alkaline pH_i contain increased activity of actin-severing proteins cofilin and Arp2/3, along with increased focal adhesion turnover rates [4, 100]. Degradation of the ECM in the local environment removes the barrier keeping them in place. Indeed, enhanced invasive and metastatic capacity has been detected upon exposure to acidosis [101–105]. Moreover, in tumors this invasion was directed toward the areas of the tumor with the most acidic pH [106]. Furthermore, the tumor-stroma interface of breast tumors was recently found to be the most acidic region of the tumor [107]. Interestingly, and in agreement with other studies, regions of hypoxia contained areas of acidosis, but not all acidic regions were hypoxic [108]. Notably, these regions of acidosis correlated highly with the expression of CAIX. Numerous studies have highlighted a role for CAIX in migration, invasion, and metastasis using genetic and pharmacological approaches [93, 109–113]. Recent proteomic analyses have provided mechanistic insight into its role in migration and invasion, identifying many interactions with known critical components of both processes [94, 95]. CAIX was found to interact with the actin-remodeling protein Arp3 and integrins β1, α2, α3, α5, and α6, all critical components of lamellipodia and invadopodia [114–116]. Furthermore, CAIX was found to interact with and colocalize with the metalloprotease MMP14 in lamellipodia and invadopdia and to provide a source of H^+ required for the collagen-degrading activity of MMP14 (Fig. 1). NHE1 has also been shown to be a critical source of pH regulation within lamellipodia and invadopodia, providing H^+ for pericellular acidification and maintaining an alkaline pH_i for cytoskeletal dynamics [100, 117, 118]. As these structures lack cellular organelles they are heavily reliant upon glycolysis for energy production. Consistent with this MCT4 has been shown to localize to the lamellipodia where it interacts with β1 intergrin [119].

Together CAIX is part of a pH regulatory network controlling intra- and pericellular pH to facilitate cancer cell migration, invasion, and metastasis.

Effect of inhibiting pH regulation on response to chemotherapy

Resistance to chemotherapy is a common feature of the hypoxic, acidic tumor microenvironment [1, 47, 120]. Chemotherapies with intracelluar targets that are weak bases such as doxorubicin, paclitaxel, vinblastine, and mitoxantrone are protonated at acidic pH, which negatively impacts diffusion capacity across the membrane, an effect known as ion trapping [121]. Furthermore, any drug that is able to cross the membrane will lose its charge in the more alkaline cytoplasm leaving it free to diffuse back across the membrane out of the cell [1, 4]. In contrast, weak acids have been shown to have greater efficacy in targeting acidic microenvironments. The negatively charged weak acid species is neutralized in the acidic TME allowing diffusion across the membrane. Once across the membrane, the alkaline intracellular pH ionizes the drug preventing its diffusion back across the membrane increasing intracellular accumulation. Combining chemotherapies with drugs targeting acidification of the TME may enhance cellular uptake and increase efficacy [32, 122]. Indeed, a number of approaches are in various stages of development and these have been reviewed recently [1, 123]. Our discussion will remain focused on exploiting pH regulators to overcome extracellular acidosis and/or disrupt pH_i homeostasis to improve therapeutic efficacy.

Targeting CAIX with sulfonamide-based inhibitors has achieved success as a single agent [109, 110, 124, 125]. Perhaps more clinically relevant is the growing body of evidence demonstrating CAIX inhibition enhances the therapeutic efficacy of several chemotherapy drugs, including gemcitabine, paclitaxel, and temozolomide in xenograft models of pancreatic, breast, and brain cancer [110, 126, 127]. While the mechanism(s) by which this is occurring is not yet known, it is possible that at least in the case of paclitaxel treatment, SLC-0111 may be diminishing the acidity of the extracellular environment, while simultaneously decreasing pH_i, thus increasing intracellular drug concentration of the weak base [110].

One of the concerns about targeting pH regulation for cancer therapy is the redundancy of pH regulatory mechanisms and the fact that pH_i may not be driven low enough to negatively impact viability [1, 47]. However, if combined with therapies that impact metabolic phenotype, pH_i may be lowered enough to cross the viability threshold [32, 128, 129]. Gemcitabine and paclitaxel have been shown to trigger ROS formation and HIF1α stabilization [130]. In line with this, gemcitabine has also been shown to trigger increased glucose uptake with enhanced flux through the pentose phosphate pathway suggested as a mechanism of resistance [127]. Moreover, gemcitabine was shown to increase the metabolic capacity of PDAC cells and tumors increasing HIF1α and CAIX expression, especially in normoxic conditions. With this enhanced metabolic activity and increased target expression these tumors were more vulnerable to SLC-0111 [127]. Biomarker assessment of tumor material revealed increased membrane positivity for the surrogate marker of acidosis LAMP2 when SLC-0111 was combined with gemcitabine, suggesting therapy-induced metabolic changes were responsible for the combinatorial effect [131]. Similar observations were made in a model of glioblastoma where the combination of SLC-0111 with temozolomide was more effective than temozolomide alone [126]. Metabolomic analysis identified the combination of the two drugs induced changes in metabolite flux not observed by temozolomide alone. The combinatorial effect inhibiting a CA with temozolomide was also recently confirmed with an inhibitor to CAXII [57]. Furthermore, CAIX inhibition with a sulfamate-based inhibitor was shown to enhance the effect of cisplatin in a model of non-small-cell lung cancer [125]. However, reasons for the enhanced effect were not explored. In a similar vein, targeting pH regulation to overcome resistance to antiangioenic therapy has proven effective [28, 129].

The acidic TME has been identified as a hurdle to antitumor immunity by limiting the function of antitumor immune response [132–134]. Indeed, neutralization of extracellular pH has been shown to increase the efficacy of immunotherapies [132, 135, 136]. Increased CAIX expression has been identified in patients that fail to respond to α-PD-1 therapy and targeting its expression/activity may be used to enhance efficacy of ICB [137, 138]. Indeed, SLC-0111 has been shown to increase the efficacy of immune checkpoint blockade with α-PD-1 and α-CTLA-4 in preclinical models of malignant melanoma and basal-like breast cancer [124]. Mechanistically, SLC-0111-treated cells had a more neutral pH_e that was conducive to T-cell immune function increasing cancer cell kill.

pH and mitochondria

pH needs to be finely tuned in order to maintain survival. Intracellular pH is maintained slightly alkaline (>7.4), whereas extracellular pH of the microenvironment of a solid tumor is acidic and can range anywhere from pH 6.5–7.0. Increased utilization of glucose leads to accumulation of lactate that is extruded by MCT4. As lactate continues to build in the extracellular environment it needs to be dissipated otherwise continued outward movement of lactate will be blocked. As this situation arises, cancer cells adjust their fuel preference to combat acidosis, which leads to a decrease in glucose utilization [9, 139–141]. Decreased glucose consumption shown by ^{18}F-FDG uptake has been observed by PET in preclinical settings in areas of increased acidosis [142]. Studies in patients with lung cancer have suggested that ^{18}F-FDG uptake does not correlate with glycolysis, but instead with proliferation, suggesting utilization of glucose by alternative pathways [143]. While chronic acidosis decreases glucose consumption, it simultaneously triggers increased consumption of glutamine and fatty acids to generate flux through the TCA cycle [139, 140]. Increased glutamine uptake and utilization via HIF2α-dependent mechanisms is used to fuel fatty acid synthesis, while increased β-oxidation of exogenous fatty acids increases TCA cycle flux along with the mitochondrial pool of acetyl coA [140, 144]. The increase in mitochondrial acetyl coA leads to further rearrangement of mitochondrial metabolism through acetylation of mitochondrial enzymes. Among the acetylated are components of complex I of the ETC whereby acetylation acts as a rheostat to restrain flux and ROS generation [144]. An increase in mitochondrial metabolism is consistent with the importance of this organelle for tumorigenesis [145–148]. Acidosis-induced changes in mitochondrial morphology permit respiration in conditions when oxygen is limiting [149], although this was demonstrated in neurons and further work is needed to determine whether similar changes occur in cancer. HIF-dependent changes in complex IV (cytochrome *c* oxidase/COX) subunit composition leads to a replacement of COX4I1 (cox4 isoform 1) with COX4I2, which permits more efficient electron transfer in low oxygen. With complex IV having an apparent K_m for oxygen of 0.1%, OXPHOS should function at near anoxic levels [3]. Titration of ROS levels in the mitochondria is especially important for maintenance of cellular respiration. Iron-sulfur clusters, cofactors for a number of TCA cycle and respiratory chain enzymes, are very sensitive to oxygen levels [150]. Acidic pH has been found to impact iron-sulfur cluster binding to proteins responsible for their transport [151]. Interestingly, CAIX was found to interact with NAF-1, a protein that localizes to the mitochondria and the ER and facilitates transport of iron-sulfur clusters (Fig. 1) [94]. NAF-1 expression was found to be elevated in breast cancer and required for tumor growth in a preclinical model [152]. The protein belongs to a family of iron-sulfur cluster binding proteins with a characteristic 39 amino acid stretch known as the CDGSH domain for iron-sulfur cluster binding. The iron-sulfur clusters are bound within a pocket coordinated by 3Cys-His conformation in NAF-1. The protonation of the histidine residue influences transferability of the iron-sulfur cluster and interfering with this activity of the protein severely impeded breast tumor growth [152]. As recent findings have implicated CAIX in control of the mitochondrial network in cells [153], it is tempting to suggest that CAIX activity may facilitate iron-sulfur cluster release from NAF-1 permitting iron-sulfur shuttling to target proteins, but further experimentation is required to unravel the significance of this interaction.

Conclusions

pH regulation in cancer is critical to survival. Upregulation of membrane transporters to extrude acids or import bicarbonate is utilized to tightly control pH_i at the expense of acidifying pH_e. However, the proteins involved in maintaining pH homeostasis have become critical therapeutic targets. As the field advances, developing isoform specific molecules, with enhanced potency and selectivity for cancer-specific transporter expression will be important. The progress of existing inhibitors to CAIX (NCT03450018) and MCT1 (NCT01791595) through clinical evaluation should be informative for the next wave of lead molecules [32, 154]. In addition, as the adaptations that occur in response to acidosis in solid tumors begin to be realized, the way forward and the therapeutic opportunities that arise will become illuminated. Combinatorial strategies targeting metabolic adaptations in response to therapy should also provide promising results for cancer research in the years to come.

References

[1] Corbet C, Feron O. Tumour acidosis: from the passenger to the driver's seat. Nat Rev Cancer 2017;17(10):577–93 [Epub 2017/09/16].

[2] Kroemer G, Pouyssegur J. Tumor cell metabolism: cancer's Achilles' heel. Cancer Cell 2008;13(6):472–82 [Epub 2008/06/10].

[3] Lee P, Chandel NS, Simon MC. Cellular adaptation to hypoxia through hypoxia inducible factors and beyond. Nat Rev Mol Cell Biol 2020;21(5):268–83 [Epub 2020/03/08].

[4] Webb BA, Chimenti M, Jacobson MP, Barber DL. Dysregulated pH: a perfect storm for cancer progression. Nat Rev Cancer 2011;11(9):671–7 [Epub 2011/08/13].

[5] Whitten ST, Garcia-Moreno EB, Hilser VJ. Local conformational fluctuations can modulate the coupling between proton binding and global structural transitions in proteins. Proc Natl Acad Sci U S A 2005;102(12): 4282–7 [Epub 2005/03/16].

[6] White KA, Kisor K, Barber DL. Intracellular pH dynamics and charge-changing somatic mutations in cancer. Cancer Metastasis Rev 2019;38(1–2):17–24 [Epub 2019/04/15].

[7] Casey JR, Grinstein S, Orlowski J. Sensors and regulators of intracellular pH. Nat Rev Mol Cell Biol 2010; 11(1):50–61 [Epub 2009/12/10].

[8] McBrian MA, Behbahan IS, Ferrari R, Su T, Huang TW, Li K, Hong CS, Christofk HR, Vogelauer M, Seligson DB, Kurdistani SK. Histone acetylation regulates intracellular pH. Mol Cell 2013;49(2):310–21 [Epub 2012/12/04].

[9] Chen JL, Lucas JE, Schroeder T, Mori S, Wu J, Nevins J, Dewhirst M, West M, Chi JT. The genomic analysis of lactic acidosis and acidosis response in human cancers. PLoS Genet 2008;4(12) e1000293 [Epub 2008/12/06].

[10] Chen JL, Merl D, Peterson CW, Wu J, Liu PY, Yin H, Muoio DM, Ayer DE, West M, Chi JT. Lactic acidosis triggers starvation response with paradoxical induction of TXNIP through MondoA. PLoS Genet 2010;6(9) e1001093 [Epub 2010/09/17].

[11] Kondo A, Yamamoto S, Nakaki R, Shimamura T, Hamakubo T, Sakai J, Kodama T, Yoshida T, Aburatani H, Osawa T. Extracellular acidic pH activates the sterol regulatory element-binding protein 2 to promote tumor progression. Cell Rep 2017;18(9):2228–42 [Epub 2017/03/02].

[12] Gillies RJ, Verduzco D, Gatenby RA. Evolutionary dynamics of carcinogenesis and why targeted therapy does not work. Nat Rev Cancer 2012;12(7):487–93 [Epub 2012/06/15].

[13] Gatenby RA, Gillies RJ. A microenvironmental model of carcinogenesis. Nat Rev Cancer 2008;8(1):56–61 [Epub 2007/12/07].

[14] Marchiq I, Pouyssegur J. Hypoxia, cancer metabolism and the therapeutic benefit of targeting lactate/H symporters. J Mol Med (Berl) 2015;94:155–71 [Epub 2015/06/24].

[15] Parks SK, Cormerais Y, Marchiq I, Pouyssegur J. Hypoxia optimises tumour growth by controlling nutrient import and acidic metabolite export. Mol Asp Med 2015;47–48:3–14 [Epub 2016/01/03].

[16] Doherty JR, Cleveland JL. Targeting lactate metabolism for cancer therapeutics. J Clin Invest 2013;123(9): 3685–92 [Epub 2013/09/04].

[17] Doyen J, Trastour C, Ettore F, Peyrottes I, Toussant N, Gal J, Ilc K, Roux D, Parks SK, Ferrero JM, Pouyssegur J. Expression of the hypoxia-inducible monocarboxylate transporter MCT4 is increased in triple negative breast cancer and correlates independently with clinical outcome. Biochem Biophys Res Commun 2014;451(1):54–61 [Epub 2014/07/25].

[18] Baltazar F, Afonso J, Costa M, Granja S. Lactate beyond a waste metabolite: metabolic affairs and signaling in malignancy. Front Oncol 2020;10:231 [Epub 2020/04/08].

[19] Halestrap AP. Monocarboxylic acid transport. Compr Physiol 2013;3(4):1611–43 [Epub 2013/11/23].

[20] Updegraff BL, Zhou X, Guo Y, Padanad MS, Chen PH, Yang C, Sudderth J, Rodriguez-Tirado C, Girard L, Minna JD, Mishra P, DeBerardinis RJ, O'Donnell KA. Transmembrane protease TMPRSS11B promotes lung cancer growth by enhancing lactate export and glycolytic metabolism. Cell Rep 2018;25(8):2223–33 e2226. [Epub 2018/11/22].

[21] Tasdogan A, Faubert B, Ramesh V, Ubellacker JM, Shen B, Solmonson A, Murphy MM, Gu Z, Gu W, Martin M, Kasitinon SY, Vandergriff T, Mathews TP, Zhao Z, Schadendorf D, DeBerardinis RJ, Morrison SJ. Metabolic heterogeneity confers differences in melanoma metastatic potential. Nature 2020;577(7788):115–20 [Epub 2019/12/20].

[22] Payen VL, Hsu MY, Radecke KS, Wyart E, Vazeille T, Bouzin C, Porporato PE, Sonveaux P. Monocarboxylate transporter MCT1 promotes tumor metastasis independently of its activity as a lactate transporter. Cancer Res 2017;77(20):5591–601 [Epub 2017/08/23].

[23] Sonveaux P, Vegran F, Schroeder T, Wergin MC, Verrax J, Rabbani ZN, De Saedeleer CJ, Kennedy KM, Diepart C, Jordan BF, Kelley MJ, Gallez B, Wahl ML, Feron O, Dewhirst MW. Targeting lactate-fueled respiration selectively kills hypoxic tumor cells in mice. J Clin Invest 2008;118(12):3930–42 [Epub 2008/11/27].

[24] Pavlides S, Whitaker-Menezes D, Castello-Cros R, Flomenberg N, Witkiewicz AK, Frank PG, Casimiro MC, Wang C, Fortina P, Addya S, Pestell RG, Martinez-Outschoorn UE, Sotgia F, Lisanti MP. The reverse Warburg effect: aerobic glycolysis in cancer associated fibroblasts and the tumor stroma. Cell Cycle 2009;8(23):3984–4001 [Epub 2009/11/20].

[25] Vegran F, Boidot R, Michiels C, Sonveaux P, Feron O. Lactate influx through the endothelial cell monocarboxylate transporter MCT1 supports an NF-kappaB/IL-8 pathway that drives tumor angiogenesis. Cancer Res 2011;71(7):2550–60 [Epub 2011/02/09].

[26] Faubert B, Li KY, Cai L, Hensley CT, Kim J, Zacharias LG, Yang C, Do QN, Doucette S, Burguete D, Li H, Huet G, Yuan Q, Wigal T, Butt Y, Ni M, Torrealba J, Oliver D, Lenkinski RE, Malloy CR, Wachsmann JW, Young JD, Kernstine K, DeBerardinis RJ. Lactate metabolism in human lung tumors. Cell 2017;171(2):358–71 e359. [Epub 2017/10/07].

[27] Dovmark TH, Saccomano M, Hulikova A, Alves F, Swietach P. Connexin-43 channels are a pathway for discharging lactate from glycolytic pancreatic ductal adenocarcinoma cells. Oncogene 2017;36(32):4538–50 [Epub 2017/04/04].

[28] Pisarsky L, Bill R, Fagiani E, Dimeloe S, Goosen RW, Hagmann J, Hess C, Christofori G. Targeting metabolic symbiosis to overcome resistance to anti-angiogenic therapy. Cell Rep 2016;15(6):1161–74 [Epub 2016/05/03].

[29] Allen E, Mieville P, Warren CM, Saghafinia S, Li L, Peng MW, Hanahan D. Metabolic symbiosis enables adaptive resistance to anti-angiogenic therapy that is dependent on mTOR signaling. Cell Rep 2016;15(6):1144–60 [Epub 2016/05/03].

[30] Benjamin D, Robay D, Hindupur SK, Pohlmann J, Colombi M, El-Shemerly MY, Maira SM, Moroni C, Lane HA, Hall MN. Dual inhibition of the lactate transporters MCT1 and MCT4 is synthetic lethal with metformin due to NAD+ depletion in cancer cells. Cell Rep 2018;25(11):3047–58 e3044. [Epub 2018/12/13].

[31] Futagi Y, Kobayashi M, Narumi K, Furugen A, Iseki K. Identification of a selective inhibitor of human monocarboxylate transporter 4. Biochem Biophys Res Commun 2018;495(1):427–32 [Epub 2017/10/11].

[32] McDonald PC, Chafe SC, Dedhar S. Overcoming hypoxia-mediated tumor progression: combinatorial approaches targeting pH regulation, angiogenesis and immune dysfunction. Front Cell Dev Biol 2016;4:27 [Epub 2016/04/12].

[33] Supuran CT. Carbonic anhydrases: novel therapeutic applications for inhibitors and activators. Nat Rev Drug Discov 2008;7(2):168–81 [Epub 2008/01/03].

[34] Pastorekova S, Parkkila S, Pastorek J, Supuran CT. Carbonic anhydrases: current state of the art, therapeutic applications and future prospects. J Enzyme Inhib Med Chem 2004;19(3):199–229 [Epub 2004/10/27].

[35] Innocenti A, Pastorekova S, Pastorek J, Scozzafava A, De Simone G, Supuran CT. The proteoglycan region of the tumor-associated carbonic anhydrase isoform IX acts as anintrinsic buffer optimizing CO2 hydration at acidic pH values characteristic of solid tumors. Bioorg Med Chem Lett 2009;19(20):5825–8 [Epub 2009/09/16].

[36] Mboge MY, Mahon BP, McKenna R, Frost SC. Carbonic anhydrases: role in pH control and cancer. Metabolites 2018;8(1) [Epub 2018/03/03].

[37] Pastorek J, Pastorekova S, Callebaut I, Mornon JP, Zelnik V, Opavsky R, Zat'ovicova M, Liao S, Portetelle D, Stanbridge EJ, et al. Cloning and characterization of MN, a human tumor-associated protein with a domain homologous to carbonic anhydrase and a putative helix-loop-helix DNA binding segment. Oncogene 1994; 9(10):2877–88 [Epub 1994/10/01].

[38] Alterio V, Hilvo M, Di Fiore A, Supuran CT, Pan P, Parkkila S, Scaloni A, Pastorek J, Pastorekova S, Pedone C, Scozzafava A, Monti SM, De Simone G. Crystal structure of the catalytic domain of the tumor-associated human carbonic anhydrase IX. Proc Natl Acad Sci U S A 2009;106(38):16233–8 [Epub 2009/10/07].

[39] Tureci O, Sahin U, Vollmar E, Siemer S, Gottert E, Seitz G, Parkkila AK, Shah GN, Grubb JH, Pfreundschuh M, Sly WS. Human carbonic anhydrase XII: cDNA cloning, expression, and chromosomal localization of a carbonic anhydrase gene that is overexpressed in some renal cell cancers. Proc Natl Acad Sci U S A 1998;95(13):7608–13 [Epub 1998/06/24].

[40] Whittington DA, Waheed A, Ulmasov B, Shah GN, Grubb JH, Sly WS, Christianson DW. Crystal structure of the dimeric extracellular domain of human carbonic anhydrase XII, a bitopic membrane protein overexpressed in certain cancer tumor cells. Proc Natl Acad Sci U S A 2001;98(17):9545–50 [Epub 2001/08/09].

[41] Csaderova L, Debreova M, Radvak P, Stano M, Vrestiakova M, Kopacek J, Pastorekova S, Svastova E. The effect of carbonic anhydrase IX on focal contacts during cell spreading and migration. Front Physiol 2013;4:271 [Epub 2013/10/09].

[42] Mahon BP, Bhatt A, Socorro L, Driscoll JM, Okoh C, Lomelino CL, Mboge MY, Kurian JJ, Tu C, Agbandje-McKenna M, Frost SC, McKenna R. The structure of carbonic anhydrase IX is adapted for low-pH catalysis. Biochemistry 2016;55(33):4642–53 [Epub 2016/07/21].

[43] Wykoff CC, Beasley N, Watson PH, Campo L, Chia SK, English R, Pastorek J, Sly WS, Ratcliffe P, Harris AL. Expression of the hypoxia-inducible and tumor-associated carbonic anhydrases in ductal carcinoma in situ of the breast. Am J Pathol 2001;158(3):1011–9 [Epub 2001/03/10].

[44] Pastorek J, Pastorekova S. Hypoxia-induced carbonic anhydrase IX as a target for cancer therapy: from biology to clinical use. Semin Cancer Biol 2015;31:52–64 [Epub 2014/08/15].

[45] McDonald PC, Dedhar S. Carbonic anhydrase IX (CAIX) as a mediator of hypoxia-induced stress response in cancer cells. Subcell Biochem 2014;75:255–69 [Epub 2013/10/23].

[46] McDonald PC, Winum JY, Supuran CT, Dedhar S. Recent developments in targeting carbonic anhydrase IX for cancer therapeutics. Oncotarget 2012;3(1):84–97 [Epub 2012/02/01].

[47] Parks SK, Chiche J, Pouyssegur J. Disrupting proton dynamics and energy metabolism for cancer therapy. Nat Rev Cancer 2013;13(9):611–23 [Epub 2013/08/24].

[48] Gillies RJ, Robey I, Gatenby RA. Causes and consequences of increased glucose metabolism of cancers. J Nucl Med 2008;49(Suppl 2):24S–42S [Epub 2008/07/01].

[49] Sedlakova O, Svastova E, Takacova M, Kopacek J, Pastorek J, Pastorekova S. Carbonic anhydrase IX, a hypoxia-induced catalytic component of the pH regulating machinery in tumors. Front Physiol 2014;4:400 [Epub 2014/01/11].

[50] Gallagher FA, Sladen H, Kettunen MI, Serrao EM, Rodrigues TB, Wright A, Gill AB, McGuire S, Booth TC, Boren J, McIntyre A, Miller JL, Lee SH, Honess D, Day SE, Hu DE, Howat WJ, Harris AL, Brindle KM. Carbonic anhydrase activity monitored in vivo by hyperpolarized 13C-magnetic resonance spectroscopy demonstrates its importance for pH regulation in tumors. Cancer Res 2015;75(19):4109–18 [Epub 2015/08/08].

[51] Lee SH, McIntyre D, Honess D, Hulikova A, Pacheco-Torres J, Cerdan S, Swietach P, Harris AL, Griffiths JR. Carbonic anhydrase IX is a pH-stat that sets an acidic tumour extracellular pH in vivo. Br J Cancer 2018; 119(5):622–30 [Epub 2018/09/13].

[52] Chiche J, Ilc K, Laferriere J, Trottier E, Dayan F, Mazure NM, Brahimi-Horn MC, Pouyssegur J. Hypoxia-inducible carbonic anhydrase IX and XII promote tumor cell growth by counteracting acidosis through the regulation of the intracellular pH. Cancer Res 2009;69(1):358–68 [Epub 2009/01/02].

[53] Doyen J, Parks SK, Marcie S, Pouyssegur J, Chiche J. Knock-down of hypoxia-induced carbonic anhydrases IX and XII radiosensitizes tumor cells by increasing intracellular acidosis. Front Oncol 2012;2:199 [Epub 2013/01/15].

[54] Parks SK, Cormerais Y, Durivault J, Pouyssegur J. Genetic disruption of the pHi-regulating proteins Na+/H+ exchanger 1 (SLC9A1) and carbonic anhydrase 9 severely reduces growth of colon cancer cells. Oncotarget 2017;8(6):10225–37 [Epub 2017/01/06].

[55] Hsieh MJ, Chen KS, Chiou HL, Hsieh YS. Carbonic anhydrase XII promotes invasion and migration ability of MDA-MB-231 breast cancer cells through the p38 MAPK signaling pathway. Eur J Cell Biol 2010;89(8):598–606 [Epub 2010/05/04].

[56] Kopecka J, Campia I, Jacobs A, Frei AP, Ghigo D, Wollscheid B, Riganti C. Carbonic anhydrase XII is a new therapeutic target to overcome chemoresistance in cancer cells. Oncotarget 2015;6(9):6776–93 [Epub 2015/02/18].

[57] Salaroglio IC, Mujumdar P, Annovazzi L, Kopecka J, Mellai M, Schiffer D, Poulsen SA, Riganti C. Carbonic anhydrase XII inhibitors overcome P-glycoprotein-mediated resistance to temozolomide in glioblastoma. Mol Cancer Ther 2018;17(12):2598–609 [Epub 2018/09/27].

[58] Guerrini G, Criscuoli M, Filippi I, Naldini A, Carraro F. Inhibition of smoothened in breast cancer cells reduces CAXII expression and cell migration. J Cell Physiol 2018;233(12):9799–811 [Epub 2018/08/23].

[59] Guerrini G, Durivault J, Filippi I, Criscuoli M, Monaci S, Pouyssegur J, Naldini A, Carraro F, Parks SK. Carbonic anhydrase XII expression is linked to suppression of sonic hedgehog ligand expression in triple negative breast cancer cells. Biochem Biophys Res Commun 2019;516(2):408–13 [Epub 2019/06/22].

[60] Watson PH, Chia SK, Wykoff CC, Han C, Leek RD, Sly WS, Gatter KC, Ratcliffe P, Harris AL. Carbonic anhydrase XII is a marker of good prognosis in invasive breast carcinoma. Br J Cancer 2003;88(7):1065–70 [Epub 2003/04/03].

[61] Ilie MI, Hofman V, Ortholan C, Ammadi RE, Bonnetaud C, Havet K, Venissac N, Mouroux J, Mazure NM, Pouyssegur J, Hofman P. Overexpression of carbonic anhydrase XII in tissues from resectable non-small cell lung cancers is a biomarker of good prognosis. Int J Cancer 2011;128(7):1614–23 [Epub 2010/06/04].

[62] Haapasalo J, Hilvo M, Nordfors K, Haapasalo H, Parkkila S, Hyrskyluoto A, Rantala I, Waheed A, Sly WS, Pastorekova S, Pastorek J, Parkkila AK. Identification of an alternatively spliced isoform of carbonic anhydrase XII in diffusely infiltrating astrocytic gliomas. Neuro-Oncology 2008;10(2):131–8 [Epub 2008/03/07].

[63] Ochi F, Shiozaki A, Ichikawa D, Fujiwara H, Nakashima S, Takemoto K, Kosuga T, Konishi H, Komatsu S, Okamoto K, Kishimoto M, Marunaka Y, Otsuji E. Carbonic anhydrase XII as an independent prognostic factor in advanced Esophageal squamous cell carcinoma. J Cancer 2015;6(10):922–9 [Epub 2015/09/01].

[64] Chien MH, Ying TH, Hsieh YH, Lin CH, Shih CH, Wei LH, Yang SF. Tumor-associated carbonic anhydrase XII is linked to the growth of primary oral squamous cell carcinoma and its poor prognosis. Oral Oncol 2012; 48(5):417–23 [Epub 2011/12/17].

[65] Reshkin SJ, Bellizzi A, Caldeira S, Albarani V, Malanchi I, Poignee M, Alunni-Fabbroni M, Casavola V, Tommasino M. Na+/H+ exchanger-dependent intracellular alkalinization is an early event in malignant transformation and plays an essential role in the development of subsequent transformation-associated phenotypes. FASEB J 2000;14(14):2185–97 [Epub 2000/10/29].

[66] Aronson PS. Properties of the renal Na+-H+ exchanger. Ann N Y Acad Sci 1985;456:220–8 [Epub 1985/01/01].

[67] Cardone RA, Casavola V, Reshkin SJ. The role of disturbed pH dynamics and the Na+/H+ exchanger in metastasis. Nat Rev Cancer 2005;5(10):786–95 [Epub 2005/09/22].

[68] Hendus-Altenburger R, Wang X, Sjogaard-Frich LM, Pedraz-Cuesta E, Sheftic SR, Bendsoe AH, Page R, Kragelund BB, Pedersen SF, Peti W. Molecular basis for the binding and selective dephosphorylation of Na (+)/H(+) exchanger 1 by calcineurin. Nat Commun 2019;10(1):3489 [Epub 2019/08/04].

[69] Lagarde AE, Franchi AJ, Paris S, Pouyssegur JM. Effect of mutations affecting Na+: H+ antiport activity on tumorigenic potential of hamster lung fibroblasts. J Cell Biochem 1988;36(3):249–60 [Epub 1988/03/01].

[70] Harley W, Floyd C, Dunn T, Zhang XD, Chen TY, Hegde M, Palandoken H, Nantz MH, Leon L, Carraway 3rd KL, Lyeth B, Gorin FA. Dual inhibition of sodium-mediated proton and calcium efflux triggers non-apoptotic cell death in malignant gliomas. Brain Res 2010;1363:159–69 [Epub 2010/09/28].

[71] Neri D, Supuran CT. Interfering with pH regulation in tumours as a therapeutic strategy. Nat Rev Drug Discov 2011;10(10):767–77 [Epub 2011/09/17].

[72] Bell SM, Schreiner CM, Schultheis PJ, Miller ML, Evans RL, Vorhees CV, Shull GE, Scott WJ. Targeted disruption of the murine Nhe1 locus induces ataxia, growth retardation, and seizures. Am J Phys 1999;276(4):C788–95 [Epub 1999/04/13].

[73] Choi I, Aalkjaer C, Boulpaep EL, Boron WF. An electroneutral sodium/bicarbonate cotransporter NBCn1 and associated sodium channel. Nature 2000;405(6786):571–5 [Epub 2000/06/13].

[74] Gross E, Hawkins K, Abuladze N, Pushkin A, Cotton CU, Hopfer U, Kurtz I. The stoichiometry of the electrogenic sodium bicarbonate cotransporter NBC1 is cell-type dependent. J Physiol 2001;531(Pt 3):597–603 [Epub 2001/03/17].

[75] Gorbatenko A, Olesen CW, Boedtkjer E, Pedersen SF. Regulation and roles of bicarbonate transporters in cancer. Front Physiol 2014;5:130 [Epub 2014/05/06].

[76] Boedtkjer E, Moreira JM, Mele M, Vahl P, Wielenga VT, Christiansen PM, Jensen VE, Pedersen SF, Aalkjaer C. Contribution of Na+,HCO3(−)-cotransport to cellular pH control in human breast cancer: a role for the breast cancer susceptibility locus NBCn1 (SLC4A7). Int J Cancer 2013;132(6):1288–99 [Epub 2012/08/22].

[77] Ahmed S, Thomas G, Ghoussaini M, Healey CS, Humphreys MK, Platte R, Morrison J, Maranian M, Pooley KA, Luben R, Eccles D, Evans DG, Fletcher O, Johnson N, dos Santos Silva I, Peto J, Stratton MR, Rahman N, Jacobs K, Prentice R, Anderson GL, Rajkovic A, Curb JD, Ziegler RG, Berg CD, Buys SS, CA MC, Feigelson HS, Calle EE, Thun MJ, Diver WR, Bojesen S, Nordestgaard BG, Flyger H, Dork T, Schurmann P, Hillemanns P, Karstens JH, Bogdanova NV, Antonenkova NN, Zalutsky IV, Bermisheva M, Fedorova S, Khusnutdinova E, Kang D, Yoo KY, Noh DY, Ahn SH, Devilee P, van Asperen CJ, Tollenaar RA, Seynaeve C, Garcia-Closas M, Lissowska J, Brinton L, Peplonska B, Nevanlinna H, Heikkinen T, Aittomaki K, Blomqvist C, Hopper JL, Southey MC, Smith L, Spurdle AB, Schmidt MK, Broeks A, van Hien RR, Cornelissen S, Milne RL, Ribas G, Gonzalez-Neira A, Benitez J, Schmutzler RK, Burwinkel B, Bartram CR, Meindl A, Brauch H, Justenhoven C, Hamann U, Chang-Claude J, Hein R, Wang-Gohrke S, Lindblom A, Margolin S, Mannermaa A, Kosma VM, Kataja V, Olson JE, Wang X, Fredericksen Z, Giles GG, Severi G, Baglietto L, English DR, Hankinson SE, Cox DG, Kraft P, Vatten LJ, Hveem K, Kumle M, Sigurdson A, Doody M, Bhatti P, Alexander BH, Hooning MJ, van den Ouweland AM, Oldenburg RA, Schutte M, Hall P, Czene K, Liu J, Li Y, Cox A, Elliott G, Brock I, Reed MW, Shen CY, Yu JC, Hsu GC, Chen ST, Anton-Culver H, Ziogas A, Andrulis IL, Knight JA, Beesley J, Goode EL, Couch F, Chenevix-Trench G, Hoover RN, Ponder BA, Hunter DJ, Pharoah PD, Dunning AM, Chanock SJ, Easton DF. Newly discovered breast cancer susceptibility loci on 3p24 and 17q23.2. Nat Genet 2009; 41(5):585–90 [Epub 2009/03/31].

[78] Han W, Woo JH, Yu JH, Lee MJ, Moon HG, Kang D, Noh DY. Common genetic variants associated with breast cancer in Korean women and differential susceptibility according to intrinsic subtype. Cancer Epidemiol Biomark Prev 2011;20(5):793–8 [Epub 2011/03/19].

[79] Long J, Shu XO, Cai Q, Gao YT, Zheng Y, Li G, Li C, Gu K, Wen W, Xiang YB, Lu W, Zheng W. Evaluation of breast cancer susceptibility loci in Chinese women. Cancer Epidemiol Biomark Prev 2010;19(9):2357–65 [Epub 2010/08/12].

[80] Sueta A, Ito H, Kawase T, Hirose K, Hosono S, Yatabe Y, Tajima K, Tanaka H, Iwata H, Iwase H, Matsuo K. A genetic risk predictor for breast cancer using a combination of low-penetrance polymorphisms in a Japanese population. Breast Cancer Res Treat 2012;132(2):711–21 [Epub 2011/12/14].

[81] Lee S, Axelsen TV, Jessen N, Pedersen SF, Vahl P, Boedtkjer E. Na(+),HCO3(−)-cotransporter NBCn1 (Slc4a7) accelerates ErbB2-induced breast cancer development and tumor growth in mice. Oncogene 2018;37(41): 5569–84 [Epub 2018/06/17].

[82] Lee S, Axelsen TV, Andersen AP, Vahl P, Pedersen SF, Boedtkjer E. Disrupting Na(+), HCO(3)(−)-cotransporter NBCn1 (Slc4a7) delays murine breast cancer development. Oncogene 2016;35(16):2112–22 [Epub 2015/07/28].

[83] Hulikova A, Harris AL, Vaughan-Jones RD, Swietach P. Regulation of intracellular pH in cancer cell lines under normoxia and hypoxia. J Cell Physiol 2013;228(4):743–52 [Epub 2012/09/06].

[84] McIntyre A, Hulikova A, Ledaki I, Snell C, Singleton D, Steers G, Seden P, Jones D, Bridges E, Wigfield S, Li JL, Russell A, Swietach P, Harris AL. Disrupting hypoxia-induced bicarbonate transport acidifies tumor cells and suppresses tumor growth. Cancer Res 2016;76(13):3744–55 [Epub 2016/05/20].

[85] Parks SK, Pouyssegur J. The Na(+)/HCO3(−) co-transporter SLC4A4 plays a role in growth and migration of colon and breast cancer cells. J Cell Physiol 2015;230(8):1954–63 [Epub 2015/01/23].

[86] Boedtkjer E, Bunch L, Pedersen SF. Physiology, pharmacology and pathophysiology of the pH regulatory transport proteins NHE1 and NBCn1: similarities, differences, and implications for cancer therapy. Curr Pharm Des 2012;18(10):1345–71 [Epub 2012/03/01].

[87] Steinkamp A-D, Seling N, Lee S, Boedtkjer E, Bolm C. Synthesis of N-cyano-substituted sulfilimine and sulfoximine derivatives of S0859 and their biological evaluation as sodium bicarbonate co-transport inhibitors. Med Chem Commun 2015;6(12):2163–9.

[88] Gawenis LR, Bradford EM, Prasad V, Lorenz JN, Simpson JE, Clarke LL, Woo AL, Grisham C, Sanford LP, Doetschman T, Miller ML, Shull GE. Colonic anion secretory defects and metabolic acidosis in mice lacking the NBC1 Na+/HCO3-cotransporter. J Biol Chem 2007;282(12):9042–52 [Epub 2006/12/29].

[89] Srere PA. Complexes of sequential metabolic enzymes. Annu Rev Biochem 1987;56:89–124 [Epub 1987/01/01].

[90] Becker HM, Deitmer JW. Transport metabolons and acid/base balance in tumor cells. Cancers (Basel) 2020;12(4) [Epub 2020/04/11].

[91] Sterling D, Reithmeier RA, Casey JR. A transport metabolon. Functional interaction of carbonic anhydrase II and chloride/bicarbonate exchangers. J Biol Chem 2001;276(51):47886–94 [Epub 2001/10/19].

[92] Morgan PE, Pastorekova S, Stuart-Tilley AK, Alper SL, Casey JR. Interactions of transmembrane carbonic anhydrase, CAIX, with bicarbonate transporters. Am J Physiol Cell Physiol 2007;293(2):C738–48 [Epub 2007/07/27].

[93] Svastova E, Witarski W, Csaderova L, Kosik I, Skvarkova L, Hulikova A, Zatovicova M, Barathova M, Kopacek J, Pastorek J, Pastorekova S. Carbonic anhydrase IX interacts with bicarbonate transporters in lamellipodia and increases cell migration via its catalytic domain. J Biol Chem 2012;287(5):3392–402 [Epub 2011/12/16].

[94] Swayampakula M, McDonald PC, Vallejo M, Coyaud E, Chafe SC, Westerback A, Venkateswaran G, Shankar J, Gao G, Laurent EMN, Lou Y, Bennewith KL, Supuran CT, Nabi IR, Raught B, Dedhar S. The interactome of metabolic enzyme carbonic anhydrase IX reveals novel roles in tumor cell migration and invadopodia/MMP14-mediated invasion. Oncogene 2017;36(45):6244–61 [Epub 2017/07/12].

[95] McDonald PC, Swayampakula M, Dedhar S. Coordinated regulation of metabolic transporters and migration/invasion by carbonic anhydrase IX. Metabolites 2018;8(1) [Epub 2018/03/09].

[96] Ames S, Andring JT, McKenna R, Becker HM. CAIX forms a transport metabolon with monocarboxylate transporters in human breast cancer cells. Oncogene 2020;39(8):1710–23 [Epub 2019/11/15].

[97] Ames S, Pastorekova S, Becker HM. The proteoglycan-like domain of carbonic anhydrase IX mediates non-catalytic facilitation of lactate transport in cancer cells. Oncotarget 2018;9(46):27940–57 [Epub 2018/07/03].

[98] Noor SI, Jamali S, Ames S, Langer S, Deitmer JW, Becker HM. A surface proton antenna in carbonic anhydrase II supports lactate transport in cancer cells. elife 2018;7 [Epub 2018/05/29].

[99] Piermarini PM, Kim EY, Boron WF. Evidence against a direct interaction between intracellular carbonic anhydrase II and pure C-terminal domains of SLC4 bicarbonate transporters. J Biol Chem 2007;282(2): 1409–21 [Epub 2006/11/09].

[100] Stock C, Schwab A. Protons make tumor cells move like clockwork. Pflugers Arch 2009;458(5):981–92 [Epub 2009/05/14].

[101] Martinez-Zaguilan R, Seftor EA, Seftor RE, Chu YW, Gillies RJ, Hendrix MJ. Acidic pH enhances the invasive behavior of human melanoma cells. Clin Exp Metastasis 1996;14(2):176–86 [Epub 1996/03/01].

[102] Robey IF, Baggett BK, Kirkpatrick ND, Roe DJ, Dosescu J, Sloane BF, Hashim AI, Morse DL, Raghunand N, Gatenby RA, Gillies RJ. Bicarbonate increases tumor pH and inhibits spontaneous metastases. Cancer Res 2009;69(6):2260–8 [Epub 2009/03/12].

[103] Schlappack OK, Zimmermann A, Hill RP. Glucose starvation and acidosis: effect on experimental metastatic potential, DNA content and MTX resistance of murine tumour cells. Br J Cancer 1991;64(4):663–70 [Epub 1991/10/01].

[104] Corbet C, Bastien E, Santiago de Jesus JP, Dierge E, Martherus R, Vander Linden C, Doix B, Degavre C, Guilbaud C, Petit L, Michiels C, Dessy C, Larondelle Y, Feron O. TGFbeta2-induced formation of lipid droplets supports acidosis-driven EMT and the metastatic spreading of cancer cells. Nat Commun 2020;11(1):454 [Epub 2020/01/25].

[105] Menard JA, Christianson HC, Kucharzewska P, Bourseau-Guilmain E, Svensson KJ, Lindqvist E, Indira Chandran V, Kjellen L, Welinder C, Bengzon J, Johansson MC, Belting M. Metastasis stimulation by hypoxia and acidosis-induced extracellular lipid uptake is mediated by proteoglycan-dependent endocytosis. Cancer Res 2016;76(16):4828–40 [Epub 2016/05/21].

[106] Estrella V, Chen T, Lloyd M, Wojtkowiak J, Cornnell HH, Ibrahim-Hashim A, Bailey K, Balagurunathan Y, Rothberg JM, Sloane BF, Johnson J, Gatenby RA, Gillies RJ. Acidity generated by the tumor microenvironment drives local invasion. Cancer Res 2013;73(5):1524–35 [Epub 2013/01/05].

[107] Rohani N, Hao L, Alexis MS, Joughin BA, Krismer K, Moufarrej MN, Soltis AR, Lauffenburger DA, Yaffe MB, Burge CB, Bhatia SN, Gertler FB. Acidification of tumor at stromal boundaries drives transcriptome alterations associated with aggressive phenotypes. Cancer Res 2019;79(8):1952–66 [Epub 2019/02/14].

[108] Helmlinger G, Yuan F, Dellian M, Jain RK. Interstitial pH and pO2 gradients in solid tumors in vivo: high-resolution measurements reveal a lack of correlation. Nat Med 1997;3(2):177–82 [Epub 1997/02/01].

[109] Lou Y, McDonald PC, Oloumi A, Chia S, Ostlund C, Ahmadi A, Kyle A, Auf Dem Keller U, Leung S, Huntsman D, Clarke B, Sutherland BW, Waterhouse D, Bally M, Roskelley C, Overall CM, Minchinton A, Pacchiano F, Carta F, Scozzafava A, Touisni N, Winum JY, Supuran CT, Dedhar S. Targeting tumor hypoxia: suppression of breast tumor growth and metastasis by novel carbonic anhydrase IX inhibitors. Cancer Res 2011;71(9):3364–76 [Epub 2011/03/19].

[110] Lock FE, McDonald PC, Lou Y, Serrano I, Chafe SC, Ostlund C, Aparicio S, Winum JY, Supuran CT, Dedhar S. Targeting carbonic anhydrase IX depletes breast cancer stem cells within the hypoxic niche. Oncogene 2013;32(44):5210–9 [Epub 2012/12/05].

[111] Gieling RG, Babur M, Mamnani L, Burrows N, Telfer BA, Carta F, Winum JY, Scozzafava A, Supuran CT, Williams KJ. Antimetastatic effect of sulfamate carbonic anhydrase IX inhibitors in breast carcinoma xenografts. J Med Chem 2012;55(11):5591–600 [Epub 2012/05/25].

[112] Kajanova I, Zatovicova M, Jelenska L, Sedlakova O, Barathova M, Csaderova L, Debreova M, Lukacikova L, Grossmannova K, Labudova M, Golias T, Svastova E, Ludwig A, Muller P, Vojtesek B, Pastorek J, Pastorekova S. Impairment of carbonic anhydrase IX ectodomain cleavage reinforces tumorigenic and metastatic phenotype of cancer cells. Br J Cancer 2020;122:1590–603 [Epub 2020/03/27].

[113] Hedlund EE, McDonald PC, Nemirovsky O, Awrey S, Jensen LDE, Dedhar S. Harnessing induced essentiality: targeting carbonic anhydrase IX and angiogenesis reduces lung metastasis of triple negative breast cancer Xenografts. Cancers (Basel) 2019;11(7) [Epub 2019/07/20].

[114] Huttenlocher A, Horwitz AR. Integrins in cell migration. Cold Spring Harb Perspect Biol 2011;3(9):a005074 [Epub 2011/09/03].

[115] Eddy RJ, Weidmann MD, Sharma VP, Condeelis JS. Tumor cell Invadopodia: invasive protrusions that orchestrate metastasis. Trends Cell Biol 2017;27(8):595–607 [Epub 2017/04/17].

[116] Millard TH, Sharp SJ, Machesky LM. Signalling to actin assembly via the WASP (Wiskott-Aldrich syndrome protein)-family proteins and the Arp2/3 complex. Biochem J 2004;380(Pt 1):1–17 [Epub 2004/03/26].

[117] Magalhaes MA, Larson DR, Mader CC, Bravo-Cordero JJ, Gil-Henn H, Oser M, Chen X, Koleske AJ, Condeelis J. Cortactin phosphorylation regulates cell invasion through a pH-dependent pathway. J Cell Biol 2011;195(5):903–20 [Epub 2011/11/23].

[118] Busco G, Cardone RA, Greco MR, Bellizzi A, Colella M, Antelmi E, Mancini MT, Dell'Aquila ME, Casavola V, Paradiso A, Reshkin SJ. NHE1 promotes invadopodial ECM proteolysis through acidification of the peri-invadopodial space. FASEB J 2010;24(10):3903–15 [Epub 2010/06/16].

[119] Gallagher SM, Castorino JJ, Philp NJ. Interaction of monocarboxylate transporter 4 with beta1-integrin and its role in cell migration. Am J Phys Cell Phys 2009;296(3):C414–21 [Epub 2008/12/17].

[120] Denny WA, Wilson WR. Considerations for the design of nitrophenyl mustards as agents with selective toxicity for hypoxic tumor cells. J Med Chem 1986;29(6):879–87 [Epub 1986/06/01].

[121] Kolosenko I, Avnet S, Baldini N, Viklund J, De Milito A. Therapeutic implications of tumor interstitial acidification. Semin Cancer Biol 2017;43:119–33 [Epub 2017/02/12].

[122] Wojtkowiak JW, Verduzco D, Schramm KJ, Gillies RJ. Drug resistance and cellular adaptation to tumor acidic pH microenvironment. Mol Pharm 2011;8(6):2032–8 [Epub 2011/10/11].

[123] Pillai SR, Damaghi M, Marunaka Y, Spugnini EP, Fais S, Gillies RJ. Causes, consequences, and therapy of tumors acidosis. Cancer Metastasis Rev 2019;38(1–2):205–22 [Epub 2019/03/27].

[124] Chafe SC, McDonald PC, Saberi S, Nemirovsky O, Venkateswaran G, Burugu S, Gao D, Delaidelli A, Kyle AH, Baker JHE, Gillespie JA, Bashashati A, Minchinton AI, Zhou Y, Shah SP, Dedhar S. Targeting hypoxia-induced carbonic anhydrase IX enhances immune checkpoint blockade locally and systemically. Cancer Immunol Res 2019; [Epub 2019/05/16].

[125] Bryant JL, Gieling RG, Meredith SL, Allen TJ, Walker L, Telfer BA, Supuran CT, Williams KJ, White A. Novel carbonic anhydrase IX-targeted therapy enhances the anti-tumour effects of cisplatin in small cell lung cancer. Int J Cancer 2018;142(1):191–201 [Epub 2017/09/15].

[126] Boyd NH, Walker K, Fried J, Hackney JR, McDonald PC, Benavides GA, Spina R, Audia A, Scott SE, Libby CJ, Tran AN, Bevensee MO, Griguer C, Nozell S, Gillespie GY, Nabors B, Bhat KP, Bar EE, Darley-Usmar V, Xu B, Gordon E, Cooper SJ, Dedhar S, Hjelmeland AB. Addition of carbonic anhydrase 9 inhibitor SLC-0111 to temozolomide treatment delays glioblastoma growth in vivo. JCI Insight 2017;2(24) [Epub 2017/12/22].

[127] McDonald PC, Chafe SC, Brown WS, Saberi S, Swayampakula M, Venkateswaran G, Nemirovsky O, Gillespie JA, Karasinska JM, Kalloger SE, Supuran CT, Schaeffer DF, Bashashati A, Shah SP, Topham JT, Yapp DT, Li J, Renouf DJ, Stanger BZ, Dedhar S. Regulation of pH by carbonic anhydrase 9 mediates survival of pancreatic cancer cells with activated KRAS in response to hypoxia. Gastroenterology 2019 [Epub 2019/05/13].

[128] Vander Heiden MG, DeBerardinis RJ. Understanding the intersections between metabolism and cancer biology. Cell 2017;168(4):657–69 [Epub 2017/02/12].

[129] McIntyre A, Harris AL. Metabolic and hypoxic adaptation to anti-angiogenic therapy: a target for induced essentiality. EMBO Mol Med 2015;7(4):368–79 [Epub 2015/02/24].

[130] Samanta D, Gilkes DM, Chaturvedi P, Xiang L, Semenza GL. Hypoxia-inducible factors are required for chemotherapy resistance of breast cancer stem cells. Proc Natl Acad Sci U S A 2014;111(50):E5429–38 [Epub 2014/12/03].

[131] Damaghi M, Tafreshi NK, Lloyd MC, Sprung R, Estrella V, Wojtkowiak JW, Morse DL, Koomen JM, Bui MM, Gatenby RA, Gillies RJ. Chronic acidosis in the tumour microenvironment selects for overexpression of LAMP2 in the plasma membrane. Nat Commun 2015;6:8752 [Epub 2015/12/15].

[132] Brand A, Singer K, Koehl GE, Kolitzus M, Schoenhammer G, Thiel A, Matos C, Bruss C, Klobuch S, Peter K, Kastenberger M, Bogdan C, Schleicher U, Mackensen A, Ullrich E, Fichtner-Feigl S, Kesselring R, Mack M, Ritter U, Schmid M, Blank C, Dettmer K, Oefner PJ, Hoffmann P, Walenta S, Geissler EK, Pouyssegur J, Villunger A, Steven A, Seliger B, Schreml S, Haferkamp S, Kohl E, Karrer S, Berneburg M, Herr W, Mueller-Klieser W, Renner K, Kreutz M. LDHA-associated lactic acid production blunts tumor immunosurveillance by T and NK cells. Cell Metab 2016;24(5):657–71 [Epub 2016/09/20].

[133] Fischer K, Hoffmann P, Voelkl S, Meidenbauer N, Ammer J, Edinger M, Gottfried E, Schwarz S, Rothe G, Hoves S, Renner K, Timischl B, Mackensen A, Kunz-Schughart L, Andreesen R, Krause SW, Kreutz M. Inhibitory effect of tumor cell-derived lactic acid on human T cells. Blood 2007;109(9):3812–9 [Epub 2007/01/27].

[134] Cascone T, McKenzie JA, Mbofung RM, Punt S, Wang Z, Xu C, Williams LJ, Bristow CA, Carugo A, Peoples MD, Li L, Karpinets T, Huang L, Malu S, Creasy C, Leahey SE, Chen J, Chen Y, Pelicano H, Bernatchez C, Gopal YNV, Heffernan TP, Hu J, Wang J, Amaria RN, Garraway LA, Huang P, Yang P, Wistuba II, Woodman SE, Roszik J, Davis RE, Davies MA, Heymach JV, Hwu P, Peng W. Increased tumor glycolysis characterizes immune resistance to adoptive T cell therapy. Cell Metab 2018;27(5):977–87 e974. [Epub 2018/04/10].

[135] Pilon-Thomas S, Kodumudi KN, El-Kenawi AE, Russell S, Weber AM, Luddy K, Damaghi M, Wojtkowiak JW, Mule JJ, Ibrahim-Hashim A, Gillies RJ. Neutralization of tumor acidity improves antitumor responses to immunotherapy. Cancer Res 2016;76(6):1381–90 [Epub 2016/01/01].

[136] Renner K, Bruss C, Schnell A, Koehl G, Becker HM, Fante M, Menevse AN, Kauer N, Blazquez R, Hacker L, Decking SM, Bohn T, Faerber S, Evert K, Aigle L, Amslinger S, Landa M, Krijgsman O, Rozeman EA, Brummer C, Siska PJ, Singer K, Pektor S, Miederer M, Peter K, Gottfried E, Herr W, Marchiq I, Pouyssegur J, Roush WR, Ong S, Warren S, Pukrop T, Beckhove P, Lang SA, Bopp T, Blank CU, Cleveland JL, Oefner PJ, Dettmer K, Selby M, Kreutz M. Restricting glycolysis preserves T cell effector functions and augments checkpoint therapy. Cell Rep 2019;29(1):135–50 e139. [Epub 2019/10/03].

[137] Hugo W, Zaretsky JM, Sun L, Song C, Moreno BH, Hu-Lieskovan S, Berent-Maoz B, Pang J, Chmielowski B, Cherry G, Seja E, Lomeli S, Kong X, Kelley MC, Sosman JA, Johnson DB, Ribas A, Lo RS. Genomic and Transcriptomic features of response to anti-PD-1 therapy in metastatic melanoma. Cell 2016;165(1):35–44 [Epub 2016/03/22].

[138] Gide TN, Quek C, Menzies AM, Tasker AT, Shang P, Holst J, Madore J, Lim SY, Velickovic R, Wongchenko M, Yan Y, Lo S, Carlino MS, Guminski A, Saw RPM, Pang A, McGuire HM, Palendira U, Thompson JF, Rizos H, Silva IPD, Batten M, Scolyer RA, Long GV, Wilmott JS. Distinct immune cell populations define response to anti-PD-1 monotherapy and anti-PD-1/anti-CTLA-4 combined therapy. Cancer Cell 2019;35(2):238–55 e236. [Epub 2019/02/13].

[139] Lamonte G, Tang X, Chen JL, Wu J, Ding CK, Keenan MM, Sangokoya C, Kung HN, Ilkayeva O, Boros LG, Newgard CB, Chi JT. Acidosis induces reprogramming of cellular metabolism to mitigate oxidative stress. Cancer Metab 2013;1(1):23 [Epub 2013/12/24].

[140] Corbet C, Draoui N, Polet F, Pinto A, Drozak X, Riant O, Feron O. The SIRT1/HIF2alpha axis drives reductive glutamine metabolism under chronic acidosis and alters tumor response to therapy. Cancer Res 2014; 74(19):5507–19 [Epub 2014/08/03].

[141] Wilde BR, Ye Z, Lim TY, Ayer DE. Cellular acidosis triggers human MondoA transcriptional activity by driving mitochondrial ATP production. elife 2019;8 [Epub 2019/02/06].

[142] Turkcan S, Kiru L, Naczynski DJ, Sasportas LS, Pratx G. Lactic acid accumulation in the tumor microenvironment suppresses (18)F-FDG uptake. Cancer Res 2019;79(2):410–9 [Epub 2018/12/05].

[143] Kernstine KH, Faubert B, Do QN, Rogers TJ, Hensley CT, Cai L, Torrealba J, Oliver D, Wachsmann JW, Lenkinski RE, Malloy CR, Deberardinis RJ. Does tumor FDG-PET avidity represent enhanced glycolytic metabolism in non-small cell lung cancer? Ann Thorac Surg 2020;109(4):1019–25 [Epub 2019/12/18].

[144] Corbet C, Pinto A, Martherus R, Santiago de Jesus JP, Polet F, Feron O. Acidosis drives the reprogramming of fatty acid metabolism in cancer cells through changes in mitochondrial and histone acetylation. Cell Metab 2016;24(2):311–23 [Epub 2016/08/11].

[145] Morais R, Zinkewich-Peotti K, Parent M, Wang H, Babai F, Zollinger M. Tumor-forming ability in athymic nude mice of human cell lines devoid of mitochondrial DNA. Cancer Res 1994;54(14):3889–96 [Epub 1994/07/15].

[146] Cavalli LR, Varella-Garcia M, Liang BC. Diminished tumorigenic phenotype after depletion of mitochondrial DNA. Cell Growth Differ 1997;8(11):1189–98 [Epub 1998/01/07].

[147] Weinberg F, Hamanaka R, Wheaton WW, Weinberg S, Joseph J, Lopez M, Kalyanaraman B, Mutlu GM, Budinger GR, Chandel NS. Mitochondrial metabolism and ROS generation are essential for Kras-mediated tumorigenicity. Proc Natl Acad Sci U S A 2010;107(19):8788–93 [Epub 2010/04/28].

[148] Tan AS, Baty JW, Dong LF, Bezawork-Geleta A, Endaya B, Goodwin J, Bajzikova M, Kovarova J, Peterka M, Yan B, Pesdar EA, Sobol M, Filimonenko A, Stuart S, Vondrusova M, Kluckova K, Sachaphibulkij K, Rohlena J, Hozak P, Truksa J, Eccles D, Haupt LM, Griffiths LR, Neuzil J, Berridge MV. Mitochondrial genome acquisition restores respiratory function and tumorigenic potential of cancer cells without mitochondrial DNA. Cell Metab 2015;21(1):81–94 [Epub 2015/01/08].

[149] Khacho M, Tarabay M, Patten D, Khacho P, MacLaurin JG, Guadagno J, Bergeron R, Cregan SP, Harper ME, Park DS, Slack RS. Acidosis overrides oxygen deprivation to maintain mitochondrial function and cell survival. Nat Commun 2014;5:3550 [Epub 2014/04/02].

[150] Rouault TA. Mammalian iron-sulphur proteins: novel insights into biogenesis and function. Nat Rev Mol Cell Biol 2015;16(1):45–55 [Epub 2014/11/27].

[151] Tamir S, Paddock ML, Darash-Yahana-Baram M, Holt SH, Sohn YS, Agranat L, Michaeli D, Stofleth JT, Lipper CH, Morcos F, Cabantchik IZ, Onuchic JN, Jennings PA, Mittler R, Nechushtai R. Structure-function analysis of NEET proteins uncovers their role as key regulators of iron and ROS homeostasis in health and disease. Biochim Biophys Acta 2015;1853(6):1294–315 [Epub 2014/12/03].

[152] Darash-Yahana M, Pozniak Y, Lu M, Sohn YS, Karmi O, Tamir S, Bai F, Song L, Jennings PA, Pikarsky E, Geiger T, Onuchic JN, Mittler R, Nechushtai R. Breast cancer tumorigenicity is dependent on high expression levels of NAF-1 and the lability of its Fe-S clusters. Proc Natl Acad Sci U S A 2016;113(39):10890–5 [Epub 2016/09/14].

[153] Li Z, Jiang L, Chew SH, Hirayama T, Sekido Y, Toyokuni S. Carbonic anhydrase 9 confers resistance to ferroptosis/apoptosis in malignant mesothelioma under hypoxia. Redox Biol 2019;26:101297 [Epub 2019/08/24].

[154] McDonald PC, Chia S, Bedard PL, Chu Q, Lyle M, Tang L, Singh M, Zhang Z, Supuran CT, Renouf DJ, Dedhar S. A phase 1 study of SLC-0111, a novel inhibitor of carbonic anhydrase IX, in patients with advanced solid Tumors. Am J Clin Oncol 2020; [Epub 2020/04/07].

CHAPTER

3

Use of pH-interfering agents as chemosensitizers: Clinical studies survey

Joonas Haapasalo[a,b], *Kristiina Nordfors*[a,c], *and Seppo Parkkila*[d,e]

[a]The Hospital for Sick Children, Arthur and Sonia Labatt Brain Tumor Research Center, Toronto, ON, Canada [b]Tampere University Hospital, Unit of Neurosurgery, Tampere, Finland [c]Division of Hematology and Oncology, The Hospital for Sick Children, Toronto, ON, Canada [d]Faculty of Medicine and Health Technology, Tampere University, Tampere, Finland [e]Fimlab Ltd, Tampere University Hospital, Tampere, Finland

Abstract

It has been widely studied that cancer thrives in hypoxic conditions. The importance of tumor metabolism was already discovered in the 1930s by Otto H. Warburg, and the Warburg effect is described by the ability of cancer cells to ferment sugars independently of the oxygen levels. Yet, there is still a lack of understanding of the mechanisms behind the cancer cell adaptation and survival, e.g., by sugar fermentation with lactate production, extracellular acidification, and finally, the survival of tumor cells in a hostile microenvironment. As a result, there has been growing interest in novel pH interfering agents that could increase the antitumoral effect of chemotherapy or even prevent tumorigenesis. It has been proposed that acidic intratumoral environment/low extracellular pH is one of the key features in tumor chemoresistance. This is theorized as due to excessive amounts of H^+, which in turn leads to protonation/neutralization of chemotherapeutic agents, leading to failure to reach their intracellular targets. There is a growing body of evidence about the mechanisms in preclinical models, but data from large clinical trials are still limited. In this chapter, we introduce both recent and current studies on pH-interfering agents that have been investigated in clinical trials.

Abbreviations

AZA	acetazolamide
CA	carbonic anhydrase
CAI	carbonic anhydrase inhibitor
FDA	Food and Drug Administration
H2RA	histamine receptor-2 antagonist

https://doi.org/10.1016/B978-0-12-820701-7.00012-9

HNSCC head and neck squamous cell carcinomas
MCT monocarboxylate transporter
MGMT O6-methylguanine-DNA methyltransferase
NHE Na^+/H^+ exchanger
PPI proton pump inhibitor
SLC solute carrier
TMZ temozolomide

Conflict of interest

No potential conflicts of interest were disclosed.

Carbonic anhydrases

Carbonic anhydrases (CAs) catalyze the reversible conversion of carbon dioxide and water to bicarbonate ion and proton, and their functions have been studied for almost 100 years. These enzymes are fundamental in normal physiological processes in organisms, and their roles have been investigated extensively. At least CA II, IX, and XII have been associated with neoplastic growth in various tumors [1–8], and their role as clinical prognosticators has been shown in several cancers, e.g., in brain tumors [3, 4, 9].

There are a few already-completed or ongoing clinical trials for carbonic anhydrase inhibitors (CAIs). One of the first CAIs studied was indisulam (E7070). It was reported to be an inhibitor of cell cycle arrest in the G1 phase, but it is also a sulfonamide-type CAI suggested to affect acid-base balance, although the mechanism of the drug is not completely understood [10]. Successful in vitro and in vivo experiments in colon carcinoma with indisulam [10] led to five phase I studies assessing the use of indisulam in patients with advanced cancer, and several phase II studies, including patients with colorectal, breast, head and neck, and non-small-cell lung carcinomas, metastatic melanoma, and renal cell cancer [11–22]. These studies showed that dose-limiting toxicities were reversible neutropenia and thrombocytopenia, and maximum-tolerated dose was established at $800\,mg\,m^{-2}$ when given as a one-hour infusion once every 3 weeks. In some of these studies, indisulam was combined with other cytotoxic drugs, and it was shown to have only minor single agent activity [14, 16, 18, 19, 21]. Siegel-Lakhai and coworkers [23] studied the combination of indisulam with capecitabine and found that it was well tolerated during the first cycle of treatment. Furthermore, the results of phase II open-label study of indisulam, idarubicin, and cytarabine showed that the combination of drugs with relapsed and refractory acute myeloid leukemia is effective and largely well tolerated [24].

Acetazolamide (AZA) is a pan-CAI. The FDA-approved indications for AZA are glaucoma, idiopathic intracranial hypertension, congestive heart failure, altitude sickness, periodic paralysis, and epilepsy, and several off-label indications have been described [25]. Acute clinical doses of AZA reduce body fluid compartments, leading to a moderate isosmotic hypovolemia and metabolic acidosis [26]. Common side effects include, e.g., fatigue, nausea, vomiting, and paresthesia. Since AZA has been used for decades in various indications, the repositioning of the drug to cancer therapy would be attractive. It has thus been assessed in preclinical settings in combination with other antitumor drugs [27–31].

In clinical trials for neoplastic disease, the dose-limiting side effects of AZA are being studied in patients with malignant astrocytoma treated with temozolomide (TMZ) (https://clinicaltrials.gov/ct2/show/NCT03011671). The study population are patients who have O6-methylguanine-DNA methyltransferase (MGMT) methylated grade III - IV astrocytoma and undergo treatment with standard adjuvant TMZ after completing treatment with TMZ and ionizing radiation. The recruitment for this trial is ongoing. Furthermore, there is an ongoing phase I study with a prospective, multicenter, nonrandomized, open-label extension cohort, evaluating the safety of AZA in combination with platinum- and etoposide-based radiochemotherapy in patients with localized small-cell lung cancer (https://clinicaltrials.gov/ct2/show/NCT03467360). From these two trials there are no published results yet available.

SLC-0111 is an ureido-benzenesulfonamide, which efficiently inhibits both CA IX and CA XII. Preclinical studies have shown that SLC-0111 sensitizes cancer cells to traditional cytotoxic drugs [32, 33]. Clinical studies of SLC-0111 include a trial on patients with advanced solid tumors (https://clinicaltrials.gov/ct2/show/NCT02215850). This phase I multicenter, open-label study investigated the safety, tolerability, and pharmacokinetics of single agent SLC-0111, but no results have been published yet. Furthermore, there is an ongoing recruitment for a phase Ib study with a combination of SLC-0111 and gemcitabine for metastatic pancreatic ductal cancer in subjects positive for CA IX (https://clinicaltrials.gov/ct2/show/NCT03450018). Data will be collected on safety, tolerability, pharmacokinetics/dynamics, and tumor response.

Proton pump inhibitors

One of the popular hypotheses has been that anticancer treatment regimens could include drugs, which reduce the excess acidity of cancer cell microenvironment. Therefore, proton pump inhibitors (PPIs) have been in the frontline of these investigations. Several PPIs have been introduced to market. They are widely considered well tolerated with very few side effects, and they represent the most effective drugs for peptic diseases [34]. Previous literature includes several preclinical in vitro and in vivo trials suggesting a role for PPIs in the regulation of tumor pH homeostasis and their applicability as potential anticancer drugs. Based on data from clinical cancer trials PPIs have been proposed for both monotherapy and combination therapy with other chemotherapeutics [35–39]. In addition, clinical studies of companion animals with spontaneous tumors supported the use of PPIs as chemosensitizers in standard treatment [40] or metronomic therapy [41].

There have been few clinical trials on the use of PPIs in cancer. A translational study conducted by Ferrari et al. [39] in human osteosarcoma showed that pretreatment with a PPI drug esomeprazole increased the focal effect of chemotherapy based on methotrexate, cisplatin, and adriamycin, as expressed by percentage of tumor necrosis. This was especially seen in chondroblastic osteosarcoma. However, the trend of higher percentage of good responders in the PPI pretreatment group was not statistically significant.

The efficacy of PPI drugs in patients with metastatic breast cancer has also been investigated with combination of docetaxel and cisplatin [42]. A significant difference was observed between the patients who had taken PPI in addition to chemotherapeutics and those who had taken only chemotherapeutics with objective response rate (67.7% vs. 46.9%) and median time to progression (9.7 months vs. 8.7 months), respectively. The authors concluded that intermittent high-dose PPI enhances the antitumor effects of chemotherapy without evidence of additional toxicity, and thus validation in a multicenter, randomized, phase III trial is warranted [42].

Papagerakis et al. [43] assessed retrospectively the usage of PPIs and histamine receptor-2 antagonists (H2RA) (mono or combination therapy) and treatment outcomes in 596 patients with head and neck squamous cell carcinomas (HNSCC). Their findings suggest that the routine use of PPIs and H2RA has therapeutic benefits evaluated in Cox models for overall survival and also H2RA alone was beneficial for recurrence free survival. Since gastroesophageal reflux often plays a role in the development of HNSCC and both PPIs and H2RA are used to treat reflux disease, these results might not be generalized to other cancers.

Monocarboxylate transporters

Monocarboxylate transporters (MCTs) are members of the solute carrier (SLC) family of proteins [44]. They are expressed in a wide variety of normal tissues and take part in key metabolic pathways. Of the 14 isoforms, MCTs 1–4 have been extensively studied, because they can transport L-lactate and pyruvate across the plasma membrane. It is also well known that MCTs form metabolon systems, enhancing lactate transport, with different CA isozymes and even with noncatalytic CA-related proteins [45]. MCT inhibitors have been proposed as good candidates for antitumor drugs since MCT expression has been reported in various human tumors, including, e.g., breast, colon, bladder, and prostate carcinomas, and gliomas [46]. Both in vitro and in vivo preclinical studies have shown that ablation of MCT1 or MCT4 can halt the proliferative capability of tumor cells [47], and MCT inhibition can be augmented with metformin and phenformin used widely in diabetes mellitus [48]. Unfortunately, the efforts to develop pan-MCT- or MCT4-specific inhibitors have not been successful.

MCT inhibitors are not yet in clinical practice. There is one small-molecule MCT inhibitor specific for MCT1 (AZD3965), which is being tested in a clinical trial for advanced adult diffuse large B-cell lymphoma and Burkitt lymphoma (https://clinicaltrials.gov/ct2/show/NCT01791595). The preliminary results presented conclude that AZD3965 can be administered at effective doses (maximum tolerated dose: 20 mg p.o.). The dose-limiting toxicities include reversible changes in retinal function [49]. The investigation is ongoing and the estimated study completion date is in 2021.

Ionophores

Ionophores are ion carries, which can bind noncovalently with ions, and thus transport them across the cell membrane [50]. They are crucial in maintaining the cell homeostasis and cell survival. It has been reported that ion transport is altered in cancer and ionophores can further affect acid base balance in neoplastic cells [50, 51].

Nigericin is a drug used as an antibiotic derived from *Streptomyces hygroscopicus*. It is an antiporter of K^+/H^+ and raises the pH of acidic compartments [52]. Cancer cells have lower extracellular pH compared to intracellular pH, leading to accelerated invasion capability and cell migration. Nigericin may have a role in transporting more H^+ ions into the cells, causing an increased intracellular acidity and thereby cancer cell death, and it has been shown to sensitize nasopharyngeal carcinoma cells to cisplatin both in vitro and in vivo [52, 53]. To our knowledge there are neither clinical trials nor other comprehensive studies discussing nigericin as a potential or existing chemosensitizing drug.

Salinomycin is a polyether antibiotic isolated from *Streptomyces albus* strain that participates in the transportation of K^+ ions through cell membranes, eventually leading to increased intracellular Ca^{2+} concentration. Researchers have found that salinomycin can select the cancerous cells, and these findings resulted in several studies showing its use as an apoptosis inducer. It has also been shown to possess an inhibiting effect on drug transporters and thus it decreases chemoresistance [52]. Case reports from clinical pilot studies showed that salinomycin, administered to cancer patients intravenously at a dose of 200–250 μg/kg every second day for up to 4 weeks, was generally well tolerated, and led to partial responses or stable disease in few patients with metastatic solid tumors [54]. However, no larger clinical trials have been conducted to assess the use of salinomycin in cancer.

Obatoclax is an apoptosis-inducing compound with ionophoretic activity [55]. It induces apoptosis via modulating Bcl2 family proteins and has been shown to increase cancer cell death, especially in leukemia patients. Obatoclax mesylate has been evaluated in several independent, nonrandomized early-phase clinical trials. They have shown that obatoclax mesylate can be safely administered to patients with a variety of hematological malignancies, including acute myeloid leukemia, chronic lymphocytic leukemia, and lymphoma [55–57]. The clinical studies in phase I-II evaluated the benefit of this mechanism in chronic lymphocytic leukemia and small-cell lung cancer with carboplatin and etoposide chemotherapy [58]. Obatoclax was well tolerated when added to carboplatin/etoposide in first-line treatment of extensive-stage small-cell lung cancer, but failed to significantly improve objective response rate, progression-free survival, or overall survival [58]. At present, there are no reported ongoing clinical trials for obatoclax (www.clinicaltrials.gov).

Other pH-interfering agents

The potential role of Na^+/H^+ exchangers (NHE) and bicarbonate transporters has been widely investigated in preclinical studies. Indeed, both bicarbonate transporter and NHE-1 represent attractive targets in cancer therapy, e.g., in glioma, hepatocellular carcinoma, and breast cancer [59–62]. NHE-1 inhibition seems to sensitize cancer cells to paclitaxel [63], imatinib [64], doxorubicin [65], and cisplatin [66]. Amiloride, one of the first NHE inhibitors, has a direct antitumor and antimetastatic effect both in vitro and in vivo [67, 68]. It has safely been used as a diuretic against hypertension and congestive heart failure. Another specific NHE-1 inhibitor is cariporide. It has shown antitumor effect [69], and in addition, the ability to overcome drug resistance and metastasis [63]. There are clinical trials that indicate cariporide generally well tolerated. Nevertheless, the promising results remain up to date unproven in clinical cancer trials.

Conclusion

There is high demand for novel treatment modalities in cancer. Hypoxia and acidic cancer microenvironment are hallmarks of most cases of aggressive cancers. Therefore, it would be important to find agents that modify intratumoral pH with sufficient clinical tolerance. Previous literature describes several promising preclinical studies showing the potential of pH interfering agents as chemosensitizers. Unfortunately, we are still lacking many large clinical trials to objectively investigate both clinical efficiency and safety of the new drugs targeting pH regulation of tumors. Only the comprehensive data from large trials would finally help us to assess the real feasibility of these drugs for the help of cancer patients.

Acknowledgments

We thank Ncedile Mankahla M.D. for commenting on the manuscript. This chapter preparation was supported by the grants of Finnish Pediatric Research Foundation (JH, KN), Pediatric Cancer Research Foundation Väre (Lasten syöpäsäätiö Väre) (JH), Sigrid Juselius Foundation (KN), Jane & Aatos Erkko Foundation (SP), and Academy of Finland (SP).

References

[1] Ivanov S, Liao SY, Ivanova A, Danilkovitch-Miagkova A, Tarasova N, Weirich G, Merrill MJ, Proescholdt MA, Oldfield EH, Lee J, Zavada J, Waheed A, Sly W, Lerman MI, Stanbridge EJ. Expression of hypoxia-inducible cell-surface transmembrane carbonic anhydrases in human cancer. Am J Pathol 2001;158:905–19.

[2] Pastorekova S, Parkkila S, Zavada J. Tumor-associated carbonic anhydrases and their clinical significance. Adv Clin Chem 2006;42:167–216.

[3] Haapasalo JA, Nordfors KM, Hilvo M, Rantala IJ, Soini Y, Parkkila AK, Pastoreková S, Pastorek J, Parkkila SM, Haapasalo HK. Expression of carbonic anhydrase IX in astrocytic tumors predicts poor prognosis. Clin Cancer Res 2006;12:473–7.

[4] Nordfors K, Haapasalo J, Korja M, Niemelä A, Laine J, Parkkila AK, Pastorekova S, Pastorek J, Waheed A, Sly WS, Parkkila S, Haapasalo H. The tumour-associated carbonic anhydrases CA II, CA IX and CA XII in a group of medulloblastomas and supratentorial primitive neuroectodermal tumours: an association of CA IX with poor prognosis. BMC Cancer 2010;10:148.

[5] Supuran CT. Carbonic anhydrase inhibition and the management of hypoxic tumors. Metabolites 2017;7:48.

[6] Nocentini A, Supuran CT. Carbonic anhydrase inhibitors as antitumor/antimetastatic agents: a patent review (2008-2018). Expert Opin Ther Pat 2018;28:729–40.

[7] Mboge MY, Mahon BP, McKenna R, Frost SC. Carbonic anhydrases: role in pH control and cancer. Metabolites 2018;8:19.

[8] Zamanova S, Shabana AM, Mondal UK, Ilies MA. Carbonic anhydrases as disease markers. Expert Opin Ther Pat 2019;29:509–33.

[9] van Kuijk SJ, Yaromina A, Houben R, Niemans R, Lambin P, Dubois LJ. Prognostic significance of carbonic anhydrase IX expression in cancer patients: a meta-analysis. Front Oncol 2016;6:69.

[10] Owa T, Yoshino H, Okauchi T, Yoshimatsu K, Ozawa Y, Sugi N, Nagasu T, Koyanagi N, Kitoh K. Discovery of novel antitumor sulfonamides targeting G1 phase of the cell cycle. J Med Chem 1999;42:3789–99.

[11] Raymond E, Fumoleau P, Roche H, Schellens JHM, Dittrich C, Punt CJA, Droz JP, Calvert AH, van Oosterom A, Wanders J, Ravic M. Combined results of 4 phase I and pharmacokinetic studies of E7070, a novel chloroindoly-sulphonamide inhibiting the activation of cdk2 and cyclin E. Clin Cancer Res 2000;6:4529s.

[12] Raymond E, ten Bokkel Huinink WW, Taïeb J, Beijnen JH, Faivre S, Wanders J, Ravic M, Fumoleau P, Armand JP, Schellens JH, European Organization for the Research and Treatment of Cancer Early Clinical Study Group. Phase I and pharmacokinetic study of E7070, a novel chloroindolyl sulfonamide cell-cycle inhibitor, administered as a one-hour infusion every three weeks in patients with advanced cancer. J Clin Oncol 2002;20:3508–21.

[13] Punt CJA, Fumoleau P, van de Walle B, Faber MN, Ravic M, Campone M. Phase I and pharmacokinetic study of E7070, a novel sulfonamide, given at a daily times five schedule in patients with solid tumors. A study by the EORTC-early clinical studies group (ECSG). Ann Oncol 2001;12:1289–93.

[14] Mainwaring PN, Van Cutsem E, Van Laethem JL, Propper D, Twelves C, Bridgewater J, Audhuy B, Carmichael J, Punt CJA, Cassidy J, Stuart N, Ravic M. A multicentre randomised phase II study of E7070 in patients with colorectal cancer who have failed 5-fluorouracil-based chemotherapy. Proc Am Soc Clin Oncol 2002;21:153a.

[15] Dittrich C, Dumez H, Calvert H, Hanauske A, Faber M, Wanders J, Yule SM, Ravic M, Fumoleau P. Phase I and pharmacokinetic study of E7070, a chloroindolyl-sulfonamide anticancer agent, administered on a weekly schedule to patients with solid tumors. Clin Cancer Res 2003;9:5195–204.

[16] Fumoleau P, Viens P, Cottu PH, De Grève J, Marty M, Delaloge S, Miles D, O'Reilly S, Wardley A, Schmid P, Ravic M, Paternoster M. A multi center phase II study of E7070, a chloroindolyl-sulfonamide anticancer agent in anthracycline and taxane pre-treated breast cancer, In: 26th San Antonio breast cancer symposium; 2003.

[17] Terret C, Zanetta S, Roche H, Schellens JHM, Faber MN, Wanders J, Ravic M, Droz JP. Phase I clinical and pharmacokinetic study of E7070, a novel sulfonamide given as a 5-day continuous infusion repeated every 3 weeks in patients with solid tumours. A study by the EORTC Early Clinical Study Group (ECSG). Eur J Cancer 2003;39:1097–104.

[18] Haddad RI, Weinstein LJ, Wieczorek TJ, Bhattacharya N, Raftopoulos H, Oster MW, Zhang X, Latham Jr. VM, Costello R, Faucher JL, DeRosa C, Yule SM, Miller LP, Loda M, Posner MR, Shapiro GI. A phase II clinical and pharmacodynamic study of E7070 in patients with metastatic, recurrent, or refractory squamous cell carcinoma of the head and neck: modulation of retinoblastoma protein phosphorylation by a novel chloroindolyl sulfonamide cell cycle inhibitor. Clin Cancer Res 2004;10:4680–7.

[19] Smyth JF, Aamdal S, Awada A, Dittrich C, Caponigro F, Schoffski P, Gore M, Lesimple T, Djurasinovic N, Baron B, Ravic M, Fumoleau P, Punt CJ. Phase II study of E7070 in patients with metastatic melanoma. Ann Oncol 2005;16:158–61.

[20] Yamada Y, Yamamoto N, Shimoyama T, Horiike A, Fujisaka Y, Takayama K, Sakamoto T, Nishioka Y, Yasuda S, Tamura T. Phase I pharmacokinetic and pharmacogenomic study of E7070 administered once every 21 days. Cancer Sci 2005;96:721–8.

[21] Talbot DC, von Pawel J, Cattell E, Yule SM, Johnston C, Zandvliet AS, Huitema ADR, Norbury CJ, Ellis P, Bosquee L, Reck M. A randomized phase II pharmacokinetic and pharmacodynamic study of indisulam as second-line therapy in patients with advanced non-small cell lung cancer. Clin Cancer Res 2007;13:1816–22.

[22] Raftopoulos H, Escudier B, Renshaw G, Lara PN, Dutcher J, Nanus D, Petrylak DP, Sacris L. A phase II multicenter study of the cyclin-dependent kinase inhibitor indisulam in patients with inoperable and/or metastatic renal cell carcinoma (RCC). J Clin Oncol—ASCO Annual Meeting Proceedings (Post-Meeting Edition) 2004;22:414s.

[23] Siegel-Lakhai WS, Zandvliet AS, Huitema AD, Tibben MM, Milano G, Girre V, Diéras V, King A, Richmond E, Wanders J, Beijnen JH, Schellens JH. A dose-escalation study of indisulam in combination with capecitabine (Xeloda) in patients with solid tumours. Br J Cancer 2008;98:1320–6.

[24] Assi R, Kantarjian HM, Kadia TM, Pemmaraju N, Jabbour E, Jain N, Daver N, Estrov Z, Uehara T, Owa T, Cortes JE, Borthakur G. Final results of a phase 2, open-label study of indisulam, idarubicin, and cytarabine in patients with relapsed or refractory acute myeloid leukemia and high-risk myelodysplastic syndrome. Cancer 2018;124:2758–65.

[25] Van Berkel MA, Elefritz JL. Evaluating off-label uses of acetazolamide. Am J Health Syst Pharm 2018;75:524–31.

[26] Brechue WF, Stager JM, Lukaski HC. Body water and electrolyte responses to acetazolamide in humans. J Appl Physiol 1990;69:1397–401.

[27] Mokhtari RB, Kumar S, Islam SS, Yazdanpanah M, Adeli K, Cutz E, Yeger H. Combination of carbonic anhydrase inhibitor, acetazolamide, and sulforaphane, reduces the viability and growth of bronchial carcinoid cell lines. BMC Cancer 2013;13:378.

[28] Said HM, Hagemann C, Carta F, Katzer A, Polat B, Staab A, Scozzafava A, Anacker J, Vince GH, Flentje M, Supuran CT. Hypoxia induced CA9 inhibitory targeting by two different sulfonamide derivatives including acetazolamide in human glioblastoma. Bioorg Med Chem 2013;21:3949–57.

[29] Islam SS, Mokhtari RB, Akbari P, Hatina J, Yeger H, Farhat WA. Simultaneous targeting of bladder tumor growth, survival, and epithelial-to-mesenchymal transition with a novel therapeutic combination of acetazolamide (AZ) and sulforaphane (SFN). Target Oncol 2016;11:209–27.

[30] Bayat Mokhtari R, Baluch N, Ka Hon Tsui M, Kumar S, Homayouni TS, Aitken K, Das B, Baruchel S, Yeger H. Acetazolamide potentiates the anti-tumor potential of HDACi, MS-275, in neuroblastoma. BMC Cancer 2017;17:156.

[31] Wu L, Bernal GM, Cahill KE, Pytel P, Fitzpatrick CA, Mashek H, Weichselbaum RR, Yamini B. BCL3 expression promotes resistance to alkylating chemotherapy in gliomas. Sci Transl Med 2018;10:448.

[32] Boyd NH, Walker K, Fried J, Hackney JR, McDonald PC, Benavides GA, Spina R, Audia A, Scott SE, Libby CJ, Tran AN, Bevensee MO, Griguer C, Nozell S, Gillespie GY, Nabors B, Bhat KP, Bar EE, Darley-Usmar V, Xu B, Gordon E, Cooper SJ, Dedhar S, Hjelmeland AB. Addition of carbonic anhydrase 9 inhibitor SLC-0111 to temozolomide treatment delays glioblastoma growth in vivo. JCI Insight 2017;2:e92928.

[33] Andreucci E, Ruzzolini J, Peppicelli S, Bianchini F, Laurenzana A, Carta F, Supuran CT, Calorini L. The carbonic anhydrase IX inhibitor SLC-0111 sensitises cancer cells to conventional chemotherapy. J Enzyme Inhib Med Chem 2019;34:117–23.

[34] Toews I, George AT, Peter JV, Kirubakaran R, Fontes LES, Ezekiel JPB, Meerpohl JJ. Interventions for preventing upper gastrointestinal bleeding in people admitted to intensive care units. Cochrane Database Syst Rev 2018;6: CD008687.

[35] Luciani F, Spada M, De Milito A, Molinari A, Rivoltini L, Montinaro A, Marra M, Lugini L, Logozzi M, Lozupone F, Federici C, Iessi E, Parmiani G, Arancia G, Belardelli F, Fais S. Effect of proton pump inhibitor pretreatment on resistance of solid tumors to cytotoxic drugs. J Natl Cancer Inst 2004;96:1702–13.

[36] Chen M, Zou X, Luo H, Cao J, Zhang X, Zhang B, Liu W. Effects and mechanisms of proton pump inhibitors as a novel chemosensitizer on human gastric adenocarcinoma (SGC7901) cells. Cell Biol Int 2009;33:1008–19.

[37] Udelnow A, Kreyes A, Ellinger S, Landfester K, Walther P, Klapperstueck T, Wohlrab J, Henne-Bruns D, Knippschild U, Wurl P. Omeprazole inhibits proliferation and modulates autophagy in pancreatic cancer cells. PLoS One 2011;6:e20143.

[38] Chen M, Huang SL, Zhang XQ, Zhang B, Zhu H, Yang VW, Zou XP. Reversal effects of pantoprazole on multidrug resistance in human gastric adenocarcinoma cells by down-regulating the V-ATPases/mTOR/HIF-1alpha/P-gp and MRP1 signaling pathway in vitro and in vivo. J Cell Biochem 2012;113:2474–87.

[39] Ferrari S, Perut F, Fagioli F, Brach Del Prever A, Meazza C, Parafioriti A, Picci P, Gambarotti M, Avnet S, Baldini N, Fais S. Proton pump inhibitor chemosensitization in human osteosarcoma: from the bench to the patients' bed. J Transl Med 2013;11:268.

[40] Spugnini EP, Baldi A, Buglioni S, Carocci F, de Bazzichini GM, Betti G, Pantaleo I, Menicagli F, Citro G, Fais S. Lansoprazole as a rescue agent in chemoresistant tumors: a phase I/II study in companion animals with spontaneously occurring tumors. J Transl Med 2011;9:221.

[41] Spugnini EP, Buglioni S, Carocci F, Francesco M, Vincenzi B, Fanciulli M, Fais S. High dose lansoprazole combined with metronomic chemotherapy: a phase I/II study in companion animals with spontaneously occurring tumors. J Transl Med 2014;12:225.

[42] Wang BY, Zhang J, Wang JL, Sun S, Wang ZH, Wang LP, Zhang QL, Lv FF, Cao EY, Shao ZM, Fais S, Hu XC. Intermittent high dose proton pump inhibitor enhances the antitumor effects of chemotherapy in metastatic breast cancer. J Exp Clin Cancer Res 2015;34:85.

[43] Papagerakis S, Bellile E, Peterson LA, Pliakas M, Balaskas K, Selman S, Hanauer D, Taylor JM, Duffy S, Wolf G. Proton pump inhibitors and histamine 2 blockers are associated with improved overall survival in patients with head and neck squamous carcinoma (HNSCC). Cancer Prev Res (Phila) 2014;712:1258–69.

[44] Felmlee MA, Jones RS, Rodriguez-Cruz V, Follman KE, Morris ME. Monocarboxylate transporters (SLC16): function, regulation, and role in health and disease. Pharmacol Rev 2020;72:466–85.

[45] Aspatwar A, Tolvanen MEE, Schneider HP, Becker HM, Narkilahti S, Parkkila S, Deitmer JW. Catalytically inactive carbonic anhydrase-related proteins enhance transport of lactate by MCT1. FEBS Open Bio 2019;9:1204–11.

[46] Pérez-Escuredo J, Van Hée VF, Sboarina M, Falces J, Payen VL, Pellerin L, Sonveaux P. Monocarboxylate transporters in the brain and in cancer. Biochim Biophys Acta 1863;2016:2481–97.

[47] Le Floch R, Chiche J, Marchiq I, Naiken T, Ilc K, Murray CM, Critchlow SE, Roux D, Simon MP, Pouysségur J. CD147 subunit of lactate/H+ symporters MCT1 and hypoxia-inducible MCT4 is critical for energetics and growth of glycolytic tumors. Proc Natl Acad Sci U S A 2011;108:16663–8.

[48] Granja S, Marchiq I, Le FR, Souto MC, Baltazar F, Pouysségur J. Disruption of BASIGIN decreases lactic acid export and sensitizes non-small cell lung cancer to biguanides independently of the LKB1 status. Oncotarget 2014;6:6708–21.

[49] Halford S, Jones P, Wedge S, Hirschberg S, Katugampola S, Veal G, Payne G, Bacon C, Potter S, Griffin M, Chenard-Poirier M, Petrides G, Holder G, Keun H, Banerji U, Plummer E. A first-in-human first-in-class (FIC) trial of the monocarboxylate transporter 1 (MCT1) inhibitor AZD3965 in patients with advanced solid tumours. J Clin Oncol 2017;35(15 Suppl):2516.
[50] Boesch M, Sopper S, Wolf D. Ionophore antibiotics as cancer stem cell-selective drugs: open questions. Oncologist 2016;21:1291–3.
[51] Prevarskaya N, Skryma R, Shuba Y. Ion channels and the hallmarks of cancer. Trends Mol Med 2010;16:107–21.
[52] Deng CC, Liang Y, Wu MS, Feng FT, Hu WR, Chen LZ, Feng QS, Bei JX, Zeng YX. Nigericin selectively targets cancer stem cells in nasopharyngeal carcinoma. Int J Biochem Cell Biol 2013;45:1997–2006.
[53] Yakisich JS, Azad N, Kaushik V, O'Doherty GA, Iyer AK. Nigericin decreases the viability of multidrug-resistant cancer cells and lung tumorspheres and potentiates the effects of cardiac glycosides. Tumour Biol 2017;39.
[54] Naujokat C, Steinhart R. Salinomycin as a drug for targeting human cancer stem cells. J Biomed Biotechnol 2012;2012:950658.
[55] Goard CA, Schimmer AD. An evidence-based review of obatoclax mesylate in the treatment of hematological malignancies. Core Evid 2013;8:15–26.
[56] Oki Y, Copeland A, Hagemeister F, Fayad LE, Fanale M, Romaguera J, Younes A. Experience with obatoclax mesylate (GX15-070), a small molecule pan-Bcl-2 family antagonist in patients with relapsed or refractory classical Hodgkin lymphoma. Blood 2012;119:2171–2.
[57] Arellano ML, Borthakur G, Berger M, Luer J, Raza A. A phase II, multicenter, open-label study of obatoclax mesylate in patients with previously untreated myelodysplastic syndromes with anemia or thrombocytopenia. Clin Lymphoma Myeloma Leuk 2014;14:534–9.
[58] Langer CJ, Albert I, Ross HJ, Kovacs P, Blakely LJ, Pajkos G, Somfay A, Zatloukal P, Kazarnowicz A, MMl M, Schreeder MT, Schnyder J, Ao-Baslock A, Pathak AK, Berger MS, GEM017 Investigators. Randomized phase II study of carboplatin and etoposide with or without obatoclax mesylate in extensive-stage small cell lung cancer. Lung Cancer 2014;85:420–8.
[59] Wong P, Kleemann HW, Tannock IF. Cytostatic potential of novel agents that inhibit the regulation of intracellular pH. Br J Cancer 2002;87:238–45.
[60] Harley W, Floyd C, Dunn T, Zhang XD, Chen TY, Hegde M, Palandoken H, Nantz MH, Leon L, Carraway 3rd KL, Lyeth B, Gorin FA. Dual inhibition of sodium-mediated proton and calcium efflux triggers non-apoptotic cell death in malignant gliomas. Brain Res 2010;1363:159–69.
[61] Yang X, Wang D, Dong W, Song Z, Dou K. Inhibition of Na(+)/H(+) exchanger 1 by 5-(N-ethyl-N-isopropyl) amiloride reduces hypoxia-induced hepatocellular carcinoma invasion and motility. Cancer Lett 2010;295:198–204.
[62] Estrella V, Chen T, Lloyd M, Wojtkowiak J, Cornnell HH, Ibrahim-Hashim A, Bailey K, Balagurunathan Y, Rothberg JM, Sloane BF, Johnson J, Gatenby RA, Gillies RJ. Acidity generated by the tumor microenvironment drives local invasion. Cancer Res 2013;73:1524–35.
[63] Reshkin SJ, Cardone RA, Harguindey S. Na+/H+ exchanger, pH regulation and cancer. Recent Pat Anticancer Drug Discov 2013;8:85–99.
[64] Chang WH, Liu TC, Yang WK, Lee CC, Lin YH, Chen TY, Chang JG. Amiloride modulates alternative splicing in leukemic cells and resensitizes Bcr-AblT315I mutant cells to imatinib. Cancer Res 2011;71:383–92.
[65] Miraglia E, Viarisio D, Riganti C, Costamagna C, Ghigo D, Bosia A. Na+/H+ exchanger activity is increased in doxorubicin-resistant human colon cancer cells and its modulation modifies the sensitivity of the cells to doxorubicin. Int J Cancer 2005;115:924–9.
[66] Lauritzen G, Jensen MB, Boedtkjer E, Dybboe R, Aalkjaer C, Nylandsted J, Pedersen SF. NBCn1 and NHE1 expression and activity in DeltaNErbB2 receptor-expressing MCF-7 breast cancer cells: contributions to pHi regulation and chemotherapy resistance. Exp Cell Res 2010;316:2538–53.
[67] Kellen JA, Mirakian A, Kolin A. Antimetastatic effect of amiloride in an animal tumour model. Anticancer Res 1988;8:1373–6.
[68] Matthews H, Ranson M, Kelso MJ. Anti-tumour/metastasis effects of the potassium-sparing diuretic amiloride: an orally active anti-cancer drug waiting for its call-of-duty? Int J Cancer 2011;129:2051–61.
[69] Harguindey S, Arranz JL, Polo Orozco JD, Rauch C, Fais S, Cardone RA, Reshkin SJ. Cariporide and other new and powerful NHE1 inhibitors as potentially selective anticancer drugs - an integral molecular/biochemical/metabolic/clinical approach after one hundred years of cancer research. J Transl Med 2013;11:282.

PART II

pH-Interfering agents as chemosensitizers

CHAPTER

4

Proton pump inhibitors and other pH-buffering agents

Alessio Nocentini[a] *and Stefano Fais*[b]

[a]NEUROFARBA Department, Section of Pharmaceutical and Nutraceutical Sciences, University of Florence, Florence, Italy [b]Department of Oncology and Molecular Medicine, Italian National Institute of Health, Rome, Italy

Abstract

By a metabolism based on sugar fermentation, cancer cells progressively acidify their microenvironment inducing a selective pressure, which leads to the development of very malignant cells entirely armed to survive in such a hostile microenvironment. Proton exchangers, among which vacuolar ATPases (V-ATPases), do not permit the intracellular acidification by a constant elimination of H^+ either outside the cells or within the internal vacuoles. This chapter describes the translational process that provided the preclinical to clinical demonstration of the potentiality of proton pump inhibitors (PPIs), exploited worldwide for peptic ulcer treatment and gastroprotection, as chemosensitizers or even antitumor agents as well.

Abbreviations

AE	anion exchanger
CA	carbonic anhydrase
CHC	α-cyano-4-hydroxycinnamate
EGF	epidermal growth factor
FASN	fatty acid synthase
HIF-1	hypoxia-inducible factor 1
IGF	insulin-like growth factor
LDHA	lactate dehydrogenase A
MCT	monocarboxylate transporter
MRP	multidrug resistance protein
mTOR	mammalian target of rapamycin
NBC	Na^+/bicarbonate cotransporter
NHE	Na^+/H^+ exchanger
PK	pyruvate kinase
PPI	proton pump inhibitor
ROS	reactive oxygen species

https://doi.org/10.1016/B978-0-12-820701-7.00011-7

SCID severe combined immunodeficient
TGFβ transforming growth factor β
TOPK T-cell-originated protein kinase
V-ATPase vacuolar ATPase

Conflict of interest

No potential conflicts of interest were disclosed.

Introduction

Cancer is the second leading cause of death globally, accounting for an estimated 9.6 million deaths, or one in six deaths, in 2018 [1]. In strong health systems countries, survival rates of many types of cancers are improving thanks to accessible early detection, quality treatment, and survivorship care. However, despite the use of multiple drug combination protocols and the development of novel targeted anticancer strategies, chemoresistance remains a primary hindrance toward cancer treatment [2]. A main hallmark of cancer is the alteration of energy metabolism that occurs in hypoxic conditions, which produces lactate accumulation and intracellular acidification [3]. Rapidly dividing cancer cells produce and release large amounts of protons into the extracellular compartment due to heightened glucose utilization, amino acid metabolism, and ATP hydrolysis [4, 5]. It can be hypothesized that the highly competitive microenvironment resulting by tumor proliferation selects the cells best adapted to survive in these hostile conditions. As a result, cancer cells overexpress many growth factors that modulate glucose metabolism like IGF [6], TGFβ, and EGF [7]. Uncontrolled tumor cell proliferation, acid production, and tissue hypoxia deriving from limited blood supply generate a highly hostile tumor microenvironment that is unsuitable for most cells. In these conditions, cancer cells have to undergo a thorough "reprogramming" that leads to the overexpression of gene products, which are involved in hypoxia and glucose metabolism, such as hypoxia-inducible factor 1 (HIF-1), Myc, p53, Ras, Akt, Src, pyruvate kinase (PK) M2, lactate dehydrogenase A (LDHA), and PK M2 [8–11]. Hence, tumor cells need to develop systems to actively extrude excess protons (Fig. 1) [12]. These mechanisms include V-ATPase, Na^+/H^+ exchanger (NHE), bicarbonate transporters (NBC and AE), monocarboxylate transporters (MCTs), and carbonic anhydrases (CAs IX and XII) [13]. Sections "V-ATPases and proton pump inhibitors" to "Inhibitors of Na^+/H^+ exchanger (NHE), Na^+/HCO_3^- co-transporter (NBC) and anion exchanger (AE)" speak on the potentiality of inhibitors of V-ATPases, MCT, NHE, NBC, and AE as chemosensitizers or even antitumor agents. A more detailed report on inhibitors of MCTs and other transporter subfamily inhibitors is provided in Chapter 7. Discussion of the inhibitors of CAs IX and XII is reported in Chapters 5 and 6.

Tumor acidity and chemoresistance

Acidity in tumor cells is caused by increase in protons generation by the anaerobic metabolism of cancer cells through the overexpression of PK M2, a dimeric form of the enzyme pyruvate kinase [14]. Hence, tumors have a significantly lower extracellular pH (~6.5–7.1) in comparison to normal tissues (7.4), due to intervention of proton efflux pumps [13].

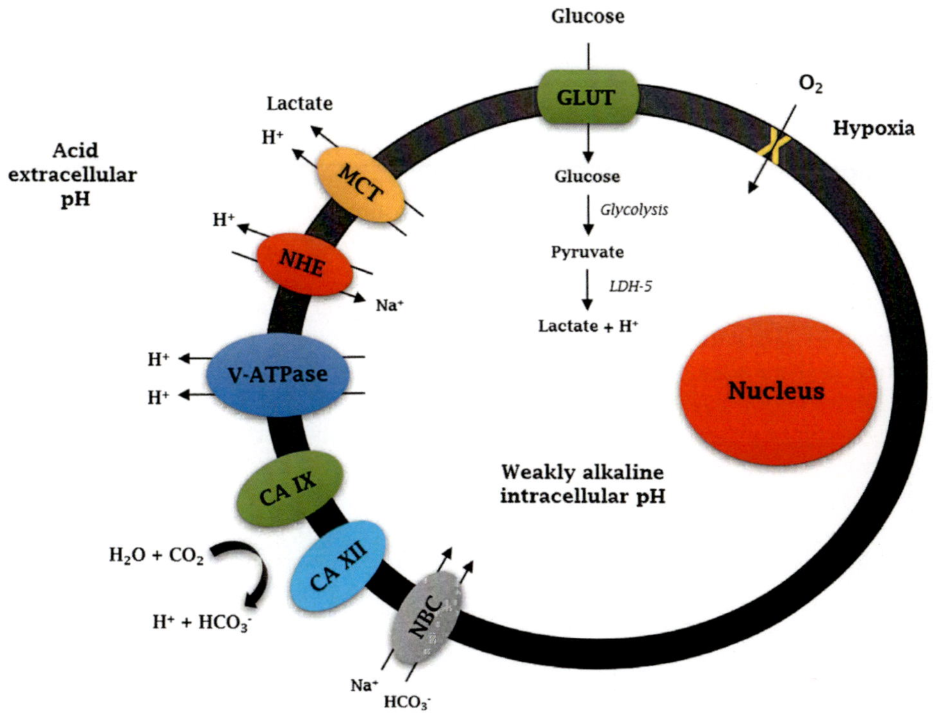

FIG. 1 Hypoxic tumor cells reprogram their metabolism to engage the glycolytic pathway that results in increased production of lactate and protons. Hence, tumor cells overexpress systems to actively extrude the excess of protons and restore a weakly alkaline intracellular pH acidifying the extracellular media. Such a pH modulation machinery is composed of monocarboxylate transporter (MCT), Na^+/H^+ exchanger (NHE), plasma membrane proton pump vacuolar ATPase (V-ATPase), carbonic anhydrases IX and XII (CA IX, CA XII), and Na^+/HCO_3^- cotransporter (NBC).

The final result of this active proton extrusion is a severe alteration of pH gradients, mainly characterized by a reversed pH gradient between the acidic extracellular microenvironment and the slightly alkaline cytosol [13]. The resulting increase in acidity induces the selection of tumor cells, which are capable of tolerating these extreme conditions leading to a more aggressive phenotype. Indeed, proton export leads inevitably to a low extracellular pH and a low endosomal and lysosomal pH, thus protecting the cell from cytosolic acidity [15, 16]. Several interdisciplinary studies have been conducted, which combine mathematical simulations and laboratory investigation, and indicate that the acidic microenvironment enhances tumor invasiveness and metastatic potential [17, 18].

A major obstacle to overcome during treatment of solid tumors is the spread of drug resistance to chemotherapy [19, 20]. A major contributor to the development of chemoresistance is the acidic tumor microenvironment (low pO_2 and pH) that attenuates and frequently abrogates therapeutic efficacy [21]. Indeed, hypoxia and altered glycolytic activity, the so-called Warburg Effect, contribute to the establishment of such a condition [21, 22]. The extracellular and endosomal and lysosomal acidic pH contribute to drug resistance both in vitro and in vivo. In fact, the reverse pH gradient occurring between the cytosolic and extracellular areas impacts negatively on the distribution, uptake, and bioavailability of weak base antineoplastic drugs [23]. Small uncharged molecules can migrate through the lipid phase of the cell membrane, which is a semipermeable barrier between the intra- and extracellular compartments [24, 25]. In contrast, charged molecules are incapable of crossing the cell membrane. The acidic extracellular space of solid tumors hinders the cellular uptake of weak bases, by a phenomenon known as "ion trapping" [21]. The ratio between ionized and nonionized forms is highly sensitive to pH changes for weak bases such as anthracyclines, anthracenediones, campothecins, and Vinca alkaloids whose uptake and efficacy are reduced [26, 27]. Also complex drugs such as cisplatin, considered weak acids in their primary structure, can be affected by a low pH. In fact, while cisplatin is partially ionized at pH6.5, the amount of ionized drug greatly increases at a pH ranging from 3 to 5 producing that the majority of cisplatin remains outside malignant melanoma cells [28].

It has been long deemed that multidrug efflux transporter P-glycoprotein (ABCB1) mediates the main mechanism of resistance within cancer cells [29]. However, it was suggested that this phenomenon may not be an essential mechanism when compared to the key role of the pH gradient that exists between cancer cells and their microenvironment. In fact, it appears that two mechanisms exist by which weak bases can be excluded from entering cancer cells: drug ionization in the acidic compartment, which reduces the ability to cross the plasma membrane [30] or by alteration of the interaction between phospholipids from the inner leaflet, which makes the bilayer membrane much firmer to drug penetration into cells [31]. What is more, such a membrane rigidity driven by the pH gradient across the membrane favors the interaction between P-glycoprotein and antineoplastic drugs.

Intriguing data were obtained with proton pump inhibitors (PPIs) with MDR-positive chemoresistant cells being resensitized to the antineoplastic effects of vinblastine by exposure to omeprazole or lansoprazole, suggesting that counteracting the acidity of the tumor microenvironment may overcome classical mechanisms of chemoresistance [32]. This, in turn, suggests that tumor acidity is the leading mechanism of tumor resistance to therapies and that counteracting acidity may represent a valuable action to increase effectiveness of existing anticancer therapies and this may be obtained by the use of PPI.

V-ATPases and proton pump inhibitors

The leading cause of tumor microenvironment acidification is the glycolytic switch of tumor metabolism coupled with reduced perfusion [2]. This condition progressively selects cells most adapted to survive in such a circumstance, where overexpression and activation of membrane-bound pH-regulating systems contribute to preventing intracellular acidification [33]. Among them, vacuolar-type H^+ ATPases (V-ATPase) are implicated in the acidification of tumor microenvironment [34–36]. V-ATPase is a complex multisubunit protein devoted to the transport of protons from the cytoplasm toward intracellular compartments and from inside to outside of the cell through the cytoplasmic membrane [37]. V-ATPases are composed of a transmembrane subunit, named V0 complex, devoted to proton transfer and a cytoplasmic portion, named V1 complex, that furnished the necessary energy for proton translocation (Fig. 2) [38].

V-ATPase is involved in multiple cellular functions, including endocytosis and activation of proteases [39], angiogenesis [40], autophagy [41], and amino acids sensing via interaction with mTOR [42]. Enhanced expression of V-ATPase confers survival and growth advantages

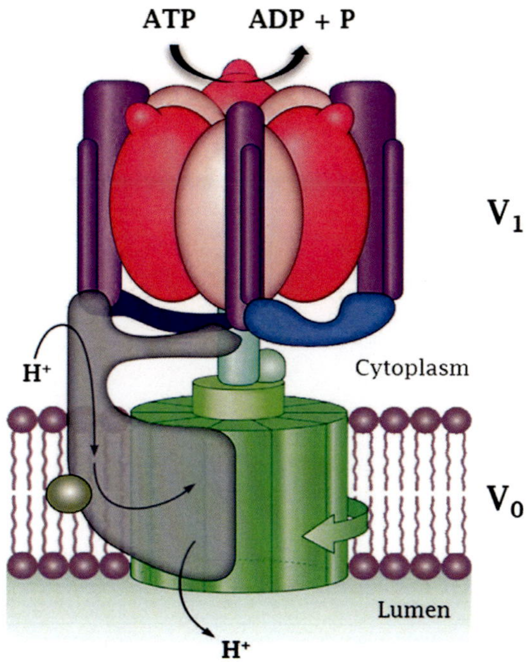

FIG. 2 Structure of eukaryotic V-ATPase.

to cancer cells [43]. In fact, the low extracellular pH also activates extracellular metalloproteinases that promote tumor cell survival, motility, and invasion, which enhanced malignant abilities. A role of V-ATPase was pointed out in tumor invasion and multidrug resistance in breast cancer [44], oral squamous cell carcinoma [45], esophageal carcinoma [46], hepatocellular and pancreatic carcinoma [47], lung carcinoma [48], sarcoma [49], and in general solid tumors [50]. Hence, inhibition of V-ATPase has become a fascinating and promising strategy to counteract proton metabolism in cancer and has been investigated in vitro and in vivo in both preclinical and clinical studies. However, as V-ATPases are ubiquitary into the body, inhibition of V-ATPases might result to be toxic [37, 39].

In this context, PPIs, that are used worldwide as potent antiacidic drugs against peptic diseases or as gastro-protectors (i.e., omeprazole, esomeprazole, lansoprazole, pantoprazole, and rabeprazole, Fig. 3), reported a huge potentiality [51]. In fact, PPIs did not show relevant systemic toxicity, even in prolonged treatments and at very high dosages [52]. PPIs specifically target gastric H^+/K^+ ATPases, but reported activity against VATPases as well [13, 37, 51]. PPI are prodrugs, which need protonation in acidic media to transform them into the active species (Fig. 4), in contrast to a plethora or weak base drugs that undergo neutralization outside the tumor cells by protonation [53]. Hence, while acidity induces tumor resistance to most drugs, PPIs exploit it to become functional [53, 54]. Preclinical investigations showed that PPIs not only sensitize tumor cells and tumors to the action of chemotherapeutics [32], but also per se exert a potent antitumor activity, by an in vivo modulation of tumor pH [55]. Furthermore, acidity represents a potent mechanism of tumor immune escape and scientific evidence has provided the proof of concept that PPIs may increase the immune reaction against tumors [56].

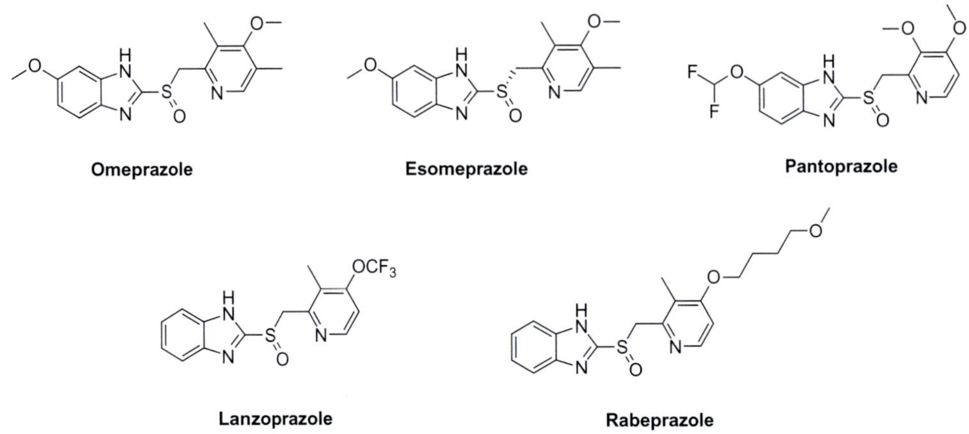

FIG. 3 Proton pump inhibitors (PPIs) in clinical use.

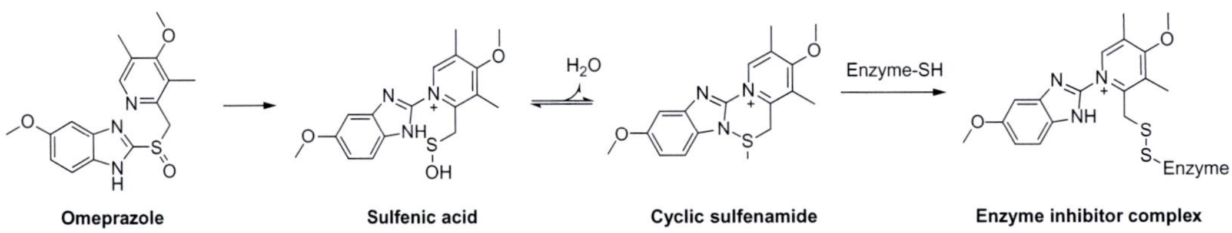

FIG. 4 Activation mechanism of omeprazole in acidic conditions.

The following sections report on these preclinical data as the background for a series of clinical investigations aimed at supporting the use of PPIs as chemosensitizers.

Preclinical data supporting the use of proton pump inhibitors in cancer treatment

The first evidence supporting the use of PPIs in cancer dates back to 2004. The pretreatment of different human tumor cell lines with PPIs was shown to improve the uptake and prolonged the withholding of some chemotherapy drugs [32]. Moreover, pretreatment of mice carrying orthotopic melanoma xenografts with PPIs promoted a marked decrease in tumor growth and the extension of survival time [32]. Additional studies extended the spectrum of PPI-responsive neoplasms: investigation of the response of B-cell lymphomas to omeprazole and esomeprazole showed for the first time that PPIs were not only efficient chemosensitizers, but also possess a clear antitumor action [55]. In fact, PPIs evidenced a cytotoxic effect against B-cell tumor derived lines and as well as against B-cell tumors xenografts [55]. A rapid production of reactive oxygen species (ROS) was detected as the intracellular mediator of the PPI action, witnessing the alkalinization of lysosomal pH, lysosomal membrane permeabilization, and cytosol acidification, and thus indicating a destabilization of the tumor acidic metabolism. This ROS-mediated, caspase-independent, alteration of the protons moving through the tumor membrane produced a proapoptotic action of the PPI that determined a delayed tumor growth in SCID mice [55]. The same authors proved that bone marrow blasts obtained from children with Acute Lymphoblastic Leukemia clearly showed to be sensible to the cytotoxic effect of PPIs [33, 55]. Likewise, PPIs were shown to exert a strong antitumor action against different human melanoma cell lines and in tumor-bearing mice [5]. In detail, human melanoma cell lines were treated with PPIs in different pH conditions. The treatment produced a marked cytotoxicity influenced by low pH and occurring by a caspase-dependent mechanism activated by rapid intracellular acidification. In vivo studies in SCID mice engrafted with human melanoma assessed the effects of PPIs on tumor growth and pH using magnetic resonance spectroscopy. The delay in tumor growth provoked by the PPI was consistently related to a neat reduction of pH gradients in tumor tissue [5]. PPIs were also evaluated against a chemoresistant gastric carcinoma cell line in vitro and in vivo, and were shown to induce the down-regulation of several metabolic pathways, including the V-ATPases/mTOR/HIF-1/P-gp and MRP1 signaling pathway and potentiated, among other agents, the antitumor effectiveness of Adriamycin in mice having gastric carcinoma xenografts [57–59]. Consistently, a study by Patel et al. showed that pantoprazole increases doxorubicin uptake and penetration through tissue in in vitro studies. Moreover, pantoprazole was found to improve doxorubicin distribution and retention and delay of breast cancer xenografts growth in mice [60]. Similar outcomes were obtained in other studies evaluating the role of tumor acidity in breast cancer and the efficacy thereby of the combination of lansoprazole/doxorubicin [61–65]. Omeprazole was shown to increase the efficacy of paclitaxel against tumor growth in orthotopic and xenograft mouse models of ovarian carcinoma [66]. The efficacy of the pretreatment with the PPI was clear in chemoresistant cancer cell lines (or xenografts) and negligible in chemosensitive cells. High-dose pantoprazole (100 mg/kg) administered intraperitoneally to mice carrying colorectal cancer halved the tumor growth compared to controls [67]. This occurred by the inhibition of T-cell-originated protein kinase (TOPK), a highly expressed enzyme responsible for promoting tumorigenesis and progression.

A new V-ATPase pump (TM9SF4) was uncovered in 2015, which is responsible for tumor aggressiveness in colorectal cancer, and inhibition thereof by PPIs was efficient in impairing tumor growth [68]. Similar outcomes arose in preclinical investigations with hepatocellular, pancreatic, prostatic, and brain tumors [66, 69–71]. Meanwhile, Azzarito et al. first evaluated that PPIs as chemosensitizers could improve the efficacy of chemotherapic agents given at suboptimal doses using low doses of paclitaxel in vitro and in vivo against melanoma. The antitumor efficacy was preserved by pretreatment with PPIs [72]. Therefore, a study identified lansoprazole as the most effective and promising agent within the PPI family in single therapy against tumor cells [73]. Indeed, lansoprazole was shown to inhibit multiple myeloma cell lines even in the absence of chemotherapic drugs [74].

The involvement of proton pumps in tumor-induced downregulation of the lymphocytic response to neoplasia and tumor resistance to chemotherapy was reported in vivo in a very aggressive murine melanoma (C57-B16) [56, 75, 76]. In this study, the H^+-rich microenvironment was shown to induce a state of anergy in tumor-infiltrating lymphocytes, which was successfully reversed by treatment with PPIs. It can be assumed that the tumor microenvironmental pH might not only impair the immune function, but induce immune cell death once immune cells try to enter the tumor mass [77]. This is further supported by the evidence that sodium bicarbonate improves the immune response against the B16 melanoma by buffering of the tumor tissue [78].

Clinical outcomes of proton pump inhibitors in association to anticancer therapies

The translation of the use of PPI from preclinical to clinical oncology occurred quite rapidly. To date, besides case reports relating long-term responses in patients treated with PPIs [79], the results of two clinical trials in human were published about osteosarcomas and metastatic breast cancer patient [34, 80]. The results showed that pretreatment with PPI increased the efficacy of neoadjuvant chemotherapy in osteosarcomas patients [34]. The inclusion of PPI pretreatment in the multidrug protocol for neoadjuvant chemotherapy in patients carrying osteosarcoma increased the feasibility of limb-sparing procedure and enhanced the overall response rate. Of note, this response was likely profoundly influenced by the acidity status, since the patients that overall experienced the best improvement in response rate were those affected by the chondroblastic variant. This subgroup of osteosarcoma shows the worse response to preoperative chemotherapy and is the most acidic of all the osteosarcoma histotypes [34]. This was the first clinical evidence that pretreatment with PPIs could be easily included in standard clinical oncology protocols leading to a clear benefit for patients having less favorable prognostic factors. This evidence was confirmed in a second trial enrolling women with metastatic breast cancer randomized to receive standard chemotherapy or chemotherapy with alkalinization [80]. The patients, experiencing PPI treatment, had the highest response rates and the longest survivals. Moreover, a maintenance protocol based only upon alkalinization showed a clear advantage to alkalinization limited to the period of chemotherapy administration [80]. This suggests including in future therapy maintenance periods of alkalinization interposed between courses of chemotherapy to aid the tumor control.

In 2014 Brana et al. investigated the efficacy of high dose i.v. pantoprazole as pretreatment of doxorubicin in patients with advanced solid tumors, identifying the dose of 240 mg as the baseline for phase II studies, showing the lack of drug-drug interaction and pointing out the increased tumor uptake of doxorubicin through biopsies [81]. In 2016 a study by Falcone et al. reported 3 cases of patients with metastatic colorectal cancer, refractory to standard chemotherapy and target therapy, treated with the combination of high dose of the PPI rabeprazole and metronomic chemotherapy, obtaining a remarkable clinical outcome [82]. In fact, such a combination of high-dose rabeprazole with metronomic capecitabine induced a long-lasting stable disease and good quality of life with tolerable side effects [82]. The same group investigated this approach in an approved clinical trial (EudraCT Number:2013-001096-20) that demonstrated that tumor alkalinization could improve the outcome of patients treated with metronomic salvage protocol by counteracting tumor chemoresistance [83, 84]. However, investigations concluded that the concomitant administration of capecitabine and PPI could actually lead to unfavorable outcome for the patients. In this context, it should be noted that Marchetti et al. had suggested to pretreat the patients with PPIs and then proceed with the administration of the chemotherapic agents to settle a microenvironment suitable (alkalinized extracellular compartment) for the antitumor intervention.

A 2019 prospective study examined the effects of rabeprazole on the pharmacokinetics of capecitabine and its metabolites [85]. It was concluded that Rabeprazole did not affect the proliferation inhibition of colon cancer cells by capecitabine. It is necessary that in the next future alkalinization might be recognized as the 5th anticancer therapy, opening a nonmainstream Avenue to clinical approach of cancer.

A metanalysis study in patients with head and neck tumor corroborated an increased response in patients receiving antiacid drugs, in particular PPIs [86]. This retrospective study unexpectedly supported the data obtained in the few available clinical trials testing the combination of PPI and chemotherapy [34, 80, 82]. Actually, two retrospective studies performed in China in an impressive number of individuals have shown that women that used PPI for treating dyspeptic diseases showed a decreased incidence of breast cancer as compared to those women that never assumed PPIs during their life [87, 88].

The series of clinical trials referring to PPI treatment in combination with chemotherapy also includes two studies in companion animals with spontaneous tumors. In the first veterinary study, PPIs showed to reverse chemoresistance in refractory tumors both hematopoietic (lymphoma) or solid (melanoma and squamous cell carcinoma) [89, 90]. In a second study PPIs acted as a response accelerator to metronomic chemotherapy, independently from the tumor histotype or the animal species affected by cancer. These studies supported the high tolerability and efficacy of this approach, and the fast evolution of cancer disease in pets, as well as the rapid and durable response, prompted the translation of the information to human patients [91, 92].

On the basis of this studies there is a general support and agreement in the "repositioning" of PPI as anticancer drugs [93–95]. In 2015 Fako et al. identified proton pump inhibitors as effective inhibitors of the thioesterase activity of human fatty acid synthase (FASN), the enzyme responsible for de novo synthesis of free fatty acids, which is upregulated in many cancers [96]. In this context, PPIs were shown to inhibit proliferation and induced apoptosis of cancer cells, while supplementation of palmitate, the end product of FASN catalysis, rescued cancer cells from PPI-induced cell death.

Monocarboxylate transporter (MCT) inhibitors

The products of glycolysis, used as metabolic intermediates, are partly extruded from the cells through the monocarboxylate transporters (MCTs), a family of proton-coupled transporters carrying lactic acid and other monocarboxylates (i.e., acetic acid) across biological membranes [97–100]. MCTs family is composed of nine different isoforms. MCT1 is the main isoform involved in entry and exit of lactate ions, whereas MCT4 was shown to be hypoxia-induced by HIF signaling [98, 99]. In 2008 Sonveaux et al. validated MCT1 as antitumor target by observing switch from lactate-fueled respiration to glycolysis upon inhibition of MCT1 by α-cyano-4-hydroxycinnamate (**CHC**, Fig. 5) in human cervix squamous carcinoma cell lines [14]. The retention of lactate in hypoxic cells prevents its usage for energy production and forces cells to metabolize glucose [14]. The MCT inhibitor **AZD3965** (Fig. 5) was shown to interfere with the cancer cell metabolism in human lymphoma and colon carcinoma cells [98]. The inhibitor increased the MCT4-dependent accumulation of intracellular lactate inhibiting both the influx and efflux of monocarboxylates. **AZD3965** also increased the levels

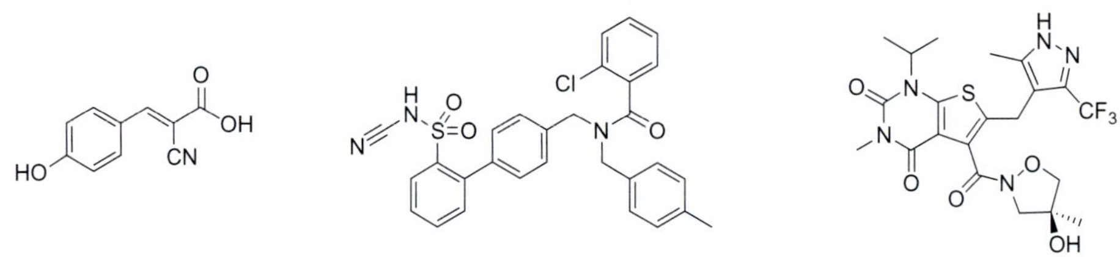

α-Cyano-4-hydroxycinnamate (CHC) S0859 AZD3965

FIG. 5 Structure of MTC inhibitors discussed in "Monocarboxylate transporter (MCT) inhibitors" section.

of Krebs cycle-related metabolites and glucose mitochondrial metabolism, enhancing oxidative pyruvate dehydrogenase and anaplerotic pyruvate carboxylase fluxes. Improved bioenergetics were also evidenced in vivo after dosing with **AZD3965** in mouse xenograft models of human lymphoma [98].

Deitmer et al. showed that **S0859** (Fig. 5), characterized as a selective inhibitor of sodium-bicarbonate cotransporters (NBC SLC4), also inhibits MCTs proving that it might act as inhibitor of anion transport with a broader spectrum than previously thought, and may also interfere with cellular metabolite uptake/release [99].

Inhibitors of Na^+/H^+ exchanger (NHE), Na^+/HCO_3^- cotransporter (NBC), and anion exchanger (AE)

As stated above, another important membrane protein for maintaining the reversed tumor pH gradient is the Na^+/H^+ exchanger 1 (NHE-1) [101, 102]. This acid extruder is the most investigated isoform among the 9 members reported so far [103]. In particular, NHE-1 was the first pH-interfering protein considered as a potential antitumor target with its activation in the early steps of carcinogenesis playing a crucial role in neoplastic evolution [101, 102]. An internal allosteric proton-binding regulatory site makes NHE-1 a pH sensor. A bulk of evidence validated the role of NHE1 in tumor progression, invasion, and metastatization, showing that it interacts with the other pH effectors, such as CA II, NBC, and AEs [104]. A study by Andersen et al. investigated the overexpression of NHE1, NBCn1, and MCT4 in mammary tumor revealing that all these membrane proteins contribute to the creation of the reversed pH gradient though via distinct but overlapping mechanisms [105]. It was shown that deletion of NHE-1 as well as its inhibition with amiloride or cariporide (Fig. 6)

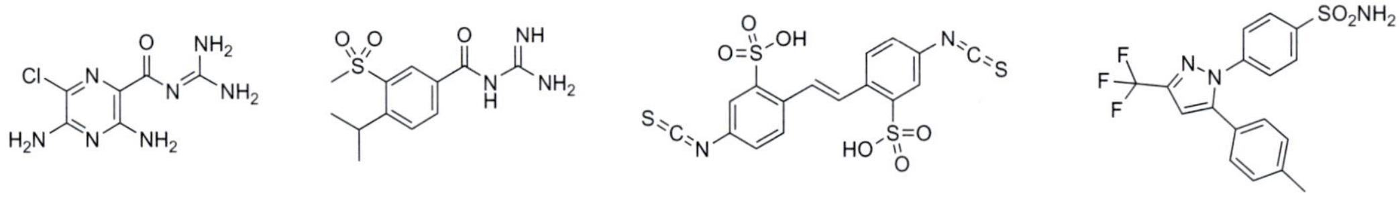

Amiloride Cariporide 4,4'-Diisothiocyano-2,2'-stilbenedisulfonic acid (DIDS) Celecoxib

FIG. 6 Structure of inhibitors discussed in "Inhibitors of Na^+/H^+ exchanger (NHE), Na^+/HCO_3^- co-transporter (NBC) and anion exchanger (AE)" section.

reduced tumor growth [106, 107]. The ubiquity of NHE-1 and the lack of isoform selectivity and significant toxicity of the most available NHE inhibitors impaired their antitumor application to date.

The lack of isoform selective inhibitors is an issue also afflicting bicarbonate transporters, another membrane protein family investigated as potential antitumor targets [108, 109]. These proteins can be divided into two main groups: the anion exchanger (AE) family, composed of three isoforms (AE1, AE2, and AE3) and the Na^+/HCO_3^- cotransporters (NBCs) which include different isoforms. AE2 expression was identified as increased in colon tumors and its suppression in cell lines led to a decreased cell proliferation, suggesting a primary role of AE2 in promoting tumor growth [110]. Among the NBCs, it was shown that NBCe1 expression varies in different cancer cell types, but its role in cancerogenesis remains to be poorly understood [108]. 4,4′-Diisothiocyano-2,2′-stilbenedisulfonate (DIDS) and other related compounds acting as bicarbonate transporters inhibitors were found to exert an antitumor activity, but also reported significant toxicity [111]. Sulfonamide containing compounds such as celecoxib (Fig. 6) also showed to inhibit AE1 transporters and exert antitumor effects [112].

Conclusions

This chapter described the translational process that provided the preclinical to clinical demonstration of the potentiality of PPIs, exploited worldwide for peptic ulcer treatment and gastroprotection, as chemosensitizers or even as antitumor agents. In fact, starting from 2004, a bulk of preclinical and clinical studies pointed out that PPIs are capable to reduce chemoresistance of cancer cell, increase uptake, and thus potentiate the antitumor effectiveness of several antitumor drugs.

References

[1] https://www.who.int/health-topics/cancer#tab=tab_1.
[2] Wilting RH, Dannenberg JH. Epigenetic mechanisms in tumorigenesis, tumor cell heterogeneity and drug resistance. Drug Resist Updat 2012;15:21–38.
[3] Chen Z, Lu W, Garcia-Prieto C, Huang P. The Warburg effect and its cancer therapeutic implications. J Bioenerg Biomembr 2007;39:267–74.
[4] Tredan O, Galmarini CM, Patel K, Tannock IF. Drug resistance and the solid tumor microenvironment. J Natl Cancer Inst 2007;99:1441–54.
[5] De Milito A, Canese R, Marino ML, Borghi M, Iero M, Villa A, et al. pH-dependent antitumor activity of proton pump inhibitors against human melanoma is mediated by inhibition of tumor acidity. Int J Cancer 2010;127:207–19.
[6] Bach LA, Headey SJ, Norton RS. IGF-binding proteins-the pieces are falling into place. Trends Endocrinol Metab 2005;16:228–34.
[7] Jiang L, Xiao L, Sugiura H, Huang X, Ali A, Kuro-O M, et al. Metabolic reprogramming during TGFβ1-induced epithelial-to-mesenchymal transition. Oncogene 2015;34:3908–16.
[8] Semenza GL. HIF-1 mediates the Warburg effect in clear cell renal carcinoma. J Bioenerg Biomembr 2007;39:231–4.
[9] Christofk HR, Vander Heiden MG, Harris MH, Ramanathan A, Gerszten RE, Wei R, et al. The M2 splice isoform of pyruvate kinase is important for cancer metabolism and tumour growth. Nature 2008;452:230–3.
[10] Hsu PP, Sabatini DM. Cancer cell metabolism: Warburg and beyond. Cell 2008;134:703–7.

[11] Popov TM, Goranova T, Stancheva G, Kaneva R, Dikov T, Chalakov I, et al. Relative quantitative expression of hypoxia-inducible factor 1α, -21α, and-31α, and vascular endothelial growth factor a in laryngeal carcinoma. Oncol Lett 2015;9:2879–85.
[12] Izumi H, Torigoe T, Ishiguchi H, Uramoto H, Yoshida Y, Tanabe M, et al. Cellular pH regulators: potentially promising molecular targets for cancer chemotherapy. Cancer Treat Rev 2003;29:541–9.
[13] Spugnini EP, Sonveaux P, Stock C, Perez-Sayans M, De Milito A, Avnet S, Garcìa AG, Harguindey S, Fais S. Proton channels and exchangers in cancer. Biochim Biophys Acta 2015;1848:2715–26.
[14] Sonveaux P, Végran F, Schroeder T, Wergin MC, Verrax J, Rabbani ZN, et al. Targeting lactate-fueled respiration selectively kills hypoxic tumor cells in mice. J Clin Invest 2008;118:3930–42.
[15] Adar Y, Stark M, Bram EE, Nowak-Sliwinska P, van den Bergh H, Szewczyk G, et al. Imidazoacridinone-dependent lysosomal photodestruction: a pharmacological Trojan horse approach to eradicate multidrug-resistant cancers. Cell Death Dis 2012;3:e293.
[16] Gotink KJ, Broxterman HJ, Labots M, de Haas RR, Dekker H, Honeywell RJ, et al. Lysosomal sequestration of sunitinib: a novel mechanism of drug resistance. Clin Cancer Res 2011;17:7337–46.
[17] Gatenby RA, Gawlinski ET, Gmitro AF, Kaylor B, Gillies R. Acid-mediated tumor invasion: a multidisciplinary study. Cancer Res 2006;66:5216–23.
[18] Martin NK, Gaffney E, Gatenby R, Maini P. Tumour-stromal interactions in acid-mediated invasion: a mathematical model. J Theor Biol 2010;267:461–70.
[19] Gonen N, Assaraf YG. Antifolates in cancer therapy: structure, activity and mechanisms of drug resistance. Drug Resist Updat 2012;15:183–210.
[20] Nobili S, Landini I, Giglioni B, Mini E. Pharmacological strategies for over-coming multidrug resistance. Curr Drug Targets 2006;7:861–79.
[21] Gillies R, Raghunand N, Garcia-Martin M, Gateby R. pH imaging. A review of pH measurement methods and applications in cancers. IEEE Eng Med Biol Mag 2004;23:57–64.
[22] Smallbone K, Gatenby RA, Maini P. Mathematical modeling of tumour acidity. J Theor Biol 2008;255:106–12.
[23] Damaghi M, Wojtkowiak JW, Gillies RJ. pH sensing and regulation in cancer. Front Physiol 2013;4:370.
[24] Wojtkowiak JW, Verduzco D, Schramm KJ, Gillies RJ. Drug resistance and cellular adaptation to tumor acidic pH microenvironment. Mol Pharm 2011;8:2032–8.
[25] Zhitomirsky B, Assaraf YG. Lysosomal sequestration of hydrophobic weak base chemotherapeutics triggers lysosomal biogenesis and lysosome-dependent cancer multidrug resistance. Oncotarget 2015;6:1143–56.
[26] Ellegaard AM, Groth-Pedersen L, Oorschot V, Klumperman J, Kirkegaard T, Nylandsted J, et al. Sunitinib and SU11652 inhibit acid sphingomyelinase, destabilize lysosomes, and inhibit multidrug resistance. Mol Cancer Ther 2013;12:2018–30.
[27] Mahoney BP, Raghunand N, Baggett B, Gillies RJ. Tumor acidity, ion trapping and chemotherapeutics. I Acid pH affects the distribution of chemotherapeutic agents in vitro. Biochem Pharmacol 2003;66:1207–18.
[28] Federici C, Petrucci F, Caimi S, Cesolini A, Logozzi M, Borghi M, et al. Exosome release and low pH belong to a framework of resistance of human melanoma cells to cisplatin. PLoS One 2014;9:e88193.
[29] Luqmani YA. Mechanisms of drug resistance in cancer chemotherapy. Med Princ Pract 2005;14:35–48.
[30] Mellor HR, Callaghan R. Accumulation and distribution of doxorubicin in tumour spheroids: the influence of acidity and expression of P-glycoprotein. Cancer Chemother Pharmacol 2011;68:1179–90.
[31] Rauch C. On the relationship between drug size, cell membrane mechanical properties and high levels of multi drug resistance: a comparison to published data. Eur Biophys J 2009;38:537–46.
[32] Luciani F, Spada M, De Milito A, Molinari A, Rivoltini L, Montinaro A, et al. Effect of proton pump inhibitor pretreatment on resistance of solid tumors to cytotoxic drugs. J Natl Cancer Inst 2004;96:1702–13.
[33] Spugnini E, Fais S. Proton pump inhibition and cancer therapeutics: a specific tumor targeting or it is a phenomenon secondary to a systemic buffering? Semin Cancer Biol 2017;43:111–8.
[34] Ferrari S, Perut F, Fagioli F, Brach Del Prever A, Meazza C, Parafioriti A, Picci P, Gambarotti M, Avnet S, Baldini N, Fais S. Proton pump inhibitor chemosensitization in human osteosarcoma: from the bench to the patients' bed. J Transl Med 2013;11:268.
[35] Barar J, Omidi Y. Dysregulated pH in tumor microenvironment checkmates cancer therapy. Bioimpacts 2013;3:149–62.
[36] Daniel C, Bell C, Burton C, Harguindey S, Reshkin SJ, Rauch C. The role of proton dynamics in the development and maintenance of multidrugresistance in cancer. Biochim Biophys Acta 2013;1832:606–17.
[37] Fais S, De Milito A, You H, Qin W. Targeting vacuolar H^{+}-ATPases as a new strategy against cancer. Cancer Res 2007;67:10627–30.

[38] Zhao J, Benlekbir S, Rubinstein JL. Electron cryomicroscopy observation of rotational states in a eukaryotic V-ATPase. Nature 2015;521:241–5.
[39] Forgac M. Vacuolar ATPases: rotary proton pumps in physiology and pathophysiology. Nat Rev Mol Cell Biol 2007;8:917–29.
[40] Rath S, Liebl J, Fürst R, Vollmar AM, Zahler S. Regulation of endothelial signaling and migration by v-ATPase. Angiogenesis 2014;17:587–601.
[41] Mijaljica D, Prescott M, Devenish RJ. V-ATPase engagement in autophagic processes. Autophagy 2011;7:666–8.
[42] Zoncu R, Bar-Peled L, Efeyan A, Wang S, Sancak Y, Sabatini DM. mTORC1senses lysosomal amino acids through an inside-out mechanism that requires the vacuolar H(+)-ATPase. Science 2011;334:678–83.
[43] von Schwarzenberg K, Wiedmann RM, Oak P, Schulz S, Zischka H, Wanner G, Efferth T, Trauner D, Vollmar AM. Mode of cell death induction by pharmacological vacuolar H^+-ATPase (V-ATPase) inhibition. J Biol Chem 2013;288:1385–96.
[44] von Schwarzenberg K, Lajtos T, Simon L, Muller R, Vereb G, Vollmar AM. V-ATPase inhibition overcomes trastuzumab resistance in breast cancer. Mol Oncol 2014;8:9–19.
[45] García-García A, Pérez-Sayáns García M, Rodríguez MJ, Antúnez-López J, Barros-Angueira F, Somoza-Martín M, Gándara-Rey JM, Aguirre-Urízar JM. Immunohistochemical localization of C1 subunit of V-ATPase (ATPase C1) in oral squamous cell cancer and normal oral mucosa. Biotech Histochem 2012;87:133–9.
[46] Huang L, Lu Q, Han Y, Li Z, Zhang Z, Li X. ABCG2/V-ATPase was associated with the drug resistance and tumor metastasis of esophageal squamous cancer cells. Diagn Pathol 2012;7:180.
[47] Chung C, Mader CC, Schmitz JC, Atladottir J, Fitchev P, Cornwell ML, Koleske AJ, Crawford SE, Gorelick F. The vacuolar-ATPase modulates matrix metalloproteinase isoforms in human pancreatic cancer. Lab Investig 2011;91:732–43.
[48] Lu Q, Lu S, Huang L, Wang T, Wan Y, Zhou CX, Zhang C, Zhang Z, Li X. The expression of V-ATPase is associated with drug resistance and pathology of non-small-cell lung cancer. Diagn Pathol 2013;8:145.
[49] Avnet S, Di Pompo G, Lemma S, Salerno M, Perut F, Bonuccelli G, Granchi D, Zini N, Baldini N. V-ATPase is a candidate therapeutic target for Ewing sarcoma. Biochim Biophys Acta 2013;1832:1105–16.
[50] Perez-Sayans M, Garcia-Garcia A, Reboiras-Lopez MD, Gandara-Vila P. Role of V-ATPases in solid tumors: importance of the subunit C. Int J Oncol 2009;34:1513–20.
[51] Mullin JM, Gabello M, Murray LJ, Farrell CP, Bellows J, Wolov KR, Kearney KR, Rudolph D, Thornton JJ. Proton pump inhibitors: actions and reactions. Drug Discov Today 2009;14:647–60.
[52] Han YM, Hahm KB, Park JM, Hong SP, Kim EH. Paradoxically augmented anti-tumorigenic action of proton pump inhibitor and GastrininAPCMin/+ intestinal polyposis model. Neoplasia 2014;16:73–83.
[53] De Milito A, Fais S. Tumor acidity, chemoresistance and proton pump inhibitors. Future Oncol 2005;1:779–86.
[54] Gerweck LE, Vijayappa S, Kozin S. Tumor pH controls the in vivo efficacy of weak acid and base chemotherapeutics. Mol Cancer Ther 2006;5:1275–9.
[55] De Milito A, Iessi E, Logozzi M, Lozupone F, Spada M, Marino ML, Federici C, Perdicchio M, Matarrese P, Lugini L, Nilsson A, Fais S. Proton pump inhibitors induce apoptosis of human B-cell tumors through a caspase-independent mechanism involving reactive oxygen species. Cancer Res 2007;67:5408–17.
[56] Calcinotto A, Filipazzi P, Grioni M, Iero M, De Milito A, Ricupito A, Cova A, Canese R, Jachetti E, Rossetti M, Huber V, Parmiani G, Generoso L, Santinami M, Borghi M, Fais S, Bellone M, Rivoltini L. Modulation of microenvironment acidity reverses anergy in human and murine tumor-infiltrating T lymphocytes. Cancer Res 2012;72:2746–56.
[57] Chen M, Huang SL, Zhang XQ, Zhang B, Zhu H, Yang VW, Zou XP. Reversal effects of pantoprazole on multidrug resistance in human gastric adenocarcinoma cells by down-regulating the V-ATPases/mTOR/HIF-1/P-gp and MRP1 signaling pathway in vitro and in vivo. J Cell Biochem 2012;113:2474–87.
[58] Yeo M, Kim DK, Kim YB, Oh TY, Lee JE, Cho SW, Kim HC, Hahm KB. Selective induction of apoptosis with proton pump inhibitor in gastric cancer cells. Clin Cancer Res 2004;10:8687–96.
[59] Yeo M, Kim DK, Park HJ, Cho SW, Cheong JY, Lee KJ. Blockage of intracellular proton extrusion with proton extrusions with proton pump inhibitor induces apoptosis in gastric cancer. Cancer Sci 2008;99:185.
[60] Patel KJ, Lee C, Tan Q, Tannock IF. Use of the proton pump inhibitor pantoprazole to modify the distribution and activity of doxorubicin: a potential strategy to improve the therapy of solid tumors. Clin Cancer Res 2013;19:6766–76.
[61] Yu M, Lee C, Wang M, Tannock IF. Influence of the proton pump inhibitor lansoprazole on distribution and activity of doxorubicin in solid tumors. Cancer Sci 2015;106:1438–47.

[62] Fan S, Niu Y, Tan N, Wu Z, Wang Y, You H, Ke R, Song J, Shen Q, Wang W, Yao G, Shu H, Lin H, Yao M, Zhang Z, Gu J, Qin W. LASS2 enhances chemosensitivity of breast cancer by counteracting acidic tumor microenvironment through inhibiting activity of V-ATPase proton pump. Oncogene 2013;32:1682–90.
[63] Glunde K, Guggino SE, Solaiyappan M, Pathak AP, Ichikawa Y, Bhujwalla ZM. Extracellular acidification alters lysosomal trafficking in human breast cancer cells. Neoplasia 2003;5:533–45.
[64] Jin UH, Lee SO, Pfent C, Safe S. The aryl hydrocarbon receptor ligand omeprazole inhibits breast cancer cell invasion and metastasis. BMC Cancer 2014;14:498.
[65] Udelnow A, Kreyes A, Ellinger S, Landfester K, Walther P, Klapperstueck T, Wohlrab J, Goh W, Sleptsova-Freidrich I, Petrovic N. Use of proton pump inhibitors as adjunct treatment for triple-negative breast cancers. An introductory study. J Pharm Pharm Sci 2014;17:439–46.
[66] Morimura T, Fujita K, Akita M, Nagashima M, Satomi A. The proton pump inhibitor inhibits cell growth and induces apoptosis in human hepatoblastoma. Pediatr Surg Int 2008;24:1087–94.
[67] Lee YY, Jeon HK, Hong JE, Cho YJ, Ryu JY, Choi JJ, Lee SH, Yoon G, Kim WY, Do IG, Kim MK, Kim TJ, Choi CH, Lee JW, Bae DS, Kim BG. Proton pump inhibitors enhance the effects of cytotoxic agents in chemoresistant epithelial ovarian carcinoma. Oncotarget 2015;6:35040–50.
[68] Lozupone F, Borghi M, Marzoli F, Azzarito T, Matarrese P, Iessi E, Venturi G, Meschini S, Canitano A, Bona R, Cara A, Fais S. TM9SF4 is a novel V-ATPase-interacting protein that modulates tumor pH alterations associated with drug resistance and invasiveness of colon cancer cells. Oncogene 2015;34:5163–74.
[69] Song J, Ge Z, Yang X, Luo Q, Wang C, You H, Ge T, Deng Y, Lin H, Cui Y, Chu W, Yao M, Zhang Z, Gu J, Fan J, Qin W. Hepatic stellate cells activated by acidic tumor microenvironment promote the metastasis of hepatocellular carcinoma via osteopontin. Cancer Lett 2015;356:713–20.
[70] Henne-Bruns D, Knippschild U, Würl P. Omeprazole inhibits proliferation and modulates autophagy in pancreatic cancer cells. PLoS One 2011;6:e20143.
[71] Ihling A, Ihling CH, Sinz A, Gekle M. Acidosis-induced changes in proteome patterns of the prostate cancer-derived tumor cell line AT-1. J Proteome Res 2015;14:3996–4004.
[72] Azzarito T, Venturi G, Cesolini A, Fais S. Lansoprazole induces sensitivity to suboptimal doses of paclitaxel in human melanoma. Cancer Lett 2015;356:697–703.
[73] Lugini L, Federici C, Borghi M, Azzarito T, Marino ML, Cesolini A, Spugnini EP, Fais S. Proton pump inhibitors while belonging to the same family of generic drugs show different anti-tumor effect. J Enzyme Inhib Med Chem 2015;28:1–8.
[74] Canitano A, Iessi E, Spugnini EP, Federici C, Fais S. Proton pump inhibitors induce a caspase-independent antitumor effect against human multiple myeloma. Cancer Lett 2016;376:278–83.
[75] Fais S, Venturi G, Gatenby B. Microenvironmental acidosis in carcinogenesis and metastases: new strategies in prevention and therapy. Cancer Metastasis Rev 2014;33:1095–108.
[76] Bellone M, Calcinotto A, Filipazzi P, De Milito A, Fais S, Rivoltini L. The acidity of the tumor microenvironment is a mechanism of immune escape that can be overcome by proton pump inhibitors. Onco Targets Ther 2013; 2:e22058.
[77] Lugini L, Matarrese P, Tinari A, Lozupone F, Federici C, Iessi E, Gentile M, Luciani F, Parmiani G, Rivoltini L, Malorni W, Fais S. Cannibalism of live lymphocytes by human metastatic but not primary melanoma cells. Cancer Res 2006;66:3629–38.
[78] Pilon-Thomas S, Kodumudi KN, El-Kenawi AE, Russell S, Weber AM, Luddy K, Damaghi M, Wojtkowiak JW, Mulé JJ, Ibrahim-Hashim A, Gillies RJ. Neutralization of tumor acidity improves antitumor responses to immunotherapy. Cancer Res 2016;76:1381–90.
[79] Sreevathsa MR, Choudhury A. Long-term survival in a patient with carcinoid syndrome receiving treatment for Zollinger-Ellison syndrome. Int Surg 2009;94:20–2.
[80] Wang BY, Zhang J, Wang JL, Sun S, Wang ZH, Wang LP, Zhang QL, Lv FF, Cao EY, Shao ZM, Fais S, Hu XC. Intermittent high dose proton pump inhibitor enhances the antitumor effects of chemotherapy in metastatic breast cancer. J Exp Clin Cancer Res 2015;34:85.
[81] Brana I, Ocana A, Chen EX, Razak AR, Haines C, Lee C, Douglas S, Wang L, Siu LL, Tannock IF, Bedard PL. A phase I trial of pantoprazole in combination with doxorubicin in patients with advanced solid tumors: evaluation of pharmacokinetics of both drugs and tissue penetration of doxorubicin. Investig New Drugs 2014;32:1269–77.
[82] Falcone R, Roberto M, D'Antonio C, Romiti A, Milano A, Onesti CE, Marchetti P, Fais S. High-doses of proton pumps inhibitors in refractory gastro-intestinal cancer: a case series and the state of art. Dig Liver Dis 2016;1:30704–6.

[83] Roberto M, Romiti A, Onesti CE, D'Antonio C, Milano A, Falcone R, Barucca V, Palombi L, Righini R, Marchetti P. A metronomic schedule as salvage chemotherapy for upper gastrointestinal tract cancer. Anti-Cancer Drugs 2016;27:106–11.
[84] Marchetti P, Milano A, D'Antonio C, Romiti A, Falcone R, Roberto M, Fais S. Association between proton pump inhibitors and metronomic capecitabine as salvage treatment for patients with advanced gastrointestinal tumors: a randomized phase II trial. Clin Colorectal Cancer 2016;15:377–80.
[85] Sekido M, Fujita K, Y3 K, Ishida H, Takahashi T, Ohkuma R, Tsunoda T, Ishikawa F, Shibanuma M, Sasaki Y. Rabeprazole intake does not affect systemic exposure to capecitabine and its metabolites, 5′-deoxy-5-fluorocytidine, 5′-deoxy-5-fluorouridine, and 5-fluorouracil. Cancer Chemother Pharmacol 2019;83:1127–35.
[86] Papagerakis S, Bellile E, Peterson LA, Pliakas M, Balaskas K, Selman S, Hanauer D, Taylor JM, Duffy S, Wolf G. Proton pump inhibitors and histamine 2 blockers are associated with improved overall survival in patients with head and neck squamous carcinoma. Cancer Prev Res (Phila) 2014;7:1258–69.
[87] Chen CH, Lee CZ, Lin YC, Kao LT, Lin HC. Negative association of proton pump inhibitors with subsequent development of breast cancer: a nationwide population-based study. J Clin Pharmacol 2019;59:350–5.
[88] Ding DC, Sung FC, Chen W, Wang JH, Lin SZ. Proton pump inhibitors reduce breast cancer risk in gastric ulcer patients: a population-based cohort study. Breast J 2020;26:474–8.
[89] Spugnini EP, Baldi A, Buglioni S, Carocci F, de Bazzichini GM, Betti G, Pantaleo I, Menicagli F, Citro G, Fais S. Lansoprazole as a rescue agent in chemoresistant tumors: a phase I/II study in companion animals with spontaneously occurring tumors. J Transl Med 2011;9:221.
[90] Spugnini EP, Buglioni S, Carocci F, Francesco M, Vincenzi B, Fanciulli M, Fais S. High dose lansoprazole combined with metronomic chemotherapy: a phase I/II study in companion animals with spontaneously occurring tumors. J Transl Med 2014;12:225.
[91] Porrello A, Cardelli P, Spugnini EP. Oncology of companion animals as a model for humans an overview of tumor histotypes. J Exp Clin Cancer Res 2006;25:97–105.
[92] Paoloni M, Khanna C. Translation of new cancer treatments from pet dogs to humans. Nat Rev Cancer 2008;8:147–56.
[93] Spugnini EP, Fais S. Drug repurposing for anticancer therapies. A lesson from proton pump inhibitors. Expert Opin Ther Pat 2020;30:15–25.
[94] Ikemura K, Hiramatsu S, Okuda M. Drug repositioning of proton pump inhibitors for enhanced efficacy and safety of cancer chemotherapy. Front Pharmacol 2017;8:911.
[95] Lu ZN, B1 T, Guo XL. Repositioning of proton pump inhibitors in cancer therapy. Cancer Chemother Pharmacol 2017;80:925–37.
[96] Fako VE, Wu X, Pflug B, Liu JY, Zhang JT. Repositioning proton pump inhibitors as anticancer drugs by targeting the thioesterase domain of human fatty acid synthase. J Med Chem 2015;58:778–84.
[97] Halestrap AP, Meredith D. The SLC16 gene family—from monocarboxylate transporters (MCTs) to aromatic amino acid transporters and beyond. Pflugers Arch - Eur J Physiol 2004;447:619–28.
[98] Beloueche-Babari M, Wantuch S, Casals Galobart T, et al. MCT1 inhibitor AZD3965 increases mitochondrial metabolism, facilitating combination therapy and noninvasive magnetic resonance spectroscopy. Cancer Res 2017;77:5913–24.
[99] Heidtmann H, Ruminot I, Becker HM, et al. Inhibition of monocarboxylate transporter by N-cyanosulphonamide S0859. Eur J Pharmacol 2015;762:344–9.
[100] Berrino E, Supuran CT. Novel approaches for designing drugs that interfere with pH regulation. Expert Opin Drug Discovery 2019;14:231–48.
[101] Counillon L, Pouyssegur J. Structure-function studies and molecular regulation of the growth factor activatable sodium- hydrogen exchanger (NHE-1). Cardiovasc Res 1995;29:147–54.
[102] Reshkin SJ, Cardone RA, Harguindey S. Na^+–H^+ exchanger, pH regulation and cancer. Recent Pat Anticancer Drug Discov 2013;8:85–99.
[103] Sardet C, Franchi A, Pouyssegur J. Molecular cloning, primary structure, and expression of the human growth factor-activatable Na^+/H^+ antiporter. Cell 1989;56:271–80.
[104] Deitmer JW, Becker HM. Transport metabolons with carbonic anhydrases. Front Physiol 2013;4:291.
[105] Andersen AP, Samsøe-Petersen J, Oernbo EK. The net acid extruders NHE1, NBCn1 and MCT4 promote mammary tumor growth through distinct but overlapping mechanisms. Int J Cancer 2018;142:2529–42.
[106] Lagarde AE, Franchi AJ, Paris S, et al. Effect of mutations affecting Na^+:H^+ antiport activity on tumorigenic potential of hamster lung fibroblasts. J Cell Biochem 1988;36:249–60.

[107] Masereel B, Pochet L, Laeckmann D. An overview of inhibitors of Na^+/H^+ exchanger. Eur J Med Chem 2003;38:547–54.
[108] Gorbatenko A, Olesen CW, Boedtkjer E, et al. Regulation and roles of bicarbonate transporters in cancer. Front Physiol 2014;5:130.
[109] Alka K, Casey JR. Bicarbonate transport in health and disease. IUBMB Life 2014;66:596–615.
[110] Song LJ, Liu RJ, Zeng Z, et al. Gastrin inhibits a novel, pathological colon cancer signaling pathway involving EGR1, AE2, and P-ERK. J Mol Med 2012;90:707–18.
[111] Jessen F, Sjøholm C, Hoffmann EK. Identification of the anion exchange protein of ehrlich cells: a kinetic analysis of the inhibitory effects of 4, 4′-diisothiocyano-2,2′-stilbene-disulfonic acid (DIDS) and labeling of membrane proteins with 3H-DIDS. J Membr Biol 1986;92:195–205.
[112] Morgan PE, Supuran CT, Casey JR. Carbonic anhydrase inhibitors that directly inhibit anion transport by the human Cl^-/HCO_3^- exchanger, AE1. Mol Membr Biol 2004;21:423–33.

CHAPTER

5

Design, synthesis, and biological evaluation of selective hCA IX inhibitors

Fabrizio Carta[a], *Daniela Vullo*[b], *and Andrea Angeli*[a,c]

[a]NEUROFARBA Department, Section of Pharmaceutical and Nutraceutical Sciences, University of Florence, Florence, Italy [b]Department of Chemistry "Ugo Schiff", University of Florence, Florence, Italy [c]Centre of Advanced Research in Bionanoconjugates and Biopolymers Department, "Petru Poni" Institute of Macromolecular Chemistry, Iasi, Romania

Abstract

Carbonic Anhydrases (CAs; EC 4.2.1.1) are well-known zinc metalloproteins involved in the catalysis of a very simple but essential physiological reaction: carbon dioxide hydration to bicarbonate and proton. These enzymes are of clinical relevance in cancer therapy and among the 15 isoforms known in humans, CA IX, almost exclusively overexpressed in hypoxic tumors, is involved with CA XII in tumorigenesis. Inhibition of the tumor associated isoform IX is now considered as a pertinent approach for the development of new therapies against hypoxic tumors. Different efforts and new families of pharmacological agents were developed that specifically target CA IX over the ubiquitous cytosolic off-target isoforms CA I and CA II. The aim of this chapter is to give a comprehensive update on the reported discoveries in the field of CA IX inhibitors with an emphasis on the new families of compounds, which reach in vivo/preclinical studies for their potential in cancer therapy.

Abbreviations

CA	carbonic anhydrase
CAI	carbonic anhydrase inhibitor
CBP	CREB-binding protein
Df	Desferrioxamine
DOTA	1,4,7,10-tetraazacyclododecane-1,4,7,10-tetraacetic acid
EDC	1-ethyl-3-(3-dimethylaminopropyl)carbodiimide
EPA	Epacadostat
FIH-1	factor inhibiting HIF-1

https://doi.org/10.1016/B978-0-12-820701-7.00014-2

GI	gastrointestinal
HIF-1	hypoxia-inducible factor 1
IC	intracellular domain
IDO1	indoleamine-2,3-dioxygenase 1
mAbs	monoclonal antibodies
NHS	*N*-hydroxysuccinimide
PG	proteoglycan-like domain
PHDs	prolyl hydroxylase domain proteins
***p*-SCN-Bn-DTPA**	*p*-isothiocyanate benzyl diethylenetriaminepentaacetic acid
SAC	saccharin
scFv	single-chain variable fragments
SIP	small immunoproteins
TM	transmembrane domain
VHL	von Hippel Lindau
ZBG	zinc-binding group

Conflict of interest

No potential conflicts of interest were disclosed.

Introduction

Human Carbonic Anhydrase (hCA) IX is a homodimeric transmembrane zinc metalloenzyme whose catalytic domain is oriented toward the extracellular milieu and functioning as the catalyst of reversible hydration of carbon dioxide to carbonic acid [1–3]. hCA IX is expressed as a 49.7-kDa protein and contains an N-terminal "exofacial" proteoglycan-like domain (PG) (59 aa), a catalytic domain (similar to CA II) (257 aa), a signal peptide domain (which is removed prior to enzyme maturation) (37 aa), a transmembrane domain (TM) (20 aa) and, finally, a C-terminal intracellular domain (25 aa) [4–6] (Fig. 1). This protein was discovered in 1992 as a cell surface protein on HeLa cervical cancer cells [7, 8] and, following its identification, was observed the presence of a catalytic domain specific for the family of CAs becoming the ninth member of the human CA family [9].

The different 15 human isoforms are present in various subcellular compartments and contribute to the production of gases, body fluids, bone resorption, and other essential biosynthetic reactions [3, 10, 11]. Excluding hCA IX, all CA isoforms are expressed mainly in specialized cells of the differentiated tissues. On the other hand, hCA IX is not expressed in the majority of normal tissues except for the stomach and gallbladder epithelia [12]. This protein is overexpressed in several kinds of hypoxic tumors making it a promising biological target [13–15]. The catalytic activity of hCA IX is very high and nearly comparable to cytosolic isoform hCA II showing a k_{cat}/K_M ($M^{-1}s^{-1}$) of 5.4×10^7, while hCA II has a k_{cat}/K_M ($M^{-1}s^{-1}$) of 1.5×10^8 [16].

CA IX expression is governed by a complex of transcription factors. Although CA IX transcription is clearly HIF-1 dependent, CA IX transcription is also regulated by extracellular acidosis, which accounts for the CA IX expression in the GI tract [17–19]. HIF-1 is a heterodimer consisting of an O_2-regulated HIF-1α subunit and a constitutively expressed HIF-1β subunit. The amount and transcriptional activity of HIF-1α subunit is regulated by two oxygen-dependent molecular switches: degradation by means of prolyl hydroxylase domain proteins (PHDs) and functional inactivation by asparaginyl hydroxylase factor inhibiting HIF-1 (FIH-1) [20]. Under hypoxic conditions, the absence of oxygen does not allow

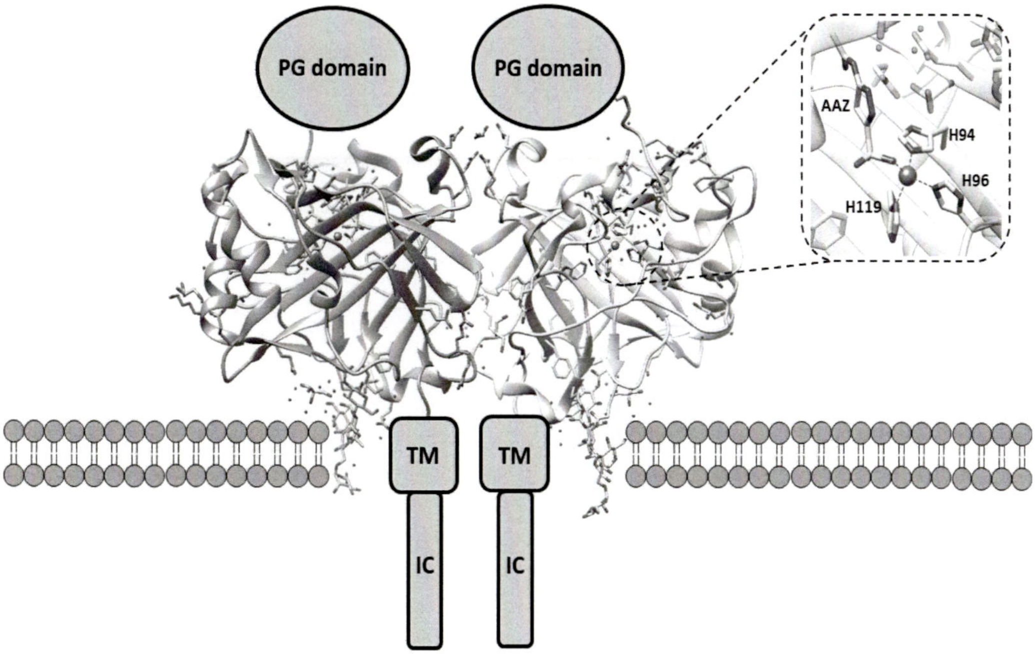

FIG. 1 Schematic diagram of hCA IX structure. Proteoglycan-like domain (PG domain), catalytic domain (from the coordinates of the CA IX crystal structure PDB ID: 3IAI), a transmembrane anchor domain (TM), and an intracellular domain (IC).

PHDs and FIH-1 to hydroxylate HIF-1α. In this way, HIF-1α accumulates in the cell, translocates to the nucleus, and dimerizes with HIF-1β forming an active HIF-1, which interacts with transcriptional coactivators p300/CBP and binds to a hypoxia response element present in the promoter of CA IX and many other hypoxia-regulated genes, activating their expression (Fig. 2) [21, 22].

Tumor cells with activated pH-regulating machinery can resist to chemo-, radio-, and immuno-therapies [23, 24]. Indeed, acidosis is a potent inhibitor of T-cell effector functions [25] and neutralization of tumor acidity can improve response to immunotherapy [26, 27]. In this context, CA IX has garnered more interest because of its limited normal expression and, to date, two main strategies have been developed for targeting it [28, 29]. The first way was the development of monoclonal antibodies and the second consists of the discovery of small molecules able to modulate preferentially CA IX activity. The most important results obtained so far in both fields are reviewed below.

Antibodies and antibody fragments which target CA IX

The most clinically investigated monoclonal antibodies (mAbs) for targeting CA IX is girentuximab (cG250) that was developed over 20 years ago [30]. This mAb binds to the extracellular proteoglycan (PG)-like domain, which is unique to CA IX compared to other isoforms and lies adjacent to the catalytic domain [30]. In addition to therapy, cG250 has also

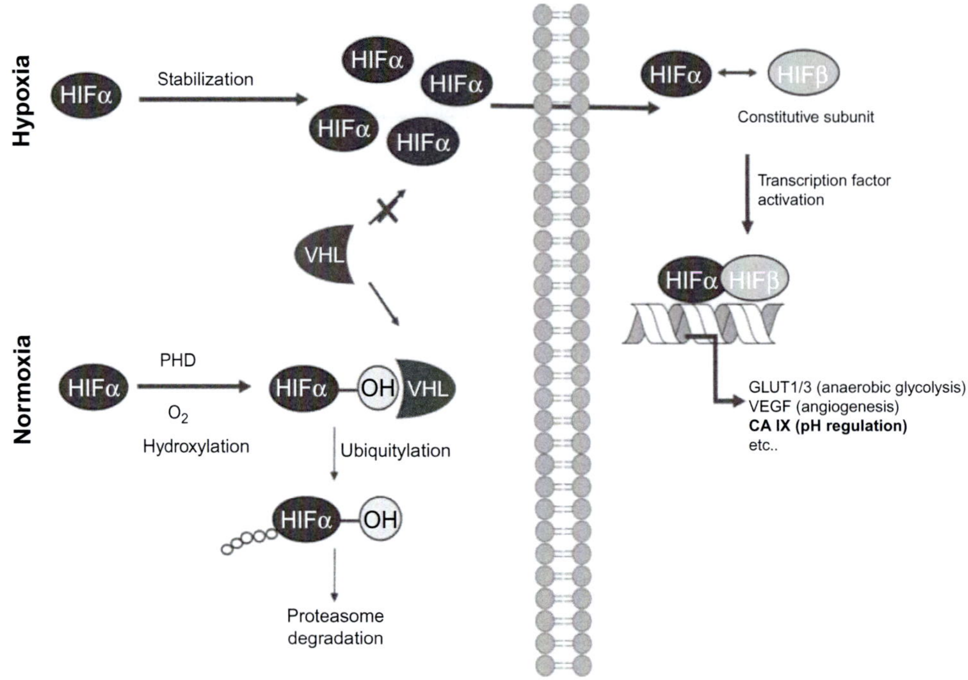

FIG. 2 Mechanism of hypoxia-induced gene expression mediated by hypoxia-inducible factor transcription factor.

been utilized as an imaging agent. The first cG250 radiolabeling, ^{131}I-labeled G250, was evaluated in two clinical trials, [31, 32]. This mAb was directly radiolabeled via the iodogen (1,3,4,6-tetrachloro-3a,6a-diphenyl glycoluril) method [33], a strong oxidizing agent, converting the radioactive iodine I^- provided via aqueous solution of NaI to an electrophilic I^+ cation. This electrophilic species reacted with tyrosine residues of the mAb and inserted in *ortho* position to the hydroxyl group (Fig. 3) [34].

However, the high-energy gamma emissions of ^{131}I and reliance on collimators for detection can lead to poor resolution images that limit the quantification [35]. Contrarily, ^{124}I is an alternative iodine radionuclide that has been used for radiolabeling cG250 [36] and enhances the sensitivity and resolution for PET imaging. This radionuclide has two potential limitations for the clinical adoption of ^{124}I-cG250: in vivo dehalogenation and the relative cost of ^{124}I production. Different approach was used for radiolabeled cG250 with different

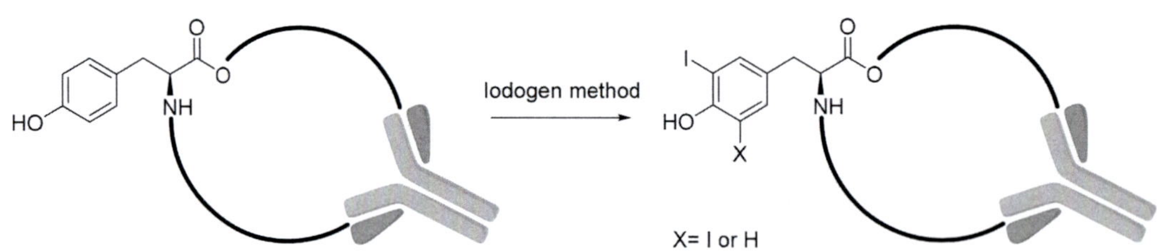

FIG. 3 Direct iodination of antibodies achieved via the iodogen method.

radionuclides such as three beta emitters (^{90}Y, ^{177}Lu, and ^{186}Re [37]) and ^{111}In and ^{89}Zr for immuno SPECT and immuno PET, respectively [38, 39]. Direct radiolabeling with most of the radionuclides mentioned above is not possible and instead an indirect approach with a host-linker moiety is needed. In this indirect labeling method, the host coordinates to the radionuclide while the linker allows covalent attachment to mAb with minimal disruption to structure. Carlin et al. [40] synthetized two low-molecular-weight fragments of cG250 (cG250-Fab and cG250-F(ab′)$_2$) and successively labeling with ^{111}In radionuclide employed an amide linkage among 1,4,7,10-tetraazacyclododecane-1,4,7,10-tetraacetic acid (DOTA) and mAb fragment (Fig. 4A).

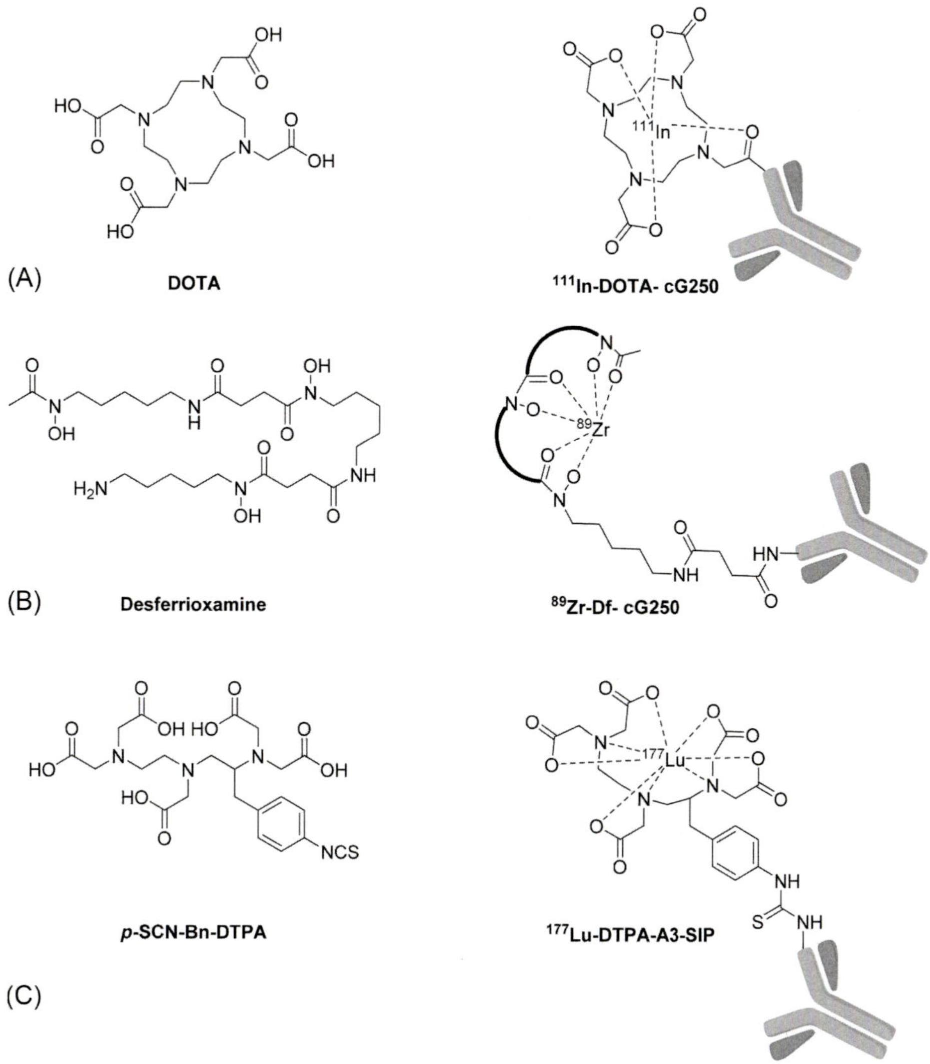

FIG. 4 Host-linker moieties that have been employed to indirectly radiolabel antibodies and antibody fragments targeting CA IX. (A) DOTA, (B) Desferrioxamine, (C) *p*-SCN-Bn-DTPA.

The F(ab′)$_2$ fragment is bivalent (two antigen-binding sites), whereas the Fab fragment is monovalent (one antigen-binding site) [41]. These fragments were conjugated to DOTA in two steps. Firstly, *N*-hydroxysuccinimide (NHS) and 1-ethyl-3-(3-dimethylaminopropyl) carbodiimide (EDC) are used to synthesize the active monoester of DOTA. Secondly, the DOTA active ester is linked to the mAb or fragment via amide bond formation through a side-chain lysine residue. The stoichiometry of incorporated DOTA moieties varied. The F (ab)$_2$ fragment was radiolabeled with ^{89}Zr from Hoeben et al. [42] using as host conjugate Desferrioxamine (Df, Fig. 4B). The labeling chemistry is very similar to ^{111}In-DOTA-cG250. Indeed, the linkage was carried out via amide bond followed radiolabeling with ^{89}Zr [43]. This radioligand was selected because its half-life (3.26 days) is similar to the clearance time of the cG250 and PET images produced using ^{89}Zr-Df-cG250 were highly superior to the ^{111}In-Df-cG250 images [43]. Ahlskog et al. characterized two humanized mAbs, A3 and CC7, identified through phage-display technology [44]. A3 and CC7 are the first fully human mAbs binding to the extracellular domain of CA IX with nanomolar affinity and were selective for CA IX over the other isoforms [44]. To facilitate faster distribution and clearance, A3 and CC7 were subcloned as single-chain variable fragments (scFv; 25 kDa) and small immunoproteins (SIP; 76 kDa). A3-SIP and CC7-SIP were radiolabeled with ^{177}Lu by covalent modification of a side-chain lysine residue of the SIP by reaction with *p*-isothiocyanate benzyl diethylenetriaminepentaacetic acid (*p*-SCN-Bn-DTPA) to form a thiourea link, followed by coordination of ^{177}Lu radionuclide to the DTPA moiety (Fig. 4C) [45]. Mass spectrometry analysis showed a stoichiometry of ~1.8 DTPA molecules per antibody molecule [44]. M75 antibody was developed for preclinical imaging of CA IX in hypoxic niches by Chrastina et al. This mAb is highly specific for CA IX [46] recognizing the extracellular proteoglycan-like domain and is useful for western blotting, immunoprecipitation, and immunohistochemistry [46]. A radiolabeled derivative has been developed for preclinical imaging of CA IX in hypoxic tumors in mouse xenograft models [47, 48] and CA IX selective accumulation of the M75 antibody labeled with ^{125}I was demonstrated in HT-29 xenografts in nude mice [47, 49]. In 2019, Stravinskiene et al. developed a new antibody (H7) against human CA IX generated after immunization of mice with either recombinant full-length extracellular domain of CA IX [50]. mAb H7 was shown to interact with the PG domain of CA IX and showed similar properties with a widely used mAb M75.

In the last years, also was developed a different approach aimed to discover specific anti CA IX inhibitors such as affibodies molecules. These proteins are engineered, based on a stable three helical bundle structure, binding their targets of interest with nano or picomolar affinities and selectivity is mediated by the permutation of 13 amino acid residues [51]. The small size (7–8 kDa) of affibodies results in a rapid localization of tumors, they are tolerant to high temperatures and extreme pHs. These features make them ideal as probes for radionuclide molecular imaging and in vivo applications [52]. The first affibody used for CA IX targeting was ZCAIX:1 reported by Honarvar et al. [53]. The affibody was radiolabelled with [^{99m}Tc]Tc(CO)$_3$ and ^{125}I via direct iodination. Among four different anti-CA IX affibody molecules, [^{99m}Tc]Tc(CO)$_3$-HE$_3$-ZCAIX:2 should provide the best imaging of CA IX-expression in disseminated cancer [53]. However, the labeling with [^{99m}Tc]Tc(CO)$_3$ required a laborious multistep procedure, which might be an obstacle for clinical translation. In order to obviate at a multistep procedure, Garousi et al. [54] selected an approach based on site-specific conjugation of DOTA at C-terminal cysteine in ZCAIX:2 with a maleimide conjugation. In the case of ^{111}In-DOTA-HE$_3$-ZCAIX:2, on the other hand, the uptake is unspecific making radiometal-labeled affibody molecules unsuitable for therapeutic targeting [54].

Small molecules which target CA IX

Sulfonamides are the most common class of CA IX inhibitors developed during the last 20 years. However, many of them act as highly efficient CA inhibitors (CAIs) in the low nano- or subnanomolar ranges, but indiscriminately inhibit most catalytically active CA isoforms with lack of selectivity [55, 56]. One of the most successful approaches for designing CA inhibitors targeting all isoforms known to date was denominated "the tail approach" [57]. This approach originally consisted in attaching different tails to scaffolds of well-known aromatic/heterocyclic sulfonamides. The nature of the "tail" moieties was very much varied, incorporating all possible moieties (lipophilic, positively charged, metal coordinating groups) as it has been observed that the tails fragments bind toward the entrance of the active site of the CA isoforms [58], a region that has the highest variability among the different isoforms, and this leads to compounds, which are able to discriminate between the various isoforms and are thus isozyme-selective CAIs [1, 2, 59]. A very good example of such "tails" is constituted by sugars. This has been denominated the "sugar approach," which led to highly successful drug design targeting many isoforms, but mostly the tumor-associated ones CA IX/XII [60]. An interesting example of this approach was achieved in 2011 by Poulsen's group developing a synthesis of carbohydrate sulfamates from monosaccharides through selective sulfamoylation of the C-6 hydroxyl group (Scheme 1) [61].

The first step, involving a deprotection of C-6 primary hydroxyl groups, followed by selective reaction with sulfamoyl chloride generated in situ, afforded the per-*O*-acylated carbohydrate sulfamates. Zemplen's conditions were employed to remove the ester-protecting groups to prepare the fully deprotected compounds [62]. The new sulfamates compounds, in particular derivatives **1** and **2**, reported here, showed a good selectivity toward CA IX than the ubiquitous CA II. Compound **1** exhibited a K_I of 82 nM for CA II and a K_I of 8.6 nM against CA IX with a selectivity for the tumor-associated isoform over 9 times. On the other hand, compound **2** had a selectivity index versus CA IX only of 1.5 but, due to their poor membrane permeability would selectively target the extracellular CA IX.

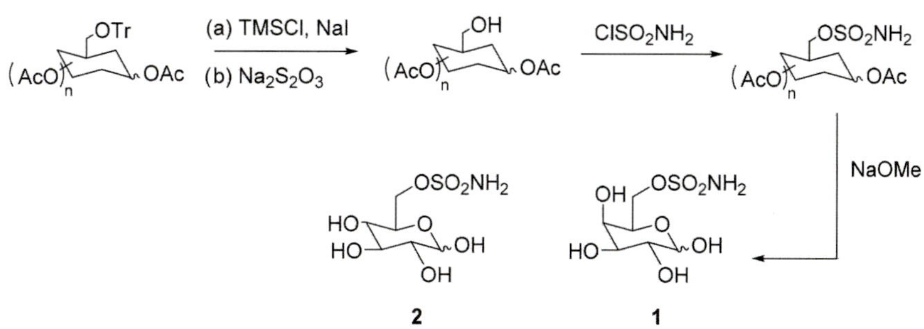

SCHEME 1 Synthesis of carbohydrate sulfamates **1** and **2**.

A second promising "tail" approach is the ureido moiety bearing benzenesulfonamide and sulfamate chemical classes [62, 63]. Ureido inhibitors have both been intensively validated in 2D (monolayer) and 3D (multicellular spheroids) cell cultures, tissue explants, and in human tumor xenograft experiments inhibiting the tumor cell proliferation, migration, and invasion when used in the low micromolar concentration range [64–66]. The tails of these inhibitors adopt rather variable orientations when bound within the enzyme cavity, which is mainly due to the flexible ureido linker between the two aromatic rings present in these class of compounds and clearly understood by X-ray crystal structure of derivatives **3** and its congeners in adduct with CA II and IX [67, 68]. The chemical structures and K_I values for CA II and CA IX of benzenesulfonamide compounds mentioned in this chapter are shown in Fig. 5.

These compounds exhibited an interesting CA IX/XII selectivity [63, 67, 68] and, to date, compound **3** (SLC-0111) is the only in clinical Phase I/II for the treatment of metastatic pancreatic ductal cancer [2]. Recently, another drug in clinical Phase was discovered by Supuran group's as effective and selective inhibitor for the tumor-associated isoforms hCA IX/XII [69, 70]. Epacadostat (EPA, Fig. 6), a selective indoleamine-2,3-dioxygenase 1 (IDO1)

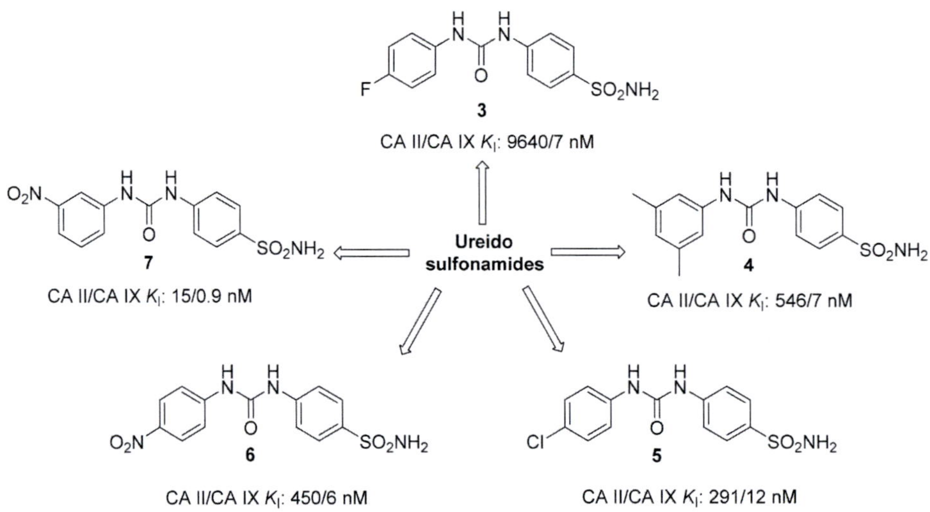

FIG. 5 Chemical structures of ureido derivatives as selective inhibitors of CA IX.

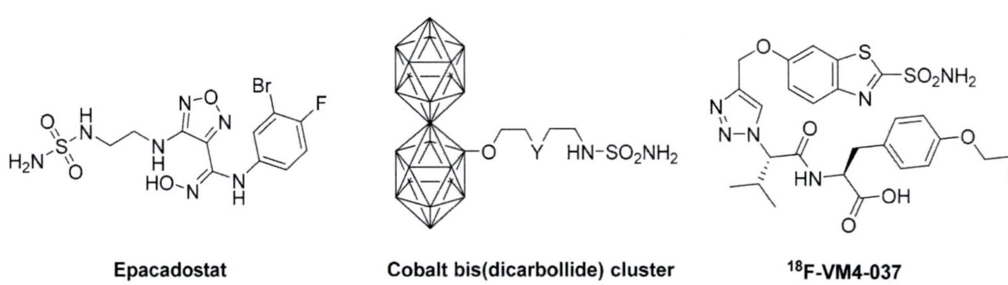

FIG. 6 Chemical structures of selective small molecule inhibitors of CA IX.

inhibitor, may exert antitumor effects also due to the potent inhibition of the tumor-associated CAs. In this scenario, EPA may exert its activity also disrupting the cellular Trp catabolism and tumor microenvironment specifically by dismantling the pH regulation machinery, which is fine regulated by the CAs IX and XII, in agreement with the tissutal acidification events as at the basis of the immune evasion mechanisms [71]. A promising field with good results is the organometallo moieties bearing benzenesulfonamide derivatives. Recently, the cobalt bis(dicarbollide) cluster ion serves as a convenient hydrophobic inorganic building block for the construction of CA IX inhibitors (Fig. 6) with CA IX/CA II selectivity index reaching a value of approximately 500 for the best inhibitor of the series, indicating specific inhibition of CA IX [72].

The only radiolabeled CA IX small-molecule inhibitor to advance into clinical trials is ^{18}F-VM4-037, an ethoxzolamide derivative (Fig. 6). At physiological pH, the free carboxylate group of ^{18}F-VM4-037 is deprotonated, introducing a negative charge that renders the compound membrane-impermeable [73]. This confers selectivity for CA IX over intracellular isoenzymes. In this context, several compounds were synthetized as imaging agents that target CA IX and a novel strategy to obtain them was reported by Thibaudeau group's with a new series of fluorinated tertiary substituted sulfonamides that showed high selectivity for CA IX and did not inhibit the CA II isoenzyme [74]. The unique selectivity of these compounds is attributed to a new but unresolved mechanism of binding.

In the continuous search of novel chemotypes, coumarins were discovered to act as a totally new class of CAIs with a particular inhibition mechanism [75]. These kinds of compounds were observed bound in hydrolyzed form (Fig. 7, compound **10**) at the entrance of the CA II active site cavity, occluding the entrance of the active site and described in detail by means of X-ray crystallographic study. It has been hypothesized that the cis-acid **9** was initially formed, which then underwent isomerization to the thermodynamically more stable

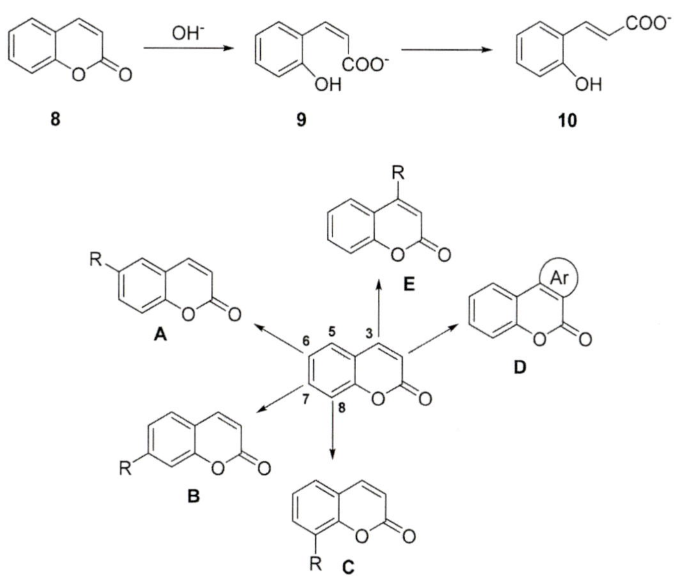

FIG. 7 CA inhibition mechanism of coumarins and different substitutions on its scaffold.

trans isomer **10**. What is really remarkable about this class of inhibitors, denominated "prodrug inhibitors" since the enzyme transforms them into the active compound, apart from their particular inhibition mechanism delineated above, is the fact that they bind in the most variable region of the CA active site.

This probably explains why coumarins are among the most isoform-selective inhibitors, with highly selective inhibitors detected against all human isoforms. For this reason, many efforts are made in the last years in order to develop several series of semisynthetic compounds with a variety of moieties substituting the coumarin ring in various positions. Indeed, it was demonstrated that substituents at carbon position C6 and C7 (Fig. 7, structures **A** and **B**) led to highly effective CA IX inhibitors [76–79]; on the other hand, it was found that substituents in C8 (structure **C**) led to increase the potency against CA XII. [80] The nature of substituents in C3 was observed crucial for the potency of inhibition, varying considerably even with very minor structural modifications [81, 82]. Many efforts were made also to discover coumarin analog scaffolds acting as highly potent and isoform-selective CAIs modifying the heterocyclic ring, as well as the eventual replacement of the oxygen atoms from the bicyclic ring system by one or two replacements [83–85]. 1,2-Benzoxathiine 2,2-dioxides, also referred as "sulfocoumarins" (Fig. 8), are bioisosters of the coumarins and, in analogy to the coumarins, their mechanism of action relies on the CA-mediated esterase activity, which cleaves the intramolecular sulfonic acid ester to afford, upon geometrical isomerization, an (*E*)-vinylsulfonic acid species represents a new chemotype acting as highly potent and isoform-selective CA IX and XII inhibitors [86–88].

A new series of homosulfocoumarins (3*H*-1,2-benzoxathiepine 2,2-dioxides) (Fig. 8) possessing various substitution patterns and moieties in the 7, 8, or 9 position of the heterocylic ring were prepared by Zalubovskis et al. with an original procedure in five steps. These classes of compounds showed selective inhibition against the tumor-associated isoform CA IX. The 8-substituted homosulfocoumarins were the most effective hCA IX/XII inhibitors followed by the 7-substituted derivatives, whereas the substitution pattern in position 9 led to less effective binders for the transmembrane, tumor-associated isoforms IX/XII [89, 90].

A novel class of selective CAIs for CA IX over the other isoforms was the cyclic sulfimide saccharin (SAC) and SAC-based derivatives (Fig. 9). This scaffold has recently showed to selectively discriminate the diverse isoforms, distinctly from classical CAIs such as primary sulfonamide [91, 92].

It is, in particular, a selective nanomolar inhibitor of CA IX, showing a weaker, micromolar inhibition of the ubiquitous CA I and II (60-fold II/IX selectivity) [93]. Its cyclic acyl sulfonamide, which is much more acidic than primary sulfonamides, functions as a zinc-binding group (ZBG) within the CA active site. Therefore, the specificity for CA IX is likely attributed to the feasibility of inhibitor movement and entry into the active site. CA IX has a larger

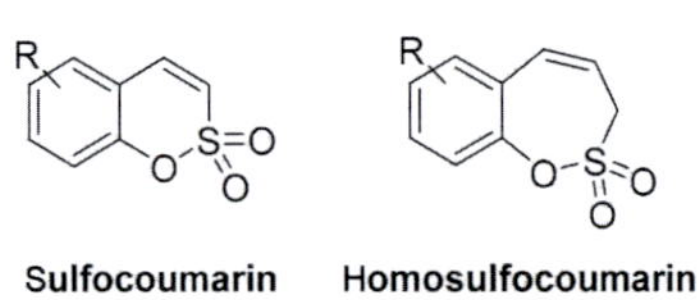

FIG. 8 Chemical structures of sulfocoumarins and homosulfocoumarins.

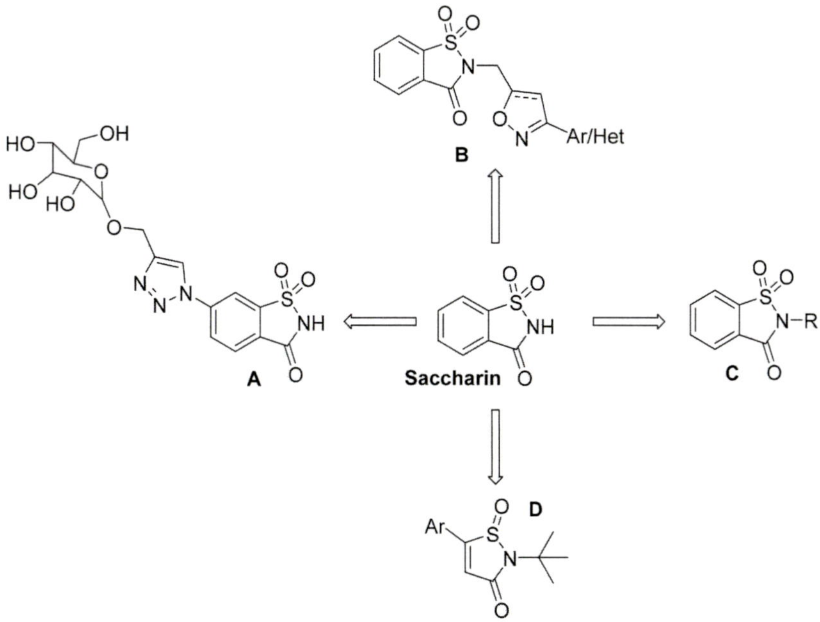

FIG. 9 Chemical structures of saccharin and saccharin derivatives.

hydrophobic pocket, along which substrate and many inhibitors travel, due to residue V131 in place of F131 in CA II. The decrease in steric hindrance of this residue at the opening of the active site facilitates the binding of bulkier, aromatic compounds such as these ureido sulfonamides. For this reason, several SAC derivatives were produced such as an addition of a triazole linker and sugar (**A**) [94, 95], isoxazoline hybrids (**B**) [96], functionalizing the sulfonamide nitrogen of saccharin [97, 98] (**C**) and synthetized the saccharin analog 5-phenylisothiazol-3(2*H*)-one-1,1-dioxide (**D**) [99].

Conclusion

CA IX is a well-known validated target for imaging and treatment of several hypoxic tumors. A large number of highly effective and selective CA IX inhibitors were developed over the last 20 years, most of them belonging to the antibodies, sulfonamide, and coumarin chemotypes. Currently, CA IX inhibitors as clinical candidates reported up to now are some sulfonamides belonging to the arylureido-benzenesulfonamide type, among them, SLC-011 recently in clinical trials Phase II results to be the most promising candidate. In addition, recent studies showing that CA IX selective inhibitors by either specific antibodies or small molecules can be conjugated to cytotoxic agents suggesting that both chemoresistance and tumorigenicity can be combatted using one single agent. Inhibition of CA IX activity has been studied in the preclinical setting and more recently in clinical trials as a mode of therapy in several types of tumor leading to accelerate the interest for the development of targeted antitumor/antimetastic drugs for the treatment of CA IX-expressing tumors.

References

[1] Supuran CT. Carbonic anhydrases: novel therapeutic applications for inhibitors and activators. Nat Rev Drug Discov 2008;7:168–81.

[2] Neri D, Supuran CT. Interfering with pH regulation in tumours as a therapeutic strategy. Nat Rev Drug Discov 2011;10:767–77.

[3] Supuran CT. Structure and function of carbonic anhydrases. Biochem J 2016;473:2023–32.

[4] Opavský R, Pastoreková S, Zelník V, Gibadulinová A, Stanbridge EJ, Závada J, Kettmann R, Pastorek J. Human MN/CA9 gene, a novel member of the carbonic anhydrase family: structure and exon to protein domain relationships. Genomics 1996;33:480–7.

[5] Alterio V, Hilvo M, Di Fiore A, Supuran CT, Pan P, Parkkila S, Scaloni A, Pastorek J, Pastorekova S, Pedone C, Scozzafava A, Monti SM, De Simone G. Crystal structure of the catalytic domain of the tumor-associated human carbonic anhydrase IX. Proc Natl Acad Sci U S A 2009;106:16233–8.

[6] De Simone G, Supuran CT. Carbonic anhydrase IX: biochemical and crystallographic characterization of a novel antitumor target. Biochim Biophys Acta 2010;1804:404–9.

[7] Pastorekova S, Zavadova Z, Kostal M, Babusikova O, Zavada J. A novel quasi-viral agent, MaTu, is a two-component system. Virology 1992;187:620–6.

[8] Zavada J, Zavadova Z, Pastorekova S, Ciampor F, Pastorek J, Zelnik V. Expression of MaTu-MN protein in human tumor cultures and in clinical specimens. Int J Cancer 1993;54:268–74.

[9] Pastorek J, Pastorekova S, Callebaut I, Mornon JP, Zelnik V, Opavsky R, Zatóvicova M, Liao S, Portetelle D, Stanbridge EJ, Zavada J, Burny A, Kettman R. Cloning and characterization of MN, a human tumor-associated protein with a domain homologous to carbonic anhydrase and a putative helix-loop-helix DNA binding segment. Oncogene 1994;9:2877–88.

[10] Bozdag M, Altamimi ASA, Vullo D, Supuran CT, Carta F. State of the art on carbonic anhydrase modulators for biomedical purposes. Curr Med Chem 2019;26:2558–73.

[11] Supuran CT. How many carbonic anhydrase inhibition mechanisms exist? J Enzyme Inhib Med Chem 2016;31:345–60.

[12] Barathova M, Takacova M, Holotnakova T, Gibadulinova A, Ohradanova A, Zatovicova M, Hulikova A, Kopacek J, Parkkila S, Supuran CT, Pastorekova S, Pastorek J. Alternative splicing variant of the hypoxia marker carbonic anhydrase IX expressed independently of hypoxia and tumour phenotype. Br J Cancer 2008;98:129–36.

[13] Supuran CT. Carbonic anhydrase inhibitors as emerging agents for the treatment and imaging of hypoxic tumors. Expert Opin Investig Drugs 2018;27:963–70.

[14] Supuran CT, Winum JY. Carbonic anhydrase IX inhibitors in cancer therapy: an update. Future Med Chem 2015;7:1407–14.

[15] Nocentini A, Supuran CT. Carbonic anhydrase inhibitors as antitumor/antimetastatic agents: a patent review (2008–2018). Expert Opin Ther Pat 2018;28:729–40.

[16] Hilvo M, Baranauskiene L, Salzano AM, Scaloni A, Matulis D, Innocenti A, Scozzafava A, Monti SM, Di Fiore A, De Simone G, Lindfors M, Jänis J, Valjakka J, Pastoreková S, Pastorek J, Kulomaa MS, Nordlund HR, Supuran CT, Parkkila S. Biochemical characterization of CA IX, one of the most active carbonic anhydrase isozymes. J Biol Chem 2008;283:27799–809.

[17] Pastorek J, Pastorekova S. Hypoxia-induced carbonic anhydrase IX as a target for cancer therapy: from biology to clinical use. Semin Cancer Biol 2015;31:52–64.

[18] Sedlakova O, Svastova E, Takacova M, Kopacek J, Pastorek J, Pastorekova S. Carbonic anhydrase IX, a hypoxia-induced catalytic component of the pH regulating machinery in tumors. Front Physiol 2014;4:400.

[19] Wykoff CC, Beasley NJ, Watson PH, Turner KJ, Pastorek J, Sibtain A, Wilson GD, Turley H, Talks KL, Maxwell PH, Pugh CW, Ratcliffe PJ, Harris AL. Hypoxia-inducible expression of tumor-associated carbonic anhydrases. Cancer Res 2000;60:7075–83.

[20] Nakazawa MS, Keith B, Simon MC. Oxygen availability and metabolic adaptations. Nat Rev Cancer 2016;16:663–73.

[21] Schofield CJ, Ratcliffe PJ. Signalling hypoxia by HIF hydroxylases. Biochem Biophys Res Commun 2005;338:617–26.

[22] Kondo K. The von Hippel-Lindau tumor suppressor gene. Exp Cell Res 2001;264:117–25.

[23] Gillies RJ, Verduzco D, Gatenby RA. Evolutionary dynamics of carcinogenesis and why targeted therapy does not work. Nat Rev Cancer 2012;12:487–93.

[24] Parks SK, Chiche J, Pouyssegur J. Disrupting proton dynamics and energy metabolism for cancer therapy. Nat Rev Cancer 2013;13:611–23.

[25] Ibrahim-Hashim A, Estrella V. Acidosis and cancer: from mechanism to neutralization. Cancer Metastasis Rev 2019;38:149–55.

[26] Pilon-Thomas S, Kodumudi KN, El-Kenawi AE, Russell S, Weber AM, Luddy K, Damaghi M, Wojtkowiak JW, Mulé JJ, Ibrahim-Hashim A, Gillies RJ. Neutralization of tumor acidity improves antitumor responses to immunotherapy. Cancer Res 2016;76:1381–90.

[27] Kouidhi S, Ben Ayed F, Benammar Elgaaied A. Targeting tumor metabolism: a new challenge to improve immunotherapy. Front Immunol 2018;9:353.

[28] Pastorekova S, Parkkila S, Parkkila AK, Opavsky R, Zelnik V, Saarnio J, Pastorek J. Carbonic anhydrase IXMN/CAIX: analysis of stomach complementaryDNA sequence and expression in human and rat alimentary tracts. Gastroenterology 1997;112:398–408.

[29] Saarnio J, Parkkila S, Parkkila AK, Waheed A, Casey MC, Zhou XY, Pastoreková S, Pastorek J, Karttunen T, Haukipuro K, Kairaluoma MI, Sly WS. Immunohistochemistry of carbonic anhydrase isozyme IX (MN/CA IX) in human gut reveals polarized expression in the epithelial cells with the highest proliferative capacity. J Histochem Cytochem 1998;46:497–504.

[30] Oosterwijk-Wakka J, Boerman O, Mulders P, Oosterwijk E. Monoclonal antibody G250 recognizing carbonic anhydrase IX in renal cell carcinoma: biological and clinical studies. In: Bukowski RM, Figlin RA, Motzer RJ, editors. Renal cell carcinoma: molecular targets and clinical applications. 2nd ed. New York: Humana Press; 2009. p. 231–47.

[31] Oosterwijk E, Bander NH, Divgi CR, Welt S, Wakka JC, Finn RD, Carswell EA, Larson SM, Warnaar SO, Fleuren GJ. Antibody localization in human renal cell carcinoma: a phase I study of monoclonal antibody G250. J Clin Oncol 1993;11:738–50.

[32] Divgi CR, Bander NH, Scott AM, O'Donoghue JA, Sgouros G, Welt S, Finn RD, Morrissey F, Capitelli P, Williams JM, Deland D, Nakhre A, Oosterwijk E, Gulec S, Graham MC, Larson SM, Old LJ. Phase I/II radioimmunotherapy trial with iodine-131-labeled monoclonal antibody G250 in metastatic renal cell carcinoma. Clin Cancer Res 1998;4:2729–39.

[33] Bailey G. The iodogen method for radiolabeling proteins. In: The protein protocols handbook. 1st ed. Humana Press; 1996. p. 673–4.

[34] Salacinski P, McLean C, Sykes J, et al. Iodination of proteins, glycoproteins, and peptides using a solid-phase oxidizing agent, 1,3,4,6-tetrachloro-3a,6a-diphenyl glycoluril (iodogen). Anal Biochem 1981;117:136–46.

[35] Rault E, Vandenberghe S, Van Holen R, De Beenhouwer J, Staelens S, Lemahieu I. Comparison of image quality of different iodine isotopes (I-123, I-124, and I-131). Cancer Biother Radiopharm 2007;22:423–30.

[36] Cascini GL, Niccoli Asabella A, Notaristefano A, Restuccia A, Ferrari C, Rubini D, Altini C, Rubini G. 124 iodine: a longer-life positron emitter isotope-new opportunities in molecular imaging. Biomed Res Int 2014;2014:672094.

[37] Brouwers AH, van Eerd JE, Frielink C, Oosterwijk E, Oyen WJ, Corstens FH, Boerman OC. Optimization of radioimmunotherapy of renal cell carcinoma: labeling of monoclonal antibody cG250 with 131I, 90Y, 177Lu, or 186Re. J Nucl Med 2004;45:327–37.

[38] Muselaers CH, Boerman OC, Oosterwijk E, Langenhuijsen JF, Oyen WJ, Mulders PF. Indium-111-labeled girentuximab immunoSPECT as a diagnostic tool in clear cell renal cell carcinoma. Eur Urol 2013;63:1101–6.

[39] Cheal SM, Punzalan B, Doran MG, Evans MJ, Osborne JR, Lewis JS, Zanzonico P, Larson SM. Pairwise comparison of 89Zr- and 124I-labeled cG250 based on positron emission tomography imaging and nonlinear immunokinetic modeling: in vivo carbonic anhydrase IX receptor binding and internalization in mouse xenografts of clear-cell renal cell carcinoma. Eur J Nucl Med Mol Imaging 2014;41:985–94.

[40] Carlin S, Khan N, Ku T, Longo VA, Larson SM, Smith-Jones PM. Molecular targeting of carbonic anhydrase IX in mice with hypoxic HT29 colorectal tumor xenografts. PLoS One 2010;5:e10857.

[41] Brouwers A, Mulders P, Oosterwijk E, Buijs W, Corstens F, Boerman O, Oyen W. Pharmacokinetics and tumour targeting of 131I-labeled F(ab')2 fragments of the chimeric monoclonal antibody G250: preclincial and clinical pilot studies. Cancer Biother Radiopharm 2004;19:466–77.

[42] Hoeben B, Kaanders J, Franssen G, et al. PET of hypoxia with 89Zr-labeled cG250-F(ab')2 in head and neck tumors. J Nucl Med 2010;51:1076–83.

[43] Brouwers A, Verel I, Van Eerd J, Visser G, Steffens M, Oosterwijk E, Corstens F, Oyen W, Van Dongen G, Boerman O. PET radioimmunoscintigraphy of renal cell cancer using 89Zr-labeled cG250 monoclonal antibody in nude rats. Cancer Biother Radiopharm 2004;19:155–63.

[44] Ahlskog JK, Schliemann C, Marlind J, Qureshi U, Ammar A, Pedley RB, Neri D. Human monoclonal antibodies targeting carbonic anhydrase IX for the molecular imaging of hypoxic regions in solid tumours. Br J Cancer 2009;101:645–57.
[45] Brechbiel M, Gansow O, Atcher R, Schlom J, Esteban J, Simpson D, Colcher D. Synthesis of 1-(p-isothiocyanatobenzyl) dericvatives of DTPA and EDTA. Antibody labeling and tumor-imaging studies. Inorg Chem 1986;25:2772–81.
[46] Li Y, Wang H, Oosterwijk E, Selman Y, Mira JC, Medrano T, Shiverick KT, Frost SC. Antibody-specific detection of CAIX in breast and prostate cancers. Biochem Biophys Res Commun 2009;386:488–92.
[47] Chrastina A, Zavada J, Parkkila S, Kaluz S, Kaluzova M, Rajcani J, Pastorek J, Pastorekova S. Biodistribution and pharmacokinetics of 125I-labeled monoclonal antibody M75 specific for carbonic anhydrase IX, an intrinsic marker of hypoxia, in nude mice xenografted with human colorectal carcinoma. Int J Cancer 2003;105:873–81.
[48] Chrastina A, Pastorekova S, Pastorek J. Immunotargeting of human cervical carcinoma xenograft expressing CA IX tumor-associated antigen by 125I-labeled M75 monoclonal antibody. Neoplasma 2003;50:13–21.
[49] Lau J, Lin KS, Bénard F. Past, present, and future: development of theranostic agents targeting carbonic anhydrase IX. Theranostics 2017;7:4322–39.
[50] Stravinskiene D, Imbrasaite A, Petrikaite V, Matulis D, Matuliene J, Zvirbliene A. New monoclonal antibodies for a selective detection of membrane-associated and soluble forms of carbonic anhydrase IX in human cell lines and biological samples. Biomolecules 2019;9:E304.
[51] Feldwisch J, Tolmachev V. Engineering of affibody molecules for therapy and diagnostics. Methods Mol Biol 2012;899:103–26.
[52] Krasniqi A, D'Huyvetter M, Devoogdt N, Frejd FY, Sörensen J, Orlova A, Keyaerts M, Tolmachev V. Same-day imaging using small proteins: clinical experience and translational prospects in oncology. J Nucl Med 2018;59:885–91.
[53] Honarvar H, Garousi J, Gunneriusson E, Hoiden-Guthenberg I, Altai M, Widstrom C, Tolmachev V, Frejd FY. Imaging of CAIX-expressing xenografts in vivo using 99mTc-HEHEHE-ZCAIX:1 affibody molecule. Int J Oncol 2015;46:513–20.
[54] Garousi J, Huizing FJ, Vorobyeva A, Mitran B, Andersson KG, Leitao CD, Frejd FY, Löfblom J, Bussink J, Orlova A, Heskamp S, Tolmachev V. Comparative evaluation of affibody- and antibody fragments-based CAIX imaging probes in mice bearing renal cell carcinoma xenografts. Sci Rep 2019;9:14907.
[55] De Simone G, Alterio V, Supuran CT. Exploiting the hydrophobic and hydrophilic binding sites for designing carbonic anhydrase inhibitors. Expert Opin Drug Discovery 2013;8:793–810.
[56] Supuran CT, Di Fiore A, Alterio V, Monti SM, De Simone G. Recent advances in structural studies of the carbonic anhydrase family: the crystal structure of human CA IX and CA XIII. Curr Pharm Des 2010;16:3246–54.
[57] Scozzafava A, Menabuoni L, Mincione F, Briganti F, Mincione G, Supuran CT. Carbonic anhydrase inhibitors. Synthesis of water-soluble, topically effective, intraocular pressure-lowering aromatic/heterocyclic sulfonamides containing cationic or anionic moieties: is the tail more important than the ring. J Med Chem 1999;42:2641–50.
[58] Alterio V, Di Fiore A, D'Ambrosio K, Supuran CT, De Simone G. Multiple binding modes of inhibitors to carbonic anhydrases: how to design specific drugs targeting 15 different isoforms? Chem Rev 2012;112:4421–68.
[59] Supuran CT, Scozzafava A, Casini A. Carbonic anhydrase inhibitors. Med Res Rev 2003;23:146–89.
[60] Winum JY, Colinas PA, Supuran CT. Glycosidic carbonic anhydrase IX inhibitors: a sweet approach against cancer. Bioorg Med Chem 2013;21:1419–26.
[61] Lopez M, Trajkovic J, Bornaghi LF, Innocenti A, Vullo D, Supuran CT, Poulsen SA. Design, synthesis, and biological evaluation of novel carbohydrate-based Sulfamates as carbonic anhydrase inhibitors. J Med Chem 2011;54:1481–9.
[62] Williams KJ, Gieling RG. Preclinical evaluation of ureidosulfamate carbonic anhydrase IX/XII inhibitors in the treatment of cancers. Int J Mol Sci 2019;20:E6080.
[63] Pacchiano F, Carta F, McDonald PC, Lou Y, Vullo D, Scozzafava A, Dedhar S, Supuran CT. Ureido-substituted benzenesulfonamides potently inhibit carbonic anhydrase IX and show antimetastatic activity in a model of breast cancer metastasis. J Med Chem 2011;54:1896–902.
[64] Lou Y, McDonald PC, Oloumi A, Chia S, Ostlund C, Ahmadi A, Kyle A, Auf dem Keller U, Leung S, Huntsman D, Clarke B, Sutherland BW, Waterhouse D, Bally M, Roskelley C, Overall CM, Minchinton A, Pacchiano F, Carta F, Scozzafava A, Touisni N, Winum JY, Supuran CT, Dedhar S. Targeting tumor

hypoxia: suppression of breast tumor growth and metastasis by novel carbonic anhydrase IX inhibitors. Cancer Res 2011;71:3364–76.

[65] Lock FE, McDonald PC, Lou Y, Serrano I, Chafe SC, Ostlund C, Aparicio S, Winum JY, Supuran CT, Dedhar S. Targeting carbonic anhydrase IX depletes breast cancer stem cells within the hypoxic niche. Oncogene 2013;32:5210–9.

[66] Swayampakula M, McDonald PC, Vallejo M, Coyaud E, Chafe SC, Westerback A, Venkateswaran G, Shankar J, Gao G, Laurent EMN, Lou Y, Bennewith KL, Supuran CT, Nabi IR, Raught B, Dedhar S. The interactome of metabolic enzyme carbonic anhydrase IX reveals novel roles in tumor cell migration and invadopodia/MMP14-mediated invasion. Oncogene 2017;36:6244–61.

[67] Pacchiano F, Aggarwal M, Avvaru BS, Robbins AH, Scozzafava A, McKenna R, Supuran CT. Selective hydrophobic pocket binding observed within the carbonic anhydrase II active site accommodate different 4-substituted-ureido-benzenesulfonamides and correlate to inhibitor potency. Chem Commun (Camb) 2010;46:8371–3.

[68] Mboge MY, Mahon BP, Lamas N, Socorro L, Carta F, Supuran CT, Frost SC, McKenna R. Structure activity study of carbonic anhydrase IX: selective inhibition with ureido-substituted benzenesulfonamides. Eur J Med Chem 2017;132:184–91.

[69] Angeli A, Peat TS, Selleri S, Saleh Alfawaz Altamimi A, Supuran CT, Carta F. X-ray crystallography of Epacadostat in adduct with carbonic anhydrase IX. Bioorg Chem 2020;97:103669.

[70] Angeli A, Ferraroni M, Nocentini A, Selleri S, Gratteri P, Supuran CT, Carta F. Polypharmacology of epacadostat: a potent and selective inhibitor of the tumor associated carbonic anhydrases IX and XII. Chem Commun (Camb) 2019;55:5720–3.

[71] Corbet C, Feron O. Tumour acidosis: from the passenger to the driver's seat. Nat Rev Cancer 2017;(10):577–93.

[72] Grüner B, Brynda J, Das V, Šícha V, Štěpánková J, Nekvinda J, Holub J, Pospíšilová K, Fábry M, Pachl P, Král V, Kugler M, Mašek V, Medvedíková M, Matějková S, Nová A, Lišková B, Gurská S, Džubák P, Hajdúch M, Řezáčová P. Metallacarborane Sulfamides: unconventional, specific, and highly selective inhibitors of carbonic anhydrase IX. J Med Chem 2019;62:9560–75.

[73] Doss M, Kolb HC, Walsh JC, Mocharla VP, Zhu Z, Haka M, Alpaugh RK, Chen DY, Yu JQ. Biodistribution and radiation dosimetry of the carbonic anhydrase IX imaging agent [(18) F]VM4-037 determined from PET/CT scans in healthy volunteers. Mol Imaging Biol 2014;16:739–46.

[74] Metayer B, Mingot A, Vullo D, Supuran CT, Thibaudeau S. New superacid synthesized (fluorinated) tertiary benzenesulfonamides acting as selective hCA IX inhibitors: toward a new mode of carbonic anhydrase inhibition by sulfonamides. Chem Commun (Camb) 2013;49:6015–7.

[75] Maresca A, Temperini C, Vu H, Pham NB, Poulsen SA, Scozzafava A, Quinn RJ, Supuran CT. Non-zinc mediated inhibition of carbonic anhydrases: coumarins are a new class of suicide inhibitors. J Am Chem Soc 2009; 131:3057–62.

[76] Melis C, Distinto S, Bianco G, Meleddu R, Cottiglia F, Fois B, Taverna D, Angius R, Alcaro S, Ortuso F, Gaspari M, Angeli A, Del Prete S, Capasso C, Supuran CT, Maccioni E. Targeting tumor associated carbonic anhydrases IX and XII: highly isozyme selective coumarin and psoralen inhibitors. ACS Med Chem Lett 2018;9:725–9.

[77] Kurt BZ, Dag A, Doğan B, Durdagi S, Angeli A, Nocentini A, Supuran CT, Sonmez F. Synthesis, biological activity and multiscale molecular modeling studies of bis-coumarins as selective carbonic anhydrase IX and XII inhibitors with effective cytotoxicity against hepatocellular carcinoma. Bioorg Chem 2019;87:838–50.

[78] Narella SG, Shaik MG, Mohammed A, Alvala M, Angeli A, Supuran CT. Synthesis and biological evaluation of coumarin-1,3,4-oxadiazole hybrids as selective carbonic anhydrase IX and XII inhibitors. Bioorg Chem 2019;87:765–72.

[79] De Luca L, Mancuso F, Ferro S, Buemi MR, Angeli A, Del Prete S, Capasso C, Supuran CT, Gitto R. Inhibitory effects and structural insights for a novel series of coumarin-based compounds that selectively target human CA IX and CA XII carbonic anhydrases. Eur J Med Chem 2018;143:276–82.

[80] Buran K, Bua S, Poli G, Önen Bayram FE, Tuccinardi T, Supuran CT. Novel 8-substituted coumarins that selectively inhibit human carbonic anhydrase IX and XII. Int J Mol Sci 2019;20:E1208.

[81] Thacker PS, Alvala M, Arifuddin M, Angeli A, Supuran CT. Design, synthesis and biological evaluation of coumarin-3-carboxamides as selective carbonic anhydrase IX and XII inhibitors. Bioorg Chem 2019;86:386–92.

[82] Bonardi A, Falsini M, Catarzi D, Varano F, Di Cesare Mannelli L, Tenci B, Ghelardini C, Angeli A, Supuran CT, Colotta V. Structural investigations on coumarins leading to chromeno[4,3-*c*]pyrazol-4-ones and pyrano[4,3-*c*]

pyrazol-4-ones: new scaffolds for the design of the tumor-associated carbonic anhydrase isoforms IX and XII. Eur J Med Chem 2018;146:47–59.

[83] Angeli A, Trallori E, Carta F, Di Cesare Mannelli L, Ghelardini C, Supuran CT. Heterocoumarins are selective carbonic anhydrase IX and XII inhibitors with cytotoxic effects against cancer cells lines. ACS Med Chem Lett 2018;9:947–51.

[84] Maresca A, Temperini C, Pochet L, Masereel B, Scozzafava A, Supuran CT. Deciphering the mechanism of carbonic anhydrase inhibition with coumarins and thiocoumarins. J Med Chem 2010;53:335–44.

[85] Ferraroni M, Carta F, Scozzafava A, Supuran CT. Thioxocoumarins show an alternative carbonic anhydrase inhibition mechanism compared to coumarins. J Med Chem 2016;59:462–73.

[86] Krasavin M, Žalubovskis R, Grandāne A, Domračeva I, Zhmurov P, Supuran CT. Sulfocoumarins as dual inhibitors of human carbonic anhydrase isoforms IX/XII and of human thioredoxin reductase. J Enzyme Inhib Med Chem 2020;35:506–10.

[87] Angapelly S, Angeli A, Khan AJ, Sri Ramya PV, Supuran CT, Arifuddin M. Synthesis and biological evaluation of 4-sulfamoylphenyl/sulfocoumarin carboxamides as selective inhibitors of carbonic anhydrase isoforms hCA II, IX, and XII. ChemMedChem 2018;13:1165–71.

[88] Grandane A, Tanc M, Žalubovskis R, Supuran CT. Synthesis of 6-aryl-substituted sulfocoumarins and investigation of their carbonic anhydrase inhibitory action. Bioorg Med Chem 2015;23:1430–6.

[89] Pustenko A, Nocentini A, Balašova A, Krasavin M, Žalubovskis R, Supuran CT. 7-Acylamino-3H-1,2-benzoxathiepine 2,2-dioxides as new isoform-selective carbonic anhydrase IX and XII inhibitors. J Enzyme Inhib Med Chem 2020;35:650–6.

[90] Pustenko A, Nocentini A, Balašova A, Alafeefy A, Krasavin M, Žalubovskis R, Supuran CT. Aryl derivatives of 3H-1,2-benzoxathiepine 2,2-dioxide as carbonic anhydrase inhibitors. J Enzyme Inhib Med Chem 2020; 35:245–54.

[91] Koruza K, Mahon BP, Blakeley MP, Ostermann A, Schrader TE, McKenna R, Knecht W, Fisher SZ. Using neutron crystallography to elucidate the basis of selective inhibition of carbonic anhydrase by saccharin and a derivative. J Struct Biol 2019;205:147–54.

[92] De Monte C, Carradori S, Secci D, D'Ascenzio M, Vullo D, Ceruso M, Supuran CT. Cyclic tertiary sulfamates: selective inhibition of the tumor-associated carbonic anhydrases IX and XII by N- and O-substituted acesulfame derivatives. Eur J Med Chem 2014;84:240–6.

[93] Köhler K, Hillebrecht A, Schulze Wischeler J, Innocenti A, Heine A, Supuran CT, Klebe G. Saccharin inhibits carbonic anhydrases: possible explanation for its unpleasant metallic aftertaste. Angew Chem Int Ed 2007;46:7697–9.

[94] Mahon BP, Hendon AM, Driscoll JM, Rankin GM, Poulsen SA, Supuran CT, McKenna R. Saccharin: a lead compound for structure-based drug design of carbonic anhydrase IX inhibitors. Bioorg Med Chem 2015;23:849–54.

[95] Murray AB, Quadri M, Li H, McKenna R, Horenstein NA. Synthesis of saccharin-glycoconjugates targeting carbonic anhydrase using a one-pot cyclization/deprotection strategy. Carbohydr Res 2019;476:65–70.

[96] D'Ascenzio M, Secci D, Carradori S, Zara S, Guglielmi P, Cirilli R, Pierini M, Poli G, Tuccinardi T, Angeli A, Supuran CT. 1,3-Dipolar cycloaddition, HPLC enantioseparation, and docking studies of saccharin/isoxazole and saccharin/isoxazoline derivatives as selective carbonic anhydrase IX and XII inhibitors. J Med Chem 2020;63:2470–88.

[97] D'Ascenzio M, Carradori S, De Monte C, Secci D, Ceruso M, Supuran CT. Design, synthesis and evaluation of *N*-substituted saccharin derivatives as selective inhibitors of tumor-associated carbonic anhydrase XII. Bioorg Med Chem 2014;22:1821–31.

[98] D'Ascenzio M, Guglielmi P, Carradori S, Secci D, Florio R, Mollica A, Ceruso M, Akdemir A, Sobolev AP, Supuran CT. Open saccharin-based secondary sulfonamides as potent and selective inhibitors of cancer-related carbonic anhydrase IX and XII isoforms. J Enzyme Inhib Med Chem 2017;32:51–9.

[99] Cornelio B, Laronze-Cochard M, Miambo R, De Grandis M, Riccioni R, Borisova B, Dontchev D, Machado C, Ceruso M, Fontana A, Supuran CT, Sapi J. 5-Arylisothiazol-3(2*H*)-one-1,(1)-(di)oxides: a new class of selective tumor-associated carbonic anhydrases (hCA IX and XII) inhibitors. Eur J Med Chem 2019;175:40–8.

CHAPTER

6

Small molecules and monoclonal antibodies as selective hCA XII inhibitors: An update

Simone Carradori[a] *and Claudiu T. Supuran*[b]

[a]Department of Pharmacy, "G. d'Annunzio" University of Chieti-Pescara, Chieti, Italy
[b]NEUROFARBA Department, Section of Pharmaceutical and Nutraceutical Sciences, University of Florence, Florence, Italy

Abstract

A plethora of membrane proteins (ion transporters, carbonic anhydrases) cooperate for the modulation of intra- and extracellular pH in hypoxic cancer cells, among which two carbonic anhydrase (CA) isoforms, CA IX and XII. Similar to human (h) hCA IX, the overexpression of hCA XII has been registered in some cancer cells and its role emerged as a marker as well as a therapeutic target for fighting chemoresistance, together with efflux pumps. In this chapter, we focus our attention on the most recent discoveries in the design of selective hCA XII inhibitors as theragnostics that could interfere with tumor pH regulation and reduce tumoral and metastatic potential. The structural requirements to achieve isoform selectivity as well as molecular modeling studies will be also discussed. Starting from in vitro assays and following to translational results we discuss available in cell-based assays, the few in vivo systems disclosed so far and their correlations with the biological (animal) data available for CA XII small molecule or antibody inhibitors.

Abbreviations

aa	amino acid
AAZ	acetazolamide
AE2	anion exchanger 2
Bn	benzyl
CAI	CA inhibitor
EGFR	epidermal growth factor receptor
EMT	epithelial mesenchymal transition
ER	estrogen receptor
ERE	estrogen-responsive enhancer
hCA	human carbonic anhydrase

https://doi.org/10.1016/B978-0-12-820701-7.00015-4

HCC	hepatocellular carcinoma
Hh	hedgehog
IDO1	indoleamine 2,3-dioxygenase-1
K_I	inhibition constant
mRNA	messenger RNA
NCI	National Cancer Institute
Pgp	P-glycoprotein
PKC	protein kinase C
RCC	renal cell carcinoma
RNA Pol II	RNA polymerase II
RNA	ribonucleic acid
ROS	reactive oxygen species
SI	selectivity index
THP	tetrahydropyran
VHL	von Hippel Lindau

Conflict of interest

No potential conflicts of interest were disclosed.

Introduction

hCA XII was first discovered in 1998 by serological screening in a human renal cell carcinoma (RCC). Along with hCA IX, hCA XII (354 aa) is characterized by a homodimeric transmembrane glycoproteic structure, with four cysteine residues establishing intramolecular disulfide bonds at the dimer interface, and an extracellular catalytic domain with GLSLS and GIILG motifs. Its gene locus relies on chromosome 15q22. Differently from hCA IX, hCA XII is usually expressed in different both normal (basolateral plasma membrane of epithelial cells of pancreas, efferent ducts, endometrium, kidneys, gastric mucosa and colon, nonpigmented epithelial cells of the eye, and salivary glands) [1] and cancer tissues (breast, RCC, colorectal, non-small-cell lung, gastrointestinal, ovarian, cervical, brain, and pancreatic). It can be slightly upregulated in tumors, but it is not under the strict control of HIF-1 as hCA IX. However, its transcription is regulated by von Hippel Lindau (VHL) protein [2] and generally associated with less aggressive tumor phenotypes and good prognosis. Both these membrane-associated CAs catalyze the reversible hydration of CO_2 directly in the extracellular space. Their catalytic activity is very efficient and the impact on the extracellular pH and maintenance of a more alkaline intracellular pH make these isozymes important as surface markers of hypoxic tumor cells [3]. Moreover, *ca IX* and *ca XII* genes silencing gave a volume reduction in xenograft tumors compared with the knockout of *ca IX* alone [4].

More recently, the role of hCA XII emerged due to the potential clinical relevance in the so-called "transport metabolon": a wide network of physical and functional interactions among membrane proteins and channels or transporters in specific cancers such as breast, glioblastoma, and hepatocellular carcinoma and all the scientific efforts were oriented to evaluate its role in relation to the prognosis. In breast cancer, hCA XII expression was demonstrated by immunohistochemistry in a large group of samples, being correlated with low tumor grade, ER (estrogen receptor) status, absence of necrosis, and negative EGFR status [5]. Enhanced expression of *ca XII* gene has been registered in breast tumors in relation to ERα upregulation via estrogens, following binding of distal estrogen-responsive enhancer (ERE) regions to ERα and successive recruitment of RNA Pol II and various activators of the steroid receptor. Furthermore, it has been demonstrated, by a combination of bioinformatics and experimental procedures, that hCA XII expression is strongly related to the epithelial phenotype controlled by EMT players, including protein kinase C (PKC). It is also regulated by hedgehog (Hh) pathway in order to control breast cancer cell proliferation and migration thus contributing to the cancer aggressiveness [6].

In hepatocellular carcinoma (HCC) specimens, mRNA levels and expression of CA XII were higher with respect to normal tissues and associated with sex, pathological grade, age, and tumor size. Its presence was described by the authors as a poor prognosis marker because of its correlation with the survival rate by the Kaplan-Meier analysis [7].

Lastly, an interesting result was reported by Salaroglio et al. who described the combined role of hCA XII and P-glycoprotein (Pgp) in mediating the temozolomide resistance in the treatment of glioblastoma [8]. The intracellular/extracellular pH regulation exerted by hCA XII influences Pgp efficiency as an efflux pump. Thus, selective hCA XII inhibitors can indirectly limit Pgp functionality and sensitize the tumor to the first-line therapy independent of specific genetic alterations. According to the Cancer Genome Atlas, CA XII and Pgp mRNA are detectable in aggressive glioblastoma patients. Furthermore, in vivo murine experiments of orthotopically implanted glioblastoma tumors demonstrated that a combination therapy of selective CA XII inhibitors and temozolomide improved the antitumor activity as well as the overall survival without exerting or exacerbating side effects.

Small molecules and monoclonal antibodies as selective hCA XII inhibitors

Differences in the active site of hCA IX and XII

Differently from hCA IX, only few molecules, which specifically inhibit hCA XII, have been developed and investigated in vitro/vivo so far [9, 10]. To design and synthesize isoform selective CA inhibitors (CAIs), different approaches can be used in Medicinal Chemistry (e.g., the tail approach, sugar approach) in order to avoid inhibition of the ubiquitous and highly expressed off-target isoforms (hCA I and II) despite the differences exploited in the active sites. Indeed, hCA I and II are characterized by a more occluded active site with respect to hCA XII, thus allowing bulky inhibitors to be accessible or a higher conformational freedom of the tail of the CAIs. Generally speaking, the active site is mostly conserved among all the CA isoforms, whereas a wider variability can be registered at a distance ≥ 5 Å from the active site zinc ion [11, 12].

The in vitro evaluation of the newly synthesized inhibitors was performed by a Stopped Flow CO_2 CO_2 hydrase assay using acetazolamide (AAZ) as the reference compound, allowing an extremely good comparison of the published data. For sake of clarity in Table 1, we report the inhibition data (K_I) and selectivity index (SI) of this potent and clinically used hCA inhibitor, in order to better compare the results relative to the other compounds recently published on this topic. It is a nanomolar inhibitor of the four most important isoforms, but characterized by a very low isoform selectivity.

TABLE 1 Inhibition data (K_I) and selectivity index (SI) of acetazolamide (AAZ) as the mostly used reference drug.

		K_I (nM)[a]				SI[b]		
		hCA I	hCA II	hCA IX	hCA XII	I/XII	II/XII	IX/XII
AAZ	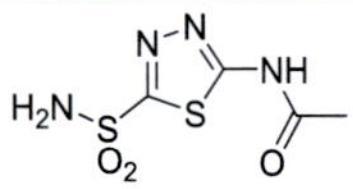	250	12.0	25.2	5.7	43.8	2.1	4.4

[a] *Mean from three different assays, by a stopped flow technique (errors were in the range of ±5%–10% of the reported values).*
[b] K_I*s ratio for the indicated enzyme isoforms. An SI > 1 indicates selectivity against hCA XII over the other isoforms.*

Recent sulfonamide derivatives as (selective) hCA XII inhibitors

Sulfonamides represent the most investigated chemical class of hCA inhibitors with their classic zinc binding group (ZBG) anchoring in their deprotonated form the prosthetic zinc(II) ion within the enzyme active site. Primary benzenesulfonamides have been explored exhaustively in the past years providing potent inhibitors characterized by nanomolar K_Is but limited isoform selectivity. This is a consequence of the direct interaction with a well-conserved region of the enzyme. Recently, Nocentini et al. proposed the introduction on the benzenesulfonamide skeleton of a nitro group at C3, which could provide an ancillary cytotoxic effect in hypoxic environments (Fig. 1A) [13]. The compounds were nanomolar inhibitors of hCA XII preferentially if an additional nitro group was introduced at C5 with respect to ureido moieties or no substitution. Despite their low SI against hCA I or II, these compounds were tested in hypoxic conditions against human colon cancer (HT-29), breast adenocarcinoma (MDA-MB-231), and prostatic cancer (PC-3) cell lines. The effects on their viability after a 48-h treatment were monitored at three discrete micromolar concentrations and compared with normoxic conditions.

Other benzenesulfonamides were designed after combination via triazole linker with chemically diverse fragments studied by differential scanning fluorimetry [14]. These hybrids showed potent hCA XII inhibitory activity in the nanomolar range, but scarce selectivity index against hCA II (Fig. 1B). Another linker exploited by Supuran's group was the hydrazido one connecting the benzenesulfonamide with substituted benzofurans (Fig. 1C) [15]. These derivatives were shown to inhibit hCA XII in the nanomolar range and some of them displayed a good isoform selectivity (SI) higher than 100. Unfortunately, the results against 60 cell lines of National Cancer Institute (NCI) did not provide a potent lead compound for further studies.

Lastly, some authors published an in-depth study about the antiproliferative effect of ureidobenzenesulfonamides (compounds **1–3**, Fig. 1D), yet reported as potent and selective dual hCA IX and XII inhibitors [16].

Other benzenesulfonamides were also functionalized via an imino linker with propargyloxy-based aryls or triazole-based coumarins (Fig. 2) [17]. The imino linker was in the para position with respect to the sulfonamide, whereas the other substituents on the second phenyl were better tolerated in the ortho position. All these derivatives were characterized by a slightly lower inhibitory activity against hCA XII, but they provided good lead

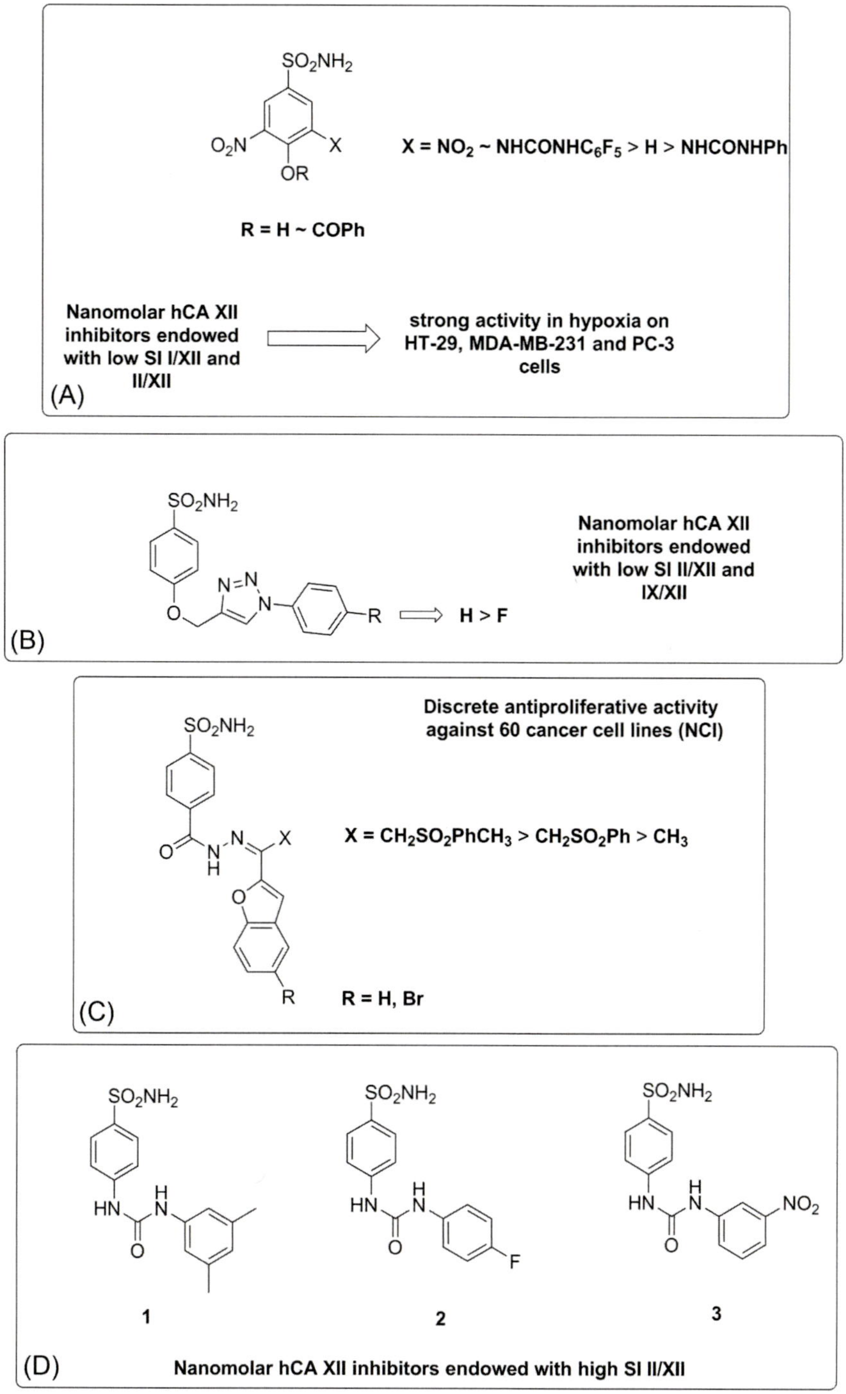

FIG. 1 Benzenesulfonamide derivatives as nanomolar hCA XII inhibitors.

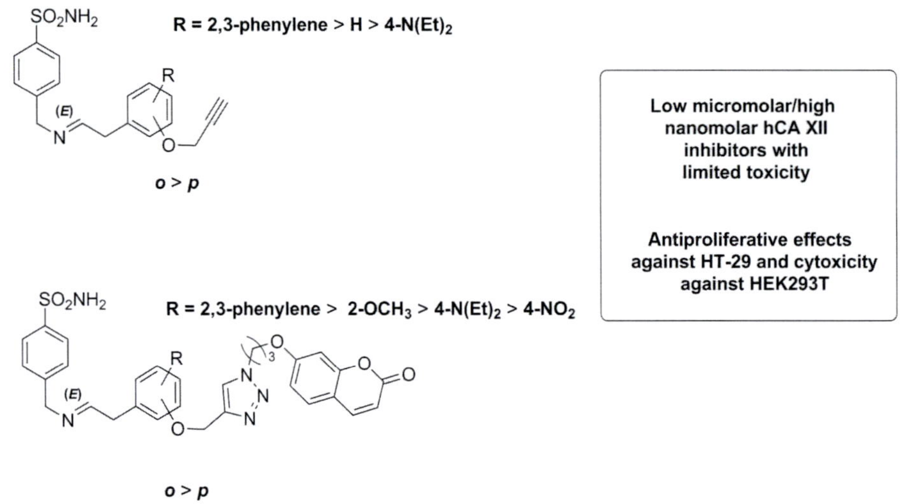

FIG. 2 Benzenesulfonamides functionalized via the imino linker.

compounds in cell-based assays revealing a selective antiproliferative effect against the tumoral HT-29 compared to the normal HEK293T cell line.

Pursuing the efforts to discover other sulfonamide-based compounds, Krasavin et al. described the hCA inhibitory activity of a large series of (hetero)aromatic sulfonamides inserting an oxadiazole as a linker (Fig. 3) [18]. These compounds were characterized by a very potent and selective hCA XII inhibitory activity and some of them were tested for their antiproliferative effects on PANC-1 and SK-MEL-2 cancer cell lines with respect to ARPE-19 normal cell line.

In addition, studies were oriented toward primary aliphatic sulfonamides starting from the picomolar inhibition exerted by the natural compound Psammaplin C against hCA XII [19]. The molecular skeleton was fractionated in three parts and modified to extend the structure-activity relationship (SAR) studies (Fig. 4). The X portion can tolerate H, Br, and OH (as in the parent compound) depending on the nature of Y group and the linker. The Y moiety is preferentially an H (no substitution), also if other protecting groups can be added (THP, Bn). The best linker was demonstrated to be a thiadiazole with respect to the aliphatic chain or a benzene as reported for compound **4**.

The most active derivative (**4**) within this large library was assessed in vitro and in vivo to establish the effects not only on the hCA XII activity, but also on the Pgp function, in association with doxorubicin for the treatment of hypoxic glioblastoma. This approach led to the discovery of a new lead compound: (i) stable in Balb/c mice plasma and mouse microsomes; (ii) able to enhance the intracellular doxorubicin concentration as the first-line therapy limiting cell viability; and (iii) able to reduce the doxorubicin-induced damage. Interestingly, in orthotopically implanted glioblastoma tumors with coexpression of hCA XII and Pgp, the administration of compound **4** with doxorubicin significantly reduced the tumor volume.

A plethora of research studies have confirmed the potential of saccharin and saccharin-based compounds in the development of potent and selective hCA XII inhibitors [20–25].

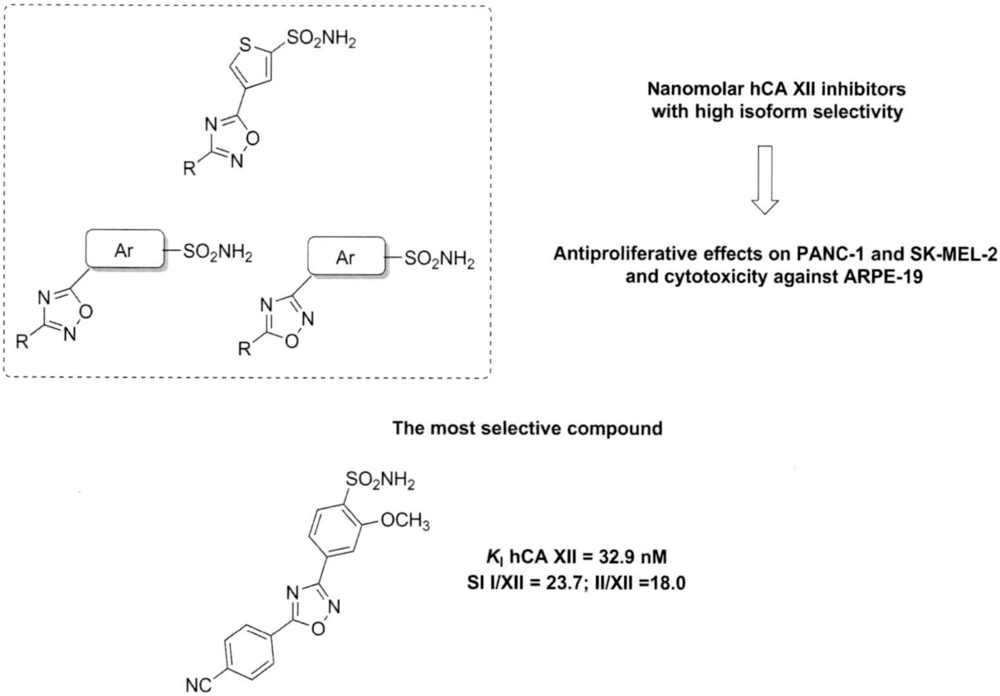

FIG. 3 (Hetero)arylsulfonamides functionalized via the oxadiazole linker.

Starting from these robust SARs, three different approaches in Medicinal Chemistry were explored: *N*-derivatization, ring expansion, and ring contraction as reported in Fig. 5. All these structures displayed a nonclassical ZBG (tertiary cyclic sulfonamide) not able to be in the reported deprotonated form and thus paving the way toward an alternative mechanism of inhibition.

The first approach led to the discovery of the most potent and selective inhibitors by means of the derivatization of the nitrogen of the saccharin skeleton [26]. These compounds were functionalized with an isoxazole or isoxazoline ring bearing an additional aryl or heteroaryl moiety and also allowed the evaluation of the impact of the stereochemistry in the enzyme recognition of the ligand. The inhibition data were further corroborated by molecular modeling and cell-based studies regarding the effects on viability and cancer cell proliferation. These compounds were strongly selective against hCA XII (SI I/XII and II/XII > 1000 for most of them) and their binding mode within the active site was described in silico depending on the interaction with the prosthetic zinc ion via one of the sulfonamide oxygen atoms. Additional H-bonds were established with essential amino acids of the enzyme and the carbonyl oxygen, whereas the isoxazol(*in*)e linker interacted positively with hydrophobic regions of the enzyme. Moreover, the two most interesting compounds were tested for their cytotoxicity on human gingival fibroblasts indicating an absence of antiproliferative effects after 24, 48, and 72 h of treatment up to 200 μM. Conversely, if associated with a subtoxic dose of

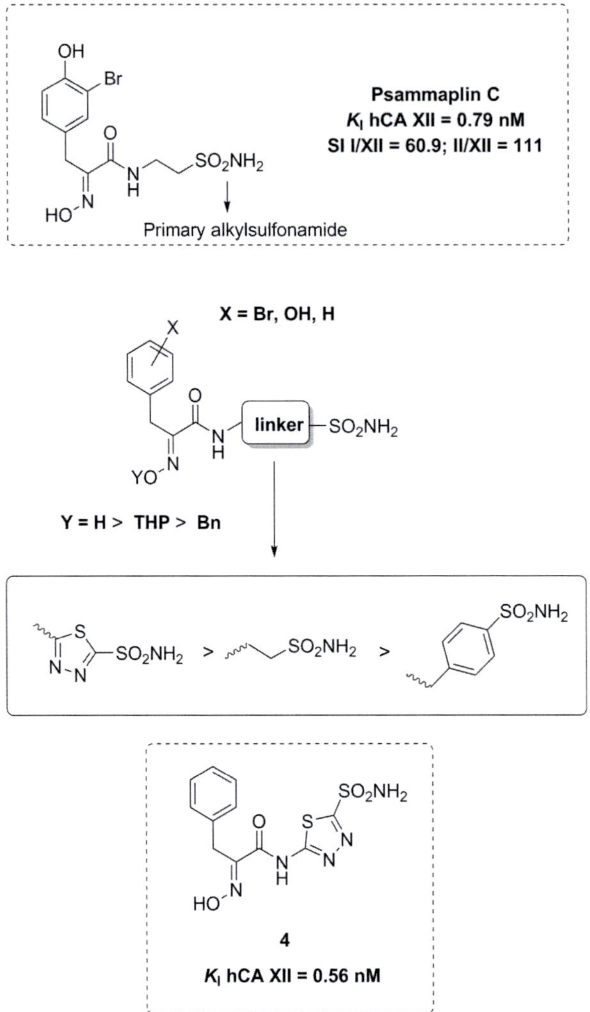

FIG. 4 Sulfonamide derivatives explored starting from Psammaplin C.

doxorubicin, they exerted a synergistic action against breast adenocarcinoma cells (MCF-7) in pseudohypoxic ($CoCl_2$-induced) conditions.

The second approach led to the design of 6-membered analogues of saccharin [27]. The compounds were endowed with potent nanomolar inhibitory activity, but higher SI was achieved only if the C7 was substituted with a COOH (compounds **5** and **6**, Fig. 6).

A combination of crystallographic studies and antiproliferative tests on three cancer cells and one normal cell line corroborated the importance of this chemical modification.

The third approach (ring contraction) offered compounds with SI I/XII and II/XII > 1000 [28]. Moreover, the oxidation state of the sulfur was important for the inhibitory activity ($SO_2 > SO > S$), regardless the C5 substituent (Cl, Ar, heterocycle). The saccharin nitrogen must

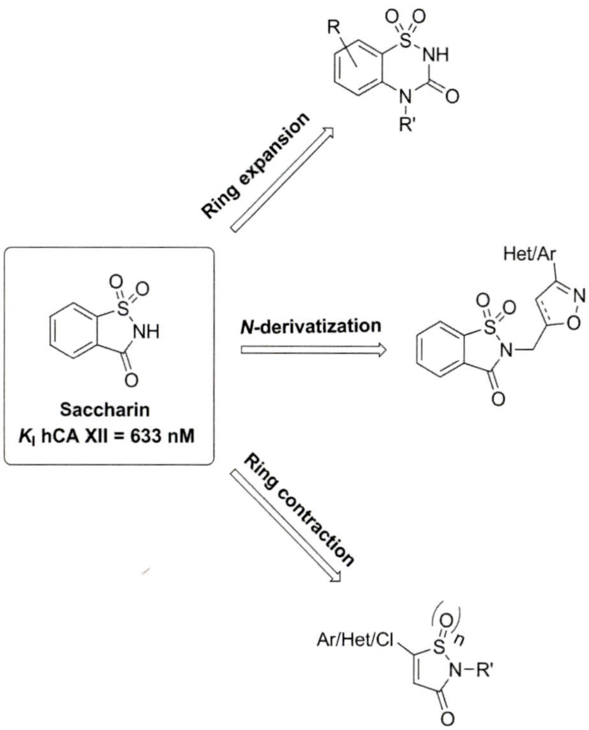

FIG. 5 The three approaches explored in the saccharin scaffold.

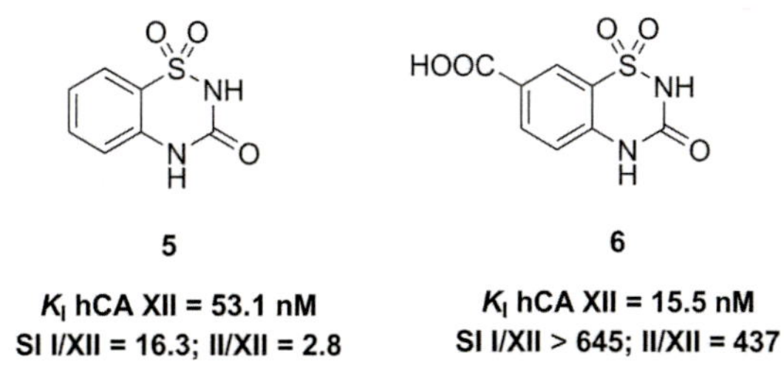

FIG. 6 The best-in-class compounds after the ring expansion of the saccharin nucleus.

be preferentially substituted to observe a better inhibitory activity in the subnanomolar range (Fig. 7).

Another chemical scaffold investigated in detail belonged to the *N*-derivatives of anthranilic acid [29]. These compounds were nanomolar inhibitors with low SIs over hCA I or II. The activity disappeared if the phthalimide was removed to add an isobenzofuran and the nitrogen must be substituted with a carboxylic acid- or ester-containing phenyl (Fig. 8).

The last example of heteroaryl sulfonamide compounds took advantage of the chelating ability of sulfonamide and nitrogen atoms in the clinically used acetazolamide structure toward different transition metal ions [30]. The authors suggested that the presence of metal

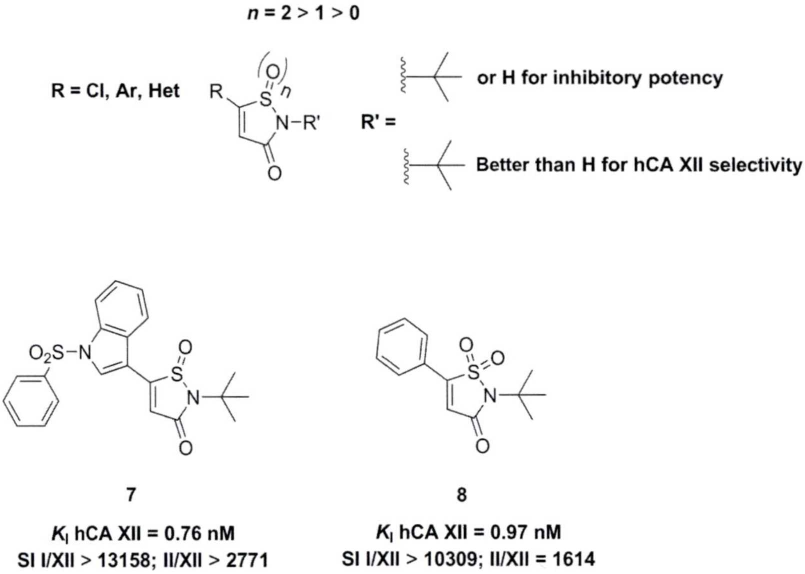

FIG. 7 Ring contraction of the saccharin nucleus.

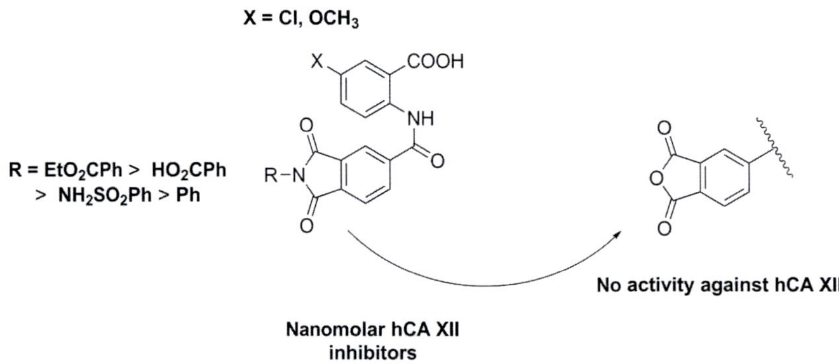

FIG. 8 Phthalimide-based hCA XII inhibitors.

ions could reinforce the pharmacological activity of the parent compound allowing promising development in the field of multivalent compounds. Due to the anticancer activity of Ru (II) complexes, the authors investigated by X-ray crystallography and in vitro assays two organoruthenium(II) derivatives of AAZ as a well-known potent hCA inhibitor (Fig. 9). Compounds **9** and **10** displayed a very potent inhibition of hCA XII (K_I values of 0.52 and 0.04 nM, respectively) and a high selectivity index of 212.5, 12, and 15.7 against hCA I, II, and IX, respectively. Moreover, they were not cytotoxic to noncancer cell lines.

Coumarins and their derivatives as hCA XII inhibitors

Among the nonconventional CA inhibitors, coumarins deserved a prominent role after the discovery of their mechanism of action. They acted as prodrugs being hydrolyzed by the

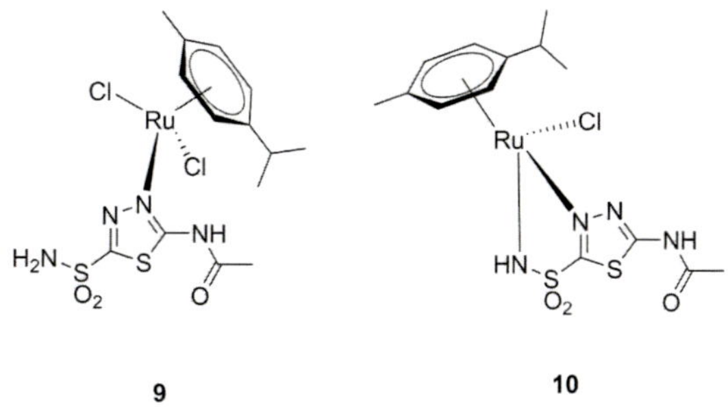

FIG. 9 Organoruthenium(II) derivatives of acetazolamide.

enzyme to cinnamic acids and then they can occlude the entrance cavity. The most important feature in this behavior is the interaction with a less conserved region of the enzyme, which can open new scenarios in the field of isoform selectivity. Coumarin-based inhibitors, and especially 7-hydroxycoumarins, were proposed and tested as promising hCA XII inhibitors after the introduction of a substituted piperidine fragment at C8 of the core nucleus (Fig. 10A) [31].

They displayed a very potent inhibitory activity oriented almost exclusively toward hCA XII isoform. Molecular modeling as well as molecular dynamics simulation studies confirmed this strong affinity predicting an H-bond between the carbonyl oxygen and a water molecule coordinating the prosthetic zinc(II) ion, other very strong H-bonds with the phenolic group with Lys97 and the accommodation of the 4-methyl into a hydrophobic pocket delimited by Val147, Leu167, and Leu225. The substituent at C8 gave van der Waals interactions with

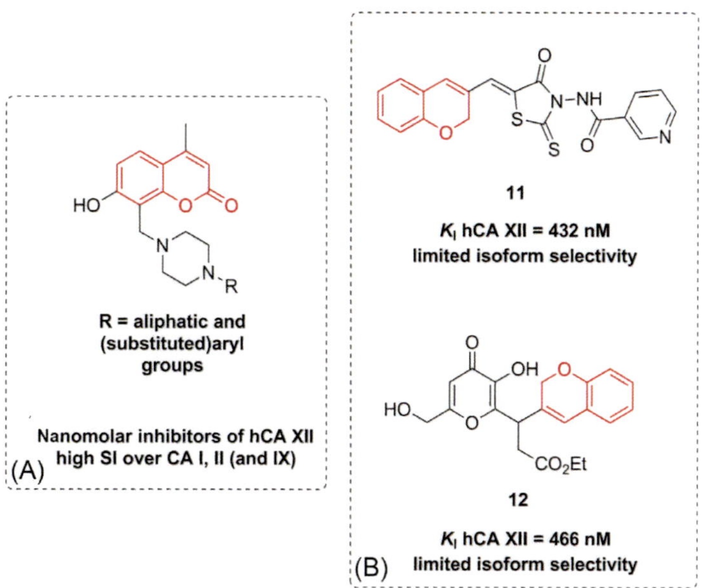

FIG. 10 7-Hydroxycoumarins and coumarin analog derivatives.

residues of the hydrophilic site of the enzyme. In order to extend the SARs within this chemical scaffold, Kartsev et al. proposed new coumarins and coumarin-like molecules (Fig. 10B) [32] characterized by a large number of chemical diversity. These derivatives were all inactive up to 10 μM against hCA I and II and potent hCA IX inhibitors. Only some of them were also shown to inhibit in the nanomolar range hCA XII (compounds **11** and **12**, Fig. 10B).

Drug repurposing and polypharmacology as innovative approaches to discover selective hCA XII inhibitors

Natural and synthetic molecules, yet approved in the clinic for other purposes, are currently candidates for repositioning in oncology as single compounds or in combination therapy [33]. Indeed, the known good safety profiles and pharmacokinetic properties may facilitate their clinical translation in cancer therapy. On the other hand, compounds, endowed with an anticancer action, can share multiple and additive mechanisms of action due to their structural features (polypharmacology).

In the past years, Supuran's research group investigated epacadostat (Fig. 11) facing clinical trials as a selective indoleamine 2,3-dioxygenase-1 (IDO1) inhibitor as a part of Immune Checkpoint Therapy [34]. This compound is able to modulate aberrant signal transduction pathways in cancer cells [35] via oral administration. A proper analysis of the chemical structure evidenced the presence of a sulfamide group (NH_2-SO_2-NH-), a bioisostere of the zinc-binding sulfonamide moiety (NH_2-SO_2-). This prompted the researchers to test its in vitro inhibition on a panel of catalytically active human carbonic anhydrase isozymes. It showed poor inhibition potencies against hCA I, II, III, IV, VA, VII, and XIV, good inhibitory activity against hCA VB, VI, and XIII, and very strong affinity and selectivity for both the tumor-associated isoforms IX and XII ($K_I = 3.0$ and 6.2 nM, respectively). More in detail, the main ligand-protein interactions were studied by means of crystallographic studies with hCA II at 1.15 Å and docking experiments with hCA XII to understand such isoform selectivity. The sulfamide moiety binds to the zinc ion and Thr199–200 residues within the hCA XII and II active sites, though in hCA XII the oxadiazole points toward the outer rim of the Leu91, Val130, and Leu199 cavity to efficiently stabilize the complex. In addition, a wide network of H-bonds is established between this heterocycle and the oxime portions with Thr91, Gln92, and Ala131. Such a result provided an additional evidence of the anticancer potential of epacadostat paving the way toward a multitargeting action (IDO1-CA XII). This important polypharmacology approach could highlight the observed antitumor effects and the

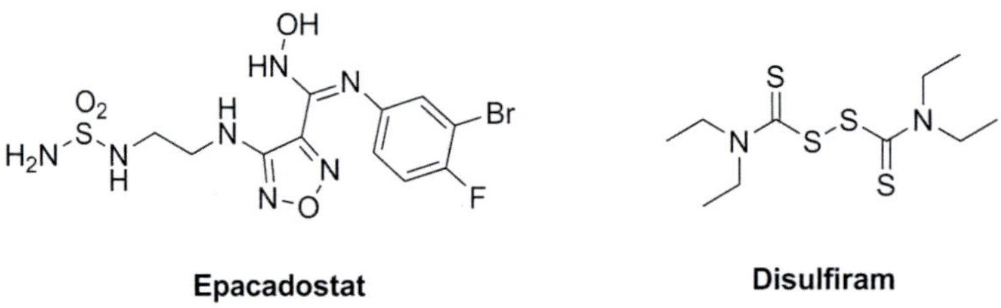

FIG. 11 New opportunities for approved drugs.

concurrent alteration of the tumor microenvironment, both useful in cancer therapy, attributed to this emerging drug.

Disulfiram (Fig. 11), well known for the treatment of alcoholism as a proteasome inhibitor, has also a strong ability to chelate copper and zinc ions leading to the generation of reactive oxygen species (ROS), matrix metalloproteinases inhibition, and cancer cell apoptosis [36]. Successively, Hwang et al. studied the effect of disulfiram on hCA XII and anion exchanger 2 (AE2) modulation to inhibit cancer cell migration. hCA XII was shown to interact with AE2 and with other ion transporters (Na^+-dependent bicarbonate transporters) in previous studies [37]. Results evidenced that disulfiram can modulate both endogenous hCA XII and AE2 in a time-dependent manner at low micromolar concentration, thus attenuating the migration and invasion of breast cancer cell lines by reduction of expression and activity of AE2, which is pathologically dysregulated in cancer [38]. Moreover, disulfiram reduced the invasion and motility of A549 cells without affecting the resting pH of the microenvironment and independent of autophagic regulation.

Monoclonal antibodies as very selective hCA XII inhibitors

CA isozymes can be detected or imaged through the development of specific monoclonal antibodies (mAbs) as an alternative approach characterized by high specificity and affinity for the target. Conversely, the use of antibodies for protein detection in vivo displays disadvantages such as bigger size, limited tissue penetrability in complex environment, nonreproducible industrial scalability, higher costs, and the development of immunological responses after administration. Keeping in mind the role exerted by hCA IX and XII in cancers, some authors proposed mAbs against these isozymes such as G250 (traded as Rencarex, Girentuximab), which was developed for the detection of hCA IX and treatment of human renal cell carcinoma [39], M75 targeting the proteoglycan domain exclusively present in hCA IX [40], EXO 610A, which binds to the catalytic domain of hCA XII, and the mAb developed by Moon et al. capable to interact with the noncatalytic region of the *N*-terminus of hCA XII [41]. More recently, another monoclonal antibody named as 6A10 has been developed to reduce the cancer cell growth in vitro and in vivo in a xenograft model of lung carcinoma [42, 43]. Its mechanism of action was oriented toward the blockade of the CA XII catalytic activity and the interference with P-glycoprotein activity, thus overcoming chemoresistance and enhancing chemosensitivity in a clinical setting [44]. It could also reduce the number of lung metastases in doxorubicin-treated mice bearing triple-negative breast cancer.

Starting from these promising premises and knowing that the use of antibody fragments could offer several advantages with respect to full IgG monoclonal antibodies (lower immunogenicity, higher tissue penetrability), a recombinant antigen-binding fragment (Fab) of 6A10 has been manufactured (Fab6A10). By means of an extensive biochemical and structural characterization, the authors revealed (i) that Fab6A10 was highly specific toward CA XII as its parent mAb in CA XII-positive human A549 lung and ASPC1 pancreatic cancer cells (ii) the crystallographic structure of the CA XII/Fab6A10 complex by interacting with rim of the active site cavity and limiting the conformational flexibility of His64; and (iii) the in vitro inhibition (K_I hCA XII = 6.6 nM, whereas K_Is values against hCA I, II and IX were > 10000 nM) and selectivity index [45].

Conclusion

The increasing knowledge about the mechanisms, which regulate pH in normal and cancer cells, has been collected in this chapter in order to better corroborate hCA XII as an attractive drug target in oncology. Medicinal chemists have disclosed new potent and selective inhibitors exploring the chemical space within the active site and enlarging the SARs within each scaffold [46, 47]. Different chemical classes of compounds were recently proposed avoiding crossinhibition with the other hCA isoforms. Sulfonamides represent the most investigated one as well as coumarins. Biological data were corroborated by molecular modeling and crystallographic studies. The most interesting compounds were also submitted to in vivo experiments.

References

[1] Whittington DA, Waheed A, Ulmasov B, Shah GN, Grubb JH, Sly WS, Christianson DW. Crystal structure of the dimeric extracellular domain of human carbonic anhydrase XII, a bitopic membrane protein overexpressed in certain cancer tumor cells. Proc Natl Acad Sci U S A 2001;98:9545–50.
[2] Supuran CT. How many carbonic anhydrase inhibition mechanisms exist? J Enzyme Inhib Med Chem 2016;31:345–60.
[3] Supuran CT. Structure and function of carbonic anhydrases. Biochem J 2016;473:2023–32.
[4] Neri D, Supuran CT. Interfering with pH regulation in tumours as a therapeutic strategy. Nat Rev Drug Discov 2011;10:767–77.
[5] Vergara D, Ravaioli S, Fonzi E, Adamo L, Damato M, Bravaccini S, Pirini F, Gaballo A, Barbano R, Pasculli B, Franck J, Fournier I, Salzet M, Maffia M. Carbonic anhydrase XII expression is modulated during epithelial mesenchymal transition and regulated through protein kinase C signaling. Int J Mol Sci 2020;21:715.
[6] Guerrini G, Durivault J, Filippi I, Criscuoli M, Monaci S, Pouyssegur J, Naldini A, Carraro F, Parks SK. Carbonic anhydrase XII expression is linked to suppression of sonic hedgehog ligand expression in triple negative breast cancer cells. Biochem Biophys Res Commun 2019;516:408–13.
[7] Gu X-F, Shi C-B, Zhao W. Prognostic value of carbonic anhydrase XII (CA XII) overexpression in hepatocellular carcinoma. Int J Clin Exp Pathol 2019;12:2173–83.
[8] Salaroglio IC, Mujumdar P, Annovazzi L, Kopecka J, Mellai M, Schiffer D, Poulsen SA, Riganti C. Carbonic anhydrase XII inhibitors overcome P-glycoprotein–mediated resistance to temozolomide in glioblastoma. Mol Cancer Ther 2018;17:2598–609.
[9] Berrino E, Supuran CT. Novel approaches for designing drugs that interfere with pH regulation. Expert Opin Drug Discov 2019;14:231–48.
[10] Alterio V, Di Fiore A, D'Ambrosio K, Supuran CT, De Simone G. Multiple binding modes of inhibitors to carbonic anhydrases: how to design specific drugs targeting 15 different isoforms? Chem Rev 2012;112:4421–68.
[11] Supuran CT. Carbonic anhydrases: novel therapeutic applications for inhibitors and activators. Nat Rev Drug Discov 2008;7:168–81.
[12] Supuran CT. Carbonic anhydrase inhibitors as emerging agents for the treatment and imaging of hypoxic tumors. Expert Opin Investig Drugs 2018;27:963–70.
[13] Nocentini A, Trallori E, Singh S, Lomelino CL, Bartolucci G, Di Cesare Mannelli L, Ghelardini C, McKenna R, Gratteri P, Supuran CT. 4-Hydroxy-3-nitro-5-ureido-benzenesulfonamides selectively target the tumor-associated carbonic anhydrase isoforms IX and XII showing hypoxia-enhanced antiproliferative profiles. J Med Chem 2018;61:10860–74.
[14] Krasavin M, Kalinin S, Zozulya S, Gryniukova A, Borysko P, Angeli A, Supuran CT. Screening of benzenesulfonamide in combination with chemically diverse fragments against carbonic anhydrase by differential scanning fluorimetry. J Enzyme Inhib Med Chem 2020;35:306–10.

[15] Abdelrahman MA, Eldehna WM, Nocentini A, Ibrahim HS, Almahli H, Abdel-Aziz HA, Abou-Seri SM, Supuran CT. Novel benzofuran-based sulphonamides as selective carbonic anhydrases IX and XII inhibitors: synthesis and in vitro biological evaluation. J Enzyme Inhib Med Chem 2020;35:298–305.

[16] Nocentini A, Supuran CT. Advances in the structural annotation of human carbonic anhydrases and impact on future drug discovery. Expert Opin Drug Discov 2019;14:1175–97.

[17] Kurt BZ, Sonmez F, Ozturk D, Akdemir A, Angeli A, Supuran CT. Synthesis of coumarin-sulfonamide derivatives and determination of their cytotoxicity, carbonic anhydrase inhibitory and molecular docking studies. Eur J Med Chem 2019;183:111702.

[18] Krasavin M, Shetnev A, Sharonova T, Baykov S, Kalinin S, Nocentini A, Sharoyko V, Poli G, Tuccinardi T, Presnukhina S, Tennikova TB, Supuran CT. Continued exploration of 1,2,4-oxadiazole periphery for carbonic anhydrase-targeting primary arene sulfonamides: discovery of subnanomolar inhibitors of membrane-bound hCA IX isoform that selectively kill cancer cells in hypoxic environment. Eur J Med Chem 2019;164:92–105.

[19] Mujumdar P, Kopecka J, Bua S, Supuran CT, Riganti C, Poulsen SA. Carbonic anhydrase XII inhibitors overcome temozolomide resistance in glioblastoma. J Med Chem 2019;62:4174–92.

[20] Ivanova J, Carta F, Vullo D, Leitans J, Kazaks A, Tars K, Žalubovskis R, Supuran CT. *N*-Substituted and ring opened saccharin derivatives selectively inhibit transmembrane, tumor-associated carbonic anhydrases IX and XII. Bioorg Med Chem 2017;25:1–7.

[21] D'Ascenzio M, Guglielmi P, Carradori S, Secci D, Florio R, Mollica A, Ceruso M, Akdemir A, Sobolev AP, Supuran CT. Open saccharin-based secondary sulfonamides as potent and selective inhibitors of cancer-related carbonic anhydrase IX and XII isoforms. J Enzyme Inhib Med Chem 2017;32:51–9.

[22] Carradori S, Secci D, De Monte C, Mollica A, Ceruso M, Akdemir A, Sobolev AP, Codispoti R, De Cosmi F, Guglielmi P, Supuran CT. A novel library of saccharin and acesulfame derivatives as potent and selective inhibitors of carbonic anhydrase IX and XII isoforms. Bioorg Med Chem 2016;24:22–5.

[23] Coviello V, Marchi B, Sartini S, Quattrini L, Marini AM, Simorini F, Taliani S, Salerno S, Orlandi P, Fioravanti A, Desidero TD, Vullo D, Da Settimo F, Supuran CT, Bocci G, La Motta C. 1,2-Benzisothiazole derivatives bearing 4-, 5-, or 6-alkyl/arylcarboxamide moieties inhibit carbonic anhydrase isoform IX (CAIX) and cell proliferation under hypoxic conditions. J Med Chem 2016;59:6547–52.

[24] Alterio V, Tanc M, Ivanova J, Zalubovskis R, Vozny I, Monti SM, Di Fiore A, De Simone G, Supuran CT. X-ray crystallographic and kinetic investigations of 6-sulfamoyl-saccharin as a carbonic anhydrase inhibitor. Org Biomol Chem 2015;13:4064–9.

[25] D'Ascenzio M, Carradori S, De Monte C, Secci D, Ceruso M, Supuran CT. Design, synthesis and evaluation of *N*-substituted saccharin derivatives as selective inhibitors of tumor-associated carbonic anhydrase XII. Bioorg Med Chem 2014;22:1821–31.

[26] D'Ascenzio M, Secci D, Carradori S, Zara S, Guglielmi P, Cirilli R, Pierini M, Poli G, Tuccinardi T, Angeli A, Supuran CT. 1,3-Dipolar cycloaddition, HPLC enantioseparation, and docking studies of saccharin/isoxazole and saccharin/isoxazoline derivatives as selective carbonic anhydrase IX and XII inhibitors. J Med Chem 2020;63:2470–88.

[27] Bua S, Lomelino CL, Murray AB, Osman SM, Alothman ZA, Bozdag M, Aziz HAA, Eldehna WM, McKenna R, Nocentini A, Supuran CT. "A sweet combination": developing saccharin and acesulfame K structures for selectively targeting the tumor-associated carbonic anhydrases IX and XII. J Med Chem 2020;63:321–33.

[28] Cornelio B, Laronze-Cochard M, Miambo R, De Grandis M, Riccioni R, Borisova B, Dontchev D, Machado C, Ceruso M, Fontana A, Supuran CT, Sapi J. 5-Arylisothiazol-3(2*H*)-one-1,(1)-(di)oxides: a new class of selective tumor-associated carbonic anhydrases (hCA IX and XII) inhibitors. Eur J Med Chem 2019;175:40–8.

[29] El-Azab AS, Abdel-Aziz AAM, Bua S, Nocentini A, AlSaif NA, Almehizia AA, Alanazi MM, Hefnawy MM, Supuran CT. New anthranilic acid-incorporating *N*-benzenesulfonamidophthalimides as potent inhibitors of carbonic anhydrases I, II, IX, and XII: synthesis, in vitro testing, and in silico assessment. Eur J Med Chem 2019;181:111573.

[30] Seršen S, Traven K, Kljun J, Turel I, Supuran CT. Organoruthenium(II) complexes of acetazolamide potently inhibit human carbonic anhydrase isoforms I, II, IX and XII. J Enzyme Inhib Med Chem 2019;34:388–93.

[31] Buran K, Bua S, Poli G, Bayram FEÖ, Tuccinardi T, Supuran CT. Novel 8-substituted coumarins that selectively inhibit human carbonic anhydrase IX and XII. Int J Mol Sci 2019;20:1208.

[32] Kartsev V, Geronikaki A, Bua S, Nocentini A, Petrou A, Lichitsky B, Frasinyuk M, Leitans J, Kazaks A, Tars K, Supuran CT. Extending the inhibition profiles of coumarin-based compounds against human carbonic anhydrases: synthesis, biological, and in silico evaluation. Molecules 2019;24:3580.

[33] Florio R, Veschi S, di Giacomo V, Pagotto S, Carradori S, Verginelli F, Cirilli R, Casulli A, Grassadonia A, Tinari N, Cataldi A, Amoroso R, Cama A, De Lellis L. The benzimidazole-based anthelmintic parbendazole: a repurposed drug candidate that synergizes with gemcitabine in pancreatic cancer. Cancer 2019;11:2042.
[34] Angeli A, Ferraroni M, Nocentini A, Selleri S, Gratteri P, Supuran CT, Carta F. Polypharmacology of epacadostat: a potent and selective inhibitor of the tumor associated carbonic anhydrases IX and XII. Chem Commun 2019;55:5720–3.
[35] Jochems C, Fantini M, Fernando RI, Kwilas AR, Donahue RN, Lepone LM, Grenga I, Kim YS, Brechbiel MW, Gulley JL, Madan RA, Heery CR, Hodge JW, Newton R, Schlom J, Tsang KY. An NK cell line (haNK) expressing high levels of granzyme and engineered to express the high affinity CD16 allele. Oncotarget 2016;25:37762.
[36] Shian SG, Kao YR, Wu FY, Wu CW. Inhibition of invasion and angiogenesis by zinc-chelating agent disulfiram. Mol Pharmacol 2003;64:1076–84.
[37] Alfarouk KO. Tumor metabolism, cancer cell transporters, and microenvironmental resistance. J Enzyme Inhib Med Chem 2016;31:859–66.
[38] Hwang S, Shin DM, Hong JH. Drug repurposing as an antitumor agent: disulfiram-mediated carbonic anhydrase 12 and anion exchanger 2 modulation to inhibit cancer cell migration. Molecules 2019;24:3409.
[39] Chamie K, Donin NM, Klöpfer P, Bevan P, Fall B, Wilhelm O, Störkel S, Said J, Gambla M, Hawkins RE, Jankilevich G, Kapoor A, Kopyltsov E, Staehler M, Taari K, Wainstein AJA, Pantuck AJ, Belldegrun AS. Adjuvant weekly girentuximab following nephrectomy for high-risk renal cell carcinoma: the ARISER randomized clinical trial. JAMA Oncol 2017;3:913–20.
[40] Pastoreková S, Závadová Z, Kostál M, Babusíková O, Závada J. A novel quasi-viral agent, MaTu, is a two-component system. Virology 1992;187:620–6.
[41] Dekaminaviciute D, Kairys V, Zilnyte M, Petrikaite V, Jogaite V, Matuliene J, Gudleviciene Z, Vullo D, Supuran CT, Zvirbliene A. Monoclonal antibodies raised against 167-180 aa sequence of human carbonic anhydrase XII inhibit its enzymatic activity. J Enzyme Inhib Med Chem 2014;29:804–10.
[42] Battke C, Kremmer E, Mysliwietz J, Gondi G, Dumitru C, Brandau S, Lang S, Vullo D, Supuran CT, Zeidler R. Generation and characterization of the first inhibitory antibody targeting tumour-associated carbonic anhydrase XII. Cancer Immunol Immunother 2011;60:649–58.
[43] Gondi G, Mysliwietz J, Hulikova A, Jen JP, Swietach P, Kremmer E, Zeidler R. Antitumor efficacy of a monoclonal antibody that inhibits the activity of cancer-associated carbonic anhydrase XII. Cancer Res 2013;73:6494–503.
[44] von Neubeck B, Gondi G, Riganti C, Pan C, Parra Damas A, Scherb H, Erturk A, Zeidler R. An inhibitory antibody targeting carbonic anhydrase XII abrogates chemoresistance and significantly reduces lung metastases in an orthotopic breast cancer model in vivo. Int J Cancer 2018;143:2065–75.
[45] Alterio V, Kellner M, Esposito D, Liesche-Starnecker F, Bua S, Supuran CT, Monti SM, Zeidler R, De Simone G. Biochemical and structural insights into carbonic anhydrase XII/Fab6A10 complex. J Mol Biol 2019;431:4910–21.
[46] Supuran CT. An update on drug interaction considerations in the therapeutic use of carbonic anhydrase inhibitors. Expert Opin Drug Metab Toxicol 2020, https://doi.org/10.1080/17425255.2020.1743679.
[47] Supuran CT. Exploring the multiple binding modes of inhibitors to carbonic anhydrases for novel drug discovery. Expert Opin Drug Discovery 2020, https://doi.org/10.1080/17460441.2020.1743676.

CHAPTER

7

Efflux pumps, NHE1, monocarboxylate transporters, and ABC transporter subfamily inhibitors

Utpal K. Mondal and Marc A. Ilies

Department of Pharmaceutical Sciences and Moulder Center for Drug Discovery Research, Temple University School of Pharmacy, Philadelphia, PA, United States

Abstract

The transporter family of proteins play an essential role in normal physiology of the human body, as well as in many pathological states, generated by genetic or acquired diseases, including many forms of cancer. The chapter details the main classes of transporters, their natural substrates, and the drugs that can be transported in/out of various cells. It also details the classes of inhibitors developed against each type of transporter, their potency, selectivity and toxicity, alone and in combination with different other drugs, in vitro and in vivo. Whenever possible, we have presented the translational efforts made to move different transporter inhibitors into clinical use, as well as the representatives currently available in clinic, their advantages and limitations.

Abbreviations

ABC ATP-binding cassette
AE1 anion exchanger 1
BCRP breast cancer resistance protein
CHC α-cyano-4-hydroxycinnamate
DBDS 4,4′-ibenzamidostilbene-2,2′-disulfonate
DIDS 4,4′-di-isothiocyanostilbene-2,2′-disulfonate
DMA 5-(*N*,*N*-dimethyl) amiloride
EIPA 5-(*N*-ethyl-*N*-isopropyl-amiloride)
ERK extracellular signal-regulated protein kinases
GSC glioblastoma stem-like cell
GSH glutathione
HMA 5-(*N*,*N*-hexamethylene) amiloride
MCT monocarboxylate transporter
MDR multidrug resistance

https://doi.org/10.1016/B978-0-12-820701-7.00017-8

MMP	matrix metallopeptidase
MNNG	*N*-methyl-*N*′-nitro-*N*-nitrosoguanidine
MRP	multidrug resistance-associated protein
NBD	nucleotide binding domain
NHE	sodium-proton exchanger
NHL	non-Hodgkin lymphoma
P-gp	P-glycoprotein
SCLC	small-cell lung cancer
SLC	solute carrier
TMD	transmembrane domain
TNF	tumor necrosis factor
uPA	urokinase-type plasminogen activator
VEGF	vascular endothelial growth factor

Conflict of interest

No potential conflicts of interest were disclosed.

Introduction

The necessity to isolate and regulate biochemical processes requires compartmentalization of the cells, which is achieved through external and internal membranes. Membranes constitute a barrier to many chemical and biochemical compounds, especially to the very hydrophilic and charged ones. In order to facilitate and regulate the flux of these compounds in and out of cells, specialized proteins evolved to perform a transport role. Thus, these drug transporters mediate drug uptake into cells (influx) and/or facilitate the export of chemical compounds, including drugs, from the cells to the external milieu (efflux). Throughout these antagonistic processes, drug transporters regulate the flux of different endogenous materials in/out of cells, thus maintaining cell homeostasis. Transporters play an essential role in the movement of metabolites, nutrients, antioxidants, neurotransmitters, hormones, and signaling molecules. They also play a central role in absorption, distribution, and excretion of drugs and impact drug efficacy, drug disposition, and drug toxicity [1,2]. It was also suggested that transporters may play a role in the remote sensing and signaling between cells.

Given their key role in movement of chemical and biochemical compounds in/out of cells, one can understand why these proteins are ubiquitously spread in almost all human tissues and organs. Transporters at the level of the gut, liver and kidneys, blood brain barrier (BBB), placenta, etc. are essential for both absorption/transfer of nutrients, as well as for elimination of toxic or potentially toxic compounds from these organs, and organisms (fetus). Therefore, mutations occurring at the level of these proteins cause genetic disorders such as cystic fibrosis, cholesterol and bile transport defects, various neurological diseases, retinal degeneration, anemia, etc. and an abnormal response to drugs used in clinic [1–3].

Moreover, in cancer and in other diseases the normal transporter function and expression is altered, allowing the tumors to develop resistance to anticancer drugs (multidrug resistance, MDR) and chemotherapy [1–6]. In what will follow we will review the main classes of transporters, emphasizing their normal biodistribution, the diseases that alter this normal distribution, and the strategies used to correct these imbalances.

Classes of transporters and efflux pumps

Drug transporters are divided into two main classes, namely, solute carrier (SLC) class and ATP-binding cassette (ABC) class. Each class contains many members, each with a specific tissue expression pattern.

Solute carrier class (SLC) transporters

Monocarboxylate transporters

Monocarboxylate transporters (MCTs) are transmembrane proteins, encoded by the SLC16 gene family, with 14 members reported in humans so far [7,8]. They are widely expressed throughout the body and play an important role in many physiological functions such as cellular metabolism, T-lymphocyte activation, spermatogenesis, pancreatic beta cell activity, and drug transport and metabolism [9,10]. Over the past decades, their expression and activities have been implicated in different pathophysiological conditions especially in cancers [7,9,11]. These proteins transport a variety of substrates across the plasma membrane and have distinct substrate specificities and affinities for each of them. Among MCT members, MCT1, MCT2, MCT3, and MCT4 are monocarboxylate transporters mainly located on the plasma membrane, where they transport different substrates such as lactate, pyruvate, and ketone bodies, along with protons, bidirectionally, depending on the concentration gradients of the substrate across the membrane [8,12,13]. L-lactate is the main substrate for these transporters, in accordance with the fundamental importance of L-stereoisomer in cellular metabolism [13]. As mentioned above, the relative affinities for different substrates differ among the MCTs. MCT2 displays the highest affinities for lactate, with a Km value of 0.5–0.75 mM, followed by MCT1 (Km = 3.5–10 mM), MCT3 (Km = 5–6 mM), and MCT4 (Km = 22–28 mM) [8,13]. A similar trend is also observed for pyruvate, with MCT2 again showing the highest affinity, followed by MCT1 and MCT4, with Km values of 0.1 mM, 1 mM, and 153 mM, respectively [14]. MCT1 and MCT2 were also characterized for their affinity toward ketone bodies in Xenopus oocytes, with MCT2 again showing higher affinity for D-β-hydroxybutyrate (Km = 1.2 mM) and acetoacetate (Km = 0.8 mM) as compared to MCT1 (Km D-β-hydroxybutyrate = 12.5 mM and Km acetoacetate = 5.5 mM) [15,16].

Besides the above-mentioned MCTs, little is known about their congeners. MCT6 was found to mediate the transport of bumetanide, a loop diuretic, while its natural substrate is unknown so far [17]. MCT7 has been reported to be a hepatic ketone body β-hydroxybutyrate transporter in zebrafishes upon fasting [18]. Loss of MCT7 was associated with hepatic steatosis in this experimental animal model, a pathological accumulation of fat, which was rescued with the introduction of MCT7 [18]. MCT8 and MCT10 were found to have affinities for thyroid hormones T3 and T4, with MCT10 also showing good affinities for aromatic amino acids [19]. MCT9 has been reported to act as a proton-coupled carnitine efflux transporter system when expressed in Xenopus oocytes incubated with [3H]-carnitine [20].

To date, no crystal structure is available for MCTs. However, topological predictions revealed that they consist of 12 transmembrane helices with intracellular N- and C- termini and a large cytosolic loop between transmembrane domains 6 and 7 [9,10,13,21].

This prediction was confirmed by proteolytic digestion studies for MCT1, done in rats. The transmembrane domains are the most conserved among the MCTs, while the loops and C-terminus are the most variable structural features. Site-directed mutagenesis studies and molecular modeling of MCT1 on the basis of *E. coli* glycerol-3-phosphate transporter (GlpT) revealed important insights about substrate binding and transporter activities [13,22,23]. Three amino acid residues, namely, Lys-38, Asp 302, and Arg 306, are particularly important for the transport of monocarboxylates and are conserved in MCT members MCT1-MCT4 and in MCT7 [13].

MCT expression in normal tissues

MCTs are ubiquitously expressed in the human body, with their abundance depending on the specific functions and metabolic requirements of the tissues (Table 1) [7,9,13]. However, their tissue distribution, subcellular localization, and regulation of expression are reported to differ significantly among the isoforms. Conspicuously, MCTs are not expressed in pancreatic β cells. This fact helps the body to avoid hypoglycemic conditions during exercise by preventing lactate-induced ATP generation and insulin secretion [10]. MCT1 was found to be most widely expressed in many tissues and organs. It is expressed in different parts of

TABLE 1 MCT isoforms, their substrates, and biodistribution.

Transporter	Substrate	Normal tissue distribution
MCT1	Lactate, pyruvate, ketone bodies, short-chain fatty acids	Ubiquitous
MCT2	Lactate, pyruvate, ketone bodies	Brain, heart, stomach, pancreas, liver, kidneys, colon, adipose tissue, testis, leukocytes
MCT3	Lactate	Retinal pigment epithelium, choroid plexus epithelium
MCT4	Lactate, ketone bodies	Brain, heart, lungs, skeletal muscles, small intestine, testis, ovaries, placenta, leucocytes, chondrocytes
MCT5		Brain, heart, lungs, skeletal muscles, liver, kidneys, ovaries, placenta
MCT6	Bumetanide	Brain, heart, lungs, skeletal muscles, liver, pancreas, kidneys, prostate, placenta
MCT7	Ketone bodies	Brain, liver, pancreas, skeletal muscles, prostate
MCT8	Thyroid hormone (T2, T3, T4)	Ubiquitous
MCT9	Carnitine	Brain, retina, breasts, kidneys, spleen, testis, ovaries, endometrium
MCT10	Thyroid hormone (T3, T4), Aromatic amino acids	Heart, skeletal muscles, small intestine, kidneys, placenta
MCT11		Retinal pigment epithelium, choroid plexus, lungs, breasts, pancreas, ovaries, skin
MCT12		Eyes, lungs, kidneys, testis
MCT13		Breasts, bone marrow, stem cells
MCT14		Brain, heart, lungs, skeletal muscles, thymus, breasts, liver, pancreas, spleen, prostate, ovaries

gastrointestinal tract such as stomach, duodenum, jejunum, ileum, cecum, colon, and rectum, with different intensities. In the gut epithelial cells, MCT1 is found to be present on both apical and basolateral membranes, where it is suggested to be involved in both the absorption of monocarboxylates and short-chain fatty acids, and in their subsequent efflux into the blood [24–31]. MCT1 is also found in red blood cells, monocytes, and T-lymphocytes [32–34]. It is involved in influx and efflux of lactate in both oxidative and glycolytic cells and thus widely expressed in red skeletal muscle fibers, white skeletal muscle fibers, cardiomyocytes, and hypoxic cells [35–37]. It mediates influx of lactate in kidney proximal convoluted tube and in liver parenchyma cells. Among the neuronal cells, it is found in astrocytes, oligodendrocytes, and microglial cells [38–41]. It is also expressed in endocrine gland such as sebaceous glands and mammary glands [42].

The highest affinity transporter, MCT2, is reported to have a narrower cellular and tissue distribution compared to MCT1 [7,9,21,43]. It is expressed in tissues that primarily depend on the uptake of monocarboxylates as fuel for oxidative phosphorylation, gluconeogenesis, or lipogenesis [44]. For instance, it is expressed in liver parenchymal cells where it transports lactate for gluconeogenesis [45]. In the kidney proximal convoluted tubes, MCT2 is involved in the efflux of lactate [45]. MCT2 is also expressed in the stomach, pancreas, neuronal cells, heart, adipose tissues, colon, and testis [44,46,47]. MCT3 is expressed in only two cell types, namely, retinal pigmented epithelial cells and choroid plexus epithelium [48–50]. Owing to a high turnover rate, MCT4 is involved in efflux of lactate from highly glycolytic cells, thus being widely expressed in glycolytic muscle fibers [47,51], astrocytes [46], and leukocytes [52]. It is also reported to be expressed in kidneys, small intestine, placenta, and Schwann cells of peripheral nervous system [47,53,54]. Other MCTs were not studied extensively for their expressions and functions.

MCT inhibitors

The idea of targeting lactic acid efflux either by pharmacological inhibition or by genetic knockdown of MCTs has been explored. Some of these earlier representative inhibitors include α-cyano-4-hydroxycinnamate (CHC), 4,4′-di-isothiocyanostilbene-2,2′-disulfonate (DIDS), and 4,4′-dibenzamidostilbene-2,2′-disulfonate (DBDS), phloretin, quercetin, and p-chloromercuribenzene sulfonate (PCMPS) (Chart 1) [9,11,55]. From these inhibitors, α-cyano-4-hydroxycinnamate (CHC) were extensively studied by several groups for anticancer applications. For example, in a study, done by Colen et al., CHC was shown to inhibit lactate efflux in malignant gliomas through MCT inhibition resulting in decreased level of glycolytic metabolites such as antioxidants taurine and glutathione and enhanced radiosensitivity in CHC-treated glioma cells, with enhanced radiation-induced cell killing [56]. The same research group demonstrated that inhibition of lactate efflux by CHC decreases invasion capacity of U87-MG and U251-MG glioma cells in vitro, causes tumor necrosis in vivo, and improves overall survival [57]. In another study, through MCT inhibition, α-cyano-4-hydroxycinnamate, quercetin, and lonidamine were found to inhibit lactate efflux in highly glycolytic human breast cancer cell lines resulting in a decrease in cell proliferation, migration, and invasion, as well as an increase in cell death. Silencing the MCT1 by using siRNA yielded similar findings, thus validating MCT1 as a promising anticancer target [55,58]. However, these compounds were found to be nonspecific inhibitors of MCTs, most of which showed only micromolar potency. For example, CHC reported to be two-fold more potent

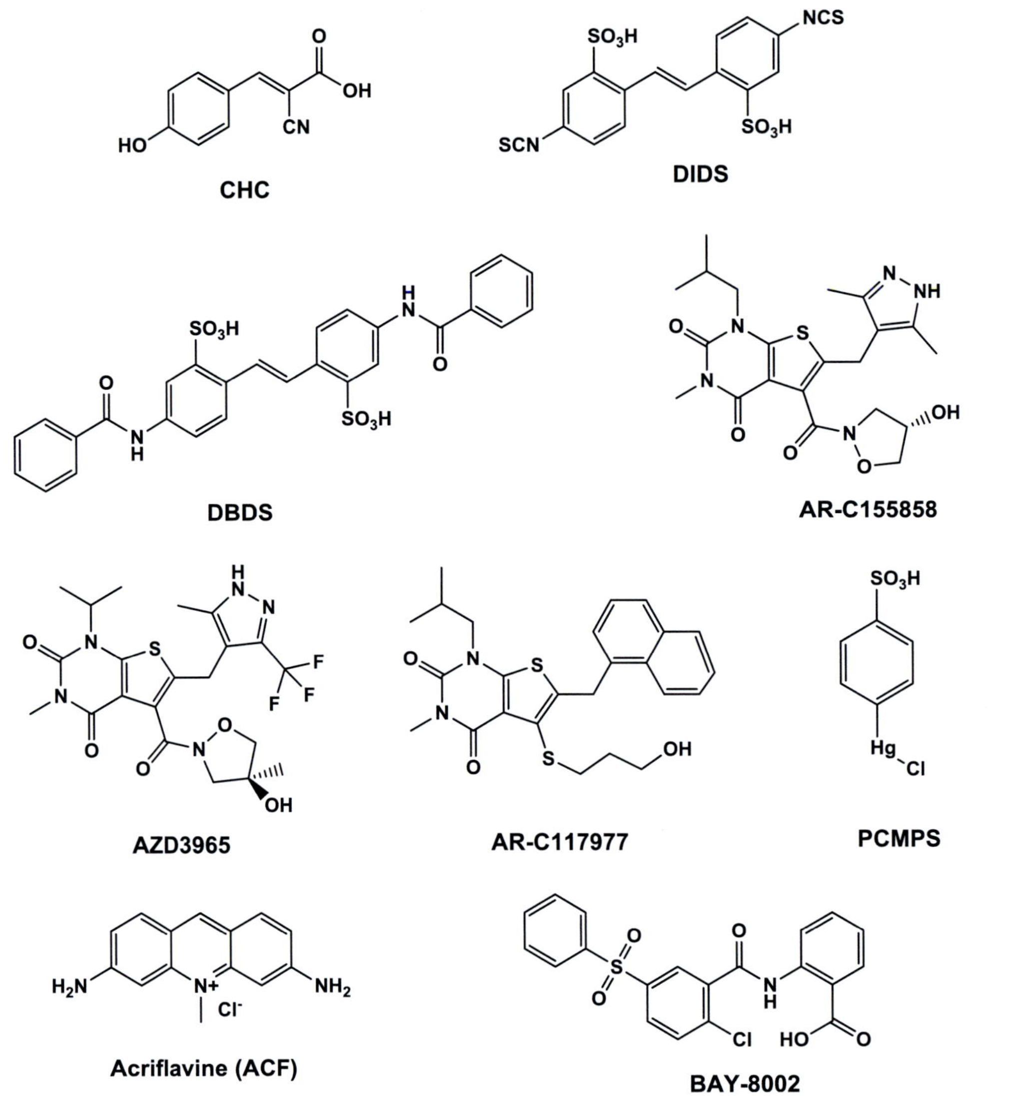

CHART 1 Representative MCT inhibitors.

inhibitors for mitochondrial pyruvate transporter than MCT1. Similarly, stilbene disulfonate derivative DIDS and DBDS were found to inhibit erythrocyte chloride/bicarbonate exchanger AE1 much more potently than MCT1 [11,59]. Therefore, more potent and specific MCT inhibitors are required to translate these inhibitors toward clinical applications.

In this context, AstraZeneca developed a new series of highly potent and specific MCT inhibitors as part of their immunosuppressive drug discovery research. These second-generation inhibitors are particularly specific to MCT1 and MCT2, with some specificity to

MCT4. Some of the representative compounds are AR-C177977, AR-C122982, AR-C117977, AR-C155858, and AZD3965 [11,60]. Among these compounds, AR-C177977, AR-C122982 were shown to suppress immune responses, with AR-C177977 being particularly efficacious in inducing donor-specific suppression and reducing both acute and chronic allograft rejection in the rats [61]. However, they exhibited quite low oral bioavailability and short plasma half-lives, which prevented their further translation into clinical use. The optimized compound, AR-C155858 (Chart 1), exhibited 2 nM potency in inhibiting MCT1-mediated L-lactate transport in rat erythrocytes. Le Floch et al. demonstrated that AR-C155858 suppresses lactate export, glycolysis, and tumor growth in vivo by inhibiting MCT1 and MCT2 in Ras-transformed fibroblasts. However, ectopic expression of MCT4 in these cells was found to compensate the effect of MCT1 and MCT2 inhibition and reverse tumorigenicity [62]. AR-C155858 was also found to potently inhibit MCT1 and MCT2 but not MCT4 in *Xenopus laevis* oocytes.

Additionally, inhibition of MCT1 was determined to be time-dependent, with binding site located at the C-terminal of MCT1 involving transmembrane domains 7–10 [63]. Further optimizations led to the development of a better MCT1 inhibitor, namely, AZD3965 (Chart 1), which showed 6-fold higher selectivity for MCT1 over MCT2 and was not active against MCT3 and MCT4. AZD3965 was extensively studied for antitumor activities [60]. It was shown to inhibit lactate efflux and result in the accumulation of glycolytic metabolites, which ultimately led to inhibition of cell proliferation in large B-cell lymphoma, non-Hodgkin lymphoma (NHL), and Burkitt's lymphoma in vitro. It also exhibited significant inhibition of tumor growth in the in vivo Raji Burkitt's lymphoma model through blocking of MCT1-mediated lactate efflux. Additionally, coadministration of AZD3965 with chemotherapeutic agent doxorubicin or rituximab showed further enhancement of cell growth inhibition and cell death compared to monotherapy treatment [64]. Based on these promising findings, AZD3965 advanced to phase I/II clinical trials in the UK for treating patients with advanced solid tumors [65]. Several other studies further confirmed the antitumor activities of AZD3965 in other cell lines, including gastric cancer and small cell lung cancer (SCLC) cell lines [66,67]. In SCLC, AZD3965 was found to be most effective under hypoxic conditions, where MCT1 was highly overexpressed along with low expression of MCT4 [67].

Recently, a novel class of selective MCT1 inhibitor was developed. The best inhibitor BAY-8002 was reported to be 5-fold selective for MCT1 compared to MCT2, with no activity observed against MCT4 as well as no off-target effects. Like AZD3965, BAY-8002 was found to bind to the same site of MCT1, as determined by the competition studies using radiolabeled-binding assays. BAY-8002 exhibited significant lactate accumulation in cells and showed significant antiproliferative effects in 11% of cell lines from 246 cell lines test against, with highest effects observed in diffuse large B-cell lymphoma (DLBCL) [68].

Surprisingly, only few MCT4 specific inhibitors have been developed so far even though MCT4 showed significant role in chemoresistance, as evidenced from different studies. Among them, a small molecule inhibitor, acriflavine (ACF, Chart 1), shown to inhibit MCT4 by blocking the binding between basigin and MCT4, required for the functional localization of MCT4 to the plasma membrane. ACF was shown to inhibit cell proliferation and self-renewal potential of several glioblastoma neurosphere cell lines in vitro, with superior activity observed under hypoxic conditions. ACF treatment also significantly inhibited tumor growth and progression in both early and late stages of the disease in mice bearing GSC-derived xenografts resulted [69].

The area is receiving significant interest and we anticipate more lead compounds to be introduced in the near future.

Na^+/H^+-exchangers (NHEs)

Sodium-proton exchangers (NHEs) are ubiquitously expressed membrane transporters, encoded by the SLC9A solute carrier gene family, with 11 human isoforms (NHE1-NHE11) identified so far [70]. They transport one intracellular H^+ in exchange for an extracellular Na^+ across the cell membrane, thereby playing an important role in pH regulation in mammalian cells. They are involved in numerous physiological processes such as regulation of cell cycle, cell volume and systemic fluid homeostasis, apoptosis, cell adhesion, migration, and proliferation [71–73].

NHE structures have not been resolved through X-ray crystallographic studies. However, topological predictions and computer modeling of hydrophilic-hydrophobic regions revealed that all the NHEs have similar structures. They predicted to have relatively conserved 11 or 12 membrane spanning domains at the N-terminus consisting of ~450–500 amino acids, which perform the Na^+ and H^+ exchange. The C-terminal domain is more variable among the isoforms having ~125–440 amino acids depending on the isoform, with multiple phosphorylation sites and ancillary protein-binding sites, and being associated with regulation of the exchange activity [71,73]. This structural information was supported by various studies utilizing a combination of different approaches such as epitopes immunolocalization [74,75], identification of glycosylation sites [76,77], protease cleavage studies [78], C-terminal truncations studies [79–81], and cysteine substitutions and accessibility studies [82] (Fig. 1).

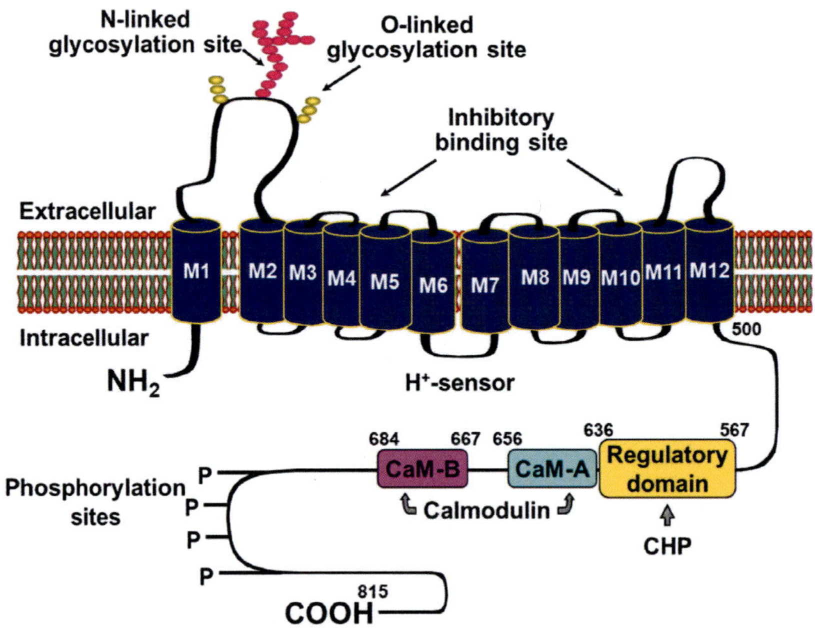

FIG. 1 Topological structural model of NHE1. *Taken from reference Gazmuri RJ, Radhakrishnan J, Ayoub IM. Sodium-hydrogen exchanger isoform-1 inhibition: a promising pharmacological intervention for resuscitation from cardiac arrest. Molecules (Basel, Switzerland) 2019;24(9):1765, with permission [83].*

Among the NHE isoforms, Na^+/H^+ exchanger isoform one (NHE1) is the most extensively characterized and studied isoform, which plays significant role in pH regulation in a wide variety of normal cells, as well as cancer cells [84,85]. It is ubiquitously expressed in most human cells except macula densa and α- and β-intercalated cells of the kidney. It is mainly localized on the cell surface, but, in some cell types, it accumulates in discrete microdomain of the plasma membranes. To further elaborate, it is found to localize on the basolateral membrane in the epithelial cells, along the border of lamellipodia in migrating fibroblasts, and the intercalated disks and t-tubules in cardiac myocytes [73,86,87].

NHE1 is reported to be involved in distinct stages of neoplastic transformation and carcinogenesis in multiple ways [88–91]. In normal cells NHE1 was found to be inactive at normal intercellular pH_i (7.0–7.1) and only become activated in response to cytosolic acidification [92]. However, in tumor cells it is hyperactive even at normal intracellular (pH_i), which leads to the increase in intracellular pH (cytosolic alkalization) and decrease in extracellular pH_e. The increase in intracellular pH was reported to be directly associated with initiation of neoplastic transformation, augmented DNA synthesis, induction of oncogene overexpression, enhanced stability of the spontaneous gene mutations, and hyperactivity of the growth factors [88,93]. All these effects contribute to the cell transformation, uncontrolled proliferation, migration, tumor formation and growth, invasion, metastasis, and chemoresistance in many cancers [28, 94–96]. Because of these multiple roles in carcinogenesis, NHE1 has been highly overexpressed in many cancers, including breast cancer [97], hepatocellular carcinoma [98], cervical cancer [99], esophageal cancer [100], and glioblastomas [101]. Therefore, inhibition of NHE1 offers a viable strategy to combat different cancer either alone or in combination with other chemotherapeutic agents. Numerous research groups explored NHE1 as an anticancer therapeutic target, with NHE1 inhibition resulting in acidification of the intracellular space, inhibition of glycolysis, reduction of cell proliferation, tumor growth and invasion, and selective apoptosis [102–104].

NHE inhibitors

There are two major classes of NHE inhibitors. First class of NHE inhibitors are pyrazine derivatives, which include amiloride, 5-(N,N dimethyl) amiloride (DMA), 5-(N-ethyl-N-isopropyl-amiloride) (EIPA), and 5-(*N,N*-hexamethylene) amiloride (HMA). The second class of NHE inhibitors are the benzoylguanidines, which include cariporide, eniporide, and HOE 694 (Chart 2) [90,105]. Few other nonamiloride-based NHE inhibitors were also developed, with improved potency and selectivity.

Amiloride was developed as a potassium-sparing diuretic in 1967, and was used for the management of high blood pressure and swelling in conjunction with other antihypertensive drugs [106]. Amiloride was reported as the first NHE1 inhibitor by Sparks et al. in 1983 [107]. In this study, amiloride was found to inhibit both tumor growth and tumor cells proliferation in the H6 hepatomas in a dose-dependent manner. Since then, amiloride was extensively investigated for antitumor, antimetastatic, and antiangiogenic agents in numerous studies [108–113]. Besides NHE inhibition, amiloride reported to reduce vascular endothelial growth factor (VEGF) expression and inhibit urokinase-type plasminogen activator (uPA), metalloproteinases and other proteases, all of which are well-known activators of cancer cells invasion and metastasis processes. Amiloride shown to inhibit proliferation and invasion of a variety of cancer cells through NHE1 inhibition such as gliomas [114], hepatocellular carcinoma [115], and both human and murine breast cancer cell lines (MCF7, MDA-MB231, and EMT6) [116].

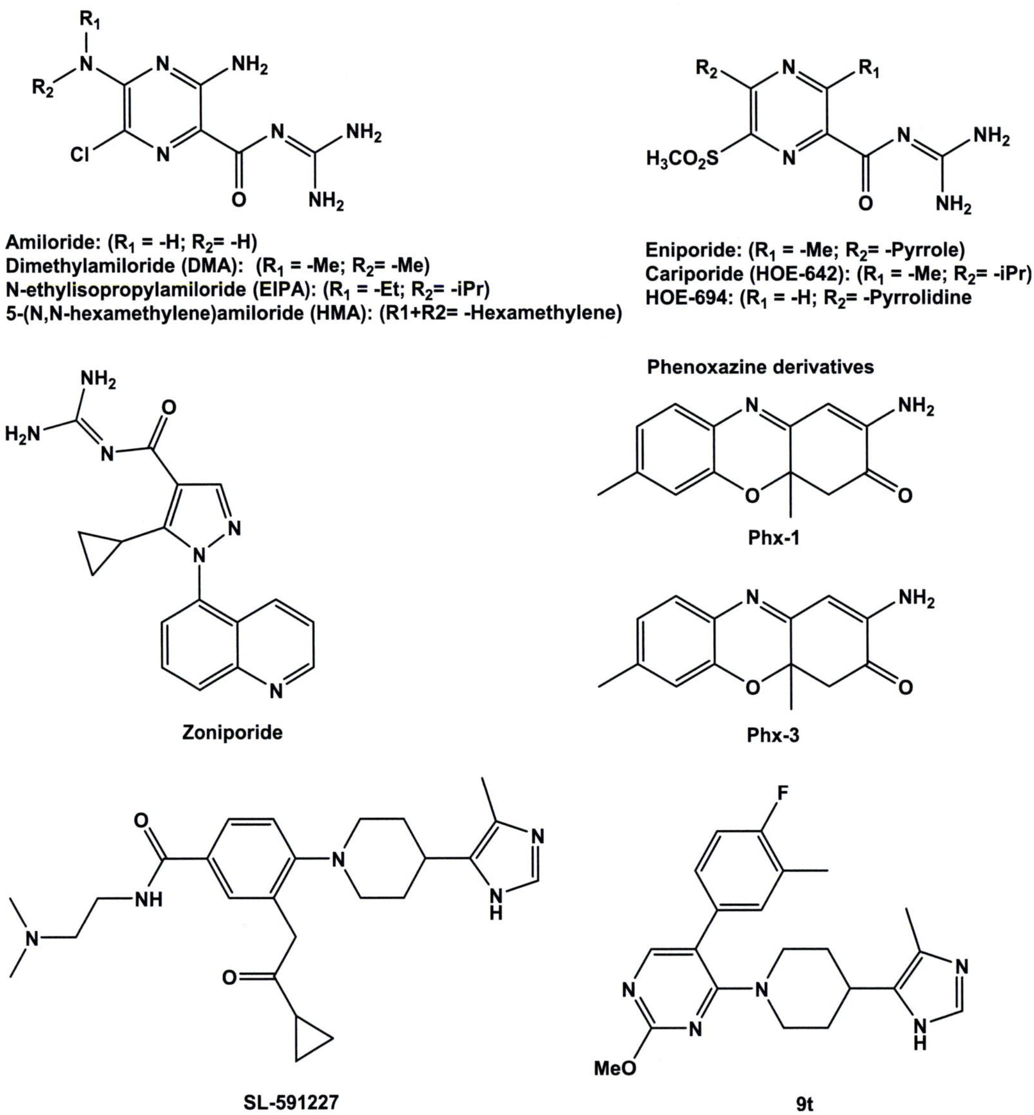

CHART 2 Representative NHE1 inhibitors available in the clinic or under development.

Several studies with amiloride on animal model of cancers have demonstrated the antitumor potential of this compound. For example, Tatsuta et al. reported the effect of amiloride on the incidence and histological features of gastric cancers in Wistar rats induced by oral administration of *N*-methyl-N′-nitro-N-nitrosoguanidine (MNNG) [117]. It was observed that s.c. injections of amiloride (5.0 mg/kg but not 2.5 mg/kg), in depot form, every other day commencing at week 25, significantly decreased the incidence of rat gastric cancers at the end of the study (week 52). However, authors didn't observe any changes in histological features of gastric cancers [117]. In a similar study, done by same research group, daily s.c. injections of amiloride (7.5 mg/kg but not 5.0 mg/kg) administered over 35 weeks were shown to decrease the incidence of azoxymethane (7.4 mg/kg)-induced colon cancers in young male Wistar rats. Like their previous study, amiloride had little or no significant influence on the histological features of colon tumors [118].

Furthermore, Evans et al. demonstrated that amiloride reduced the invasive capacity of unaltered human breast cancer cells MDA-MB-231 and MDA-MB-436 in a dose-dependent manner by using Matrigel invasion assays. Authors also observed amiloride could inhibit the cell proliferation in a nonsignificant manner [119]. In another Matrigel invasion assays, amiloride was found to decrease invasion of MATB rat mammary carcinoma cells, either administered alone or in combination with COX-2 inhibitor celecoxib [120]. Same research group also reported the dose- and time-dependent antimetastasis effects of amiloride in two different rat models of metastasis [113].

In order to improve potency and selectivity toward NHE1, first generation of amiloride scaffold-based NHE1 inhibitors were generated through the substitution of both H atoms of the 5-aminopyrazine moiety of amiloride with different R1 and R2 groups (Chart 2) [121]. This yielded more potent and specific NHE1 inhibitors such as dimethylamiloride (DMA), N-ethylisopropylamiloride (EIPA) and Hexamethylene amiloride (HMA) (Chart 2). Among these inhibitors, N-ethylisopropylamiloride (EIPA) and Hexamethylene amiloride (HMA) are 223- and 540-times more potent relative to amiloride, respectively, with less off-target diuretic and antikaliuretic effects, thus particularly promising agents for anticancer applications [122,123].

In this context, Chen et al. demonstrated that the administration of EIPA inhibits smooth muscle cells proliferation and migration through the induction of apoptosis. EIPA was shown to reduce DNA synthesis and mitochondrial respiration. These effects might be due to inhibition of both NHE activity and uPA activity [124]. In another study, EIPA exhibited suppression of invasion and metastasis of HepG2 hepatocellular carcinoma cells under hypoxic conditions. These effects were experimentally determined to act through the downregulation of MMP-2, MMP-9, and VEGF, in an ERK1/2-dependent pathway [115].

In an extensive study, amiloride and HMA were shown to significantly reduce the growth and viability of two human hepatocarcinoma cell lines and a rat hepatocarcinoma cell line. Both amiloride and HMA decreased the DNA synthesis and cell protein content of the cultures after a 24-h treatment time in a concentration-dependent manner along with decrease in intracellular pHi via NHE inhibition. Authors suggested that intracellular acidification might be the reason behind diminished cell viability through the reduction in glutathione (GSH) levels and impairment of lysosomal integrity [125]. In another study, Rich et al. demonstrated that pharmacologic doses of HMA decrease the pHi and induce apoptosis in patient leukemic cells in vitro for up to 5 h treatment time, with low toxicity observed for normal cells [103].

Coadministration of amiloride and its analogs with chemotherapeutic agents was reported to produce synergistic effects. For instance, Reshkin, et al. demonstrated that concurrent administration of NHE1 inhibitor 5-N,N-dimethylamiloride (DMA) with paclitaxel resulted in a synergistic induction of apoptosis and a reduction in the paclitaxel IC_{50} for apoptosis in MDA-MB-436 breast cancer cells compared with paclitaxel or DMA administered alone

[126]. In another study, EIPA was shown to potentiate the cytotoxic effect of doxorubicin in HT29 colon cancer cells when administered together. Authors also reported that EIPA significantly reduced the pHi value, which led to the increased the intracellular accumulation of doxorubicin [96]. Similar finding was also reported by Poon et al., where amiloride was found to synergize with doxorubicin to potentiate apoptosis in canine osteosarcoma cells in a combination treatment [127].

In an effort to inhibit NHE1 specifically, a new series of benzoylguanidine class of inhibitors was developed through the simultaneous substitution of the 6-chloro- moiety by a sulfomethyl group, along with the replacement of 2-amino group with a -H or a methyl group. The resulted inhibitors, such as HOE-694, cariporide (HOE-642), and eniporide (Chart 2), showed improved selectivity toward NHE1, with diminished off-target diuretic and antikaliuretic effects. Among these inhibitors, cariporide has been studied extensively, which showed significant suppression of invasive capacity in various cancer cell lines [91,128]. Like amiloride derivatives, cariporide was shown to reduce proliferation and induce apoptosis in cholangiocarcinoma cells due to acidification of the intracellular pH through selective inhibition of NHE1 [90]. Cariporide was also reported to reduce invasion of invasion of human tongue squamous carcinoma cells under hypoxic conditions [105].

Several nonamiloride and nonguanidine NHE inhibitors were developed in order to further improve the potency and selectivity toward NHE1, such as SL-591227, phenoxazine derivatives Phx-1 and Phx-3, and pyrimidine analog compound 9t (Chart 2). Phx-3, reported by Tomoda et al., showed excellent selectivity for NHE1 and was shown to induce apoptosis in a variety of cancer cell lines [129,130]. Compound 9t, developed by Bristol-Meyers, was reported to be 500 times more potent against NHE1 relative to cariporide, having an IC50 value of 6.5nM, with 1400 times higher selectivity toward NHE1 over NHE2 [131]. However, no further publications are available regarding anticancer activities of this compound. These studies toward more selective NHE isoform inhibitors are ongoing.

ATP-binding cassette (ABC) transporters

ATP-binding cassette (ABC) transporters belong to the superfamily of transmembrane transporter proteins, encoded by 48 human genes. Seven subfamilies (ABCA-ABCG) were identified so far, based on the sequence identity and structural homology [132–134].These transporters are composed of two domains, namely, an ATP-binding domains and transmembrane domains (TMDs). The ATP-binding domains, also called nucleotide binding domains (NBDs), consist of approximately 90–110 amino acid residues, which organize to form two highly conserved Walker motifs A and B, and a less conserved linker motif C in-between them. The TMD unit is responsible for substrate specificities and consists of six α-helices. It is not well conserved [132–134]. The functional unit of ABC transporters consists of two NBDs and two TMDs, with transmembrane domains (TMDs) binding the substrate, while two NBDs bind and hydrolyze the ATP that drive the transport across the membrane (Fig. 2) [133,135,136].

These transporters are involved in the movement across the plasma membrane of a wide variety of substrates such as carbohydrates, amino acids, lipids, ions, metabolic products, toxins, and drugs across the plasma and intracellular membranes [132,135,137]. Thus, they play important roles in physiological processes and their mutations result in various

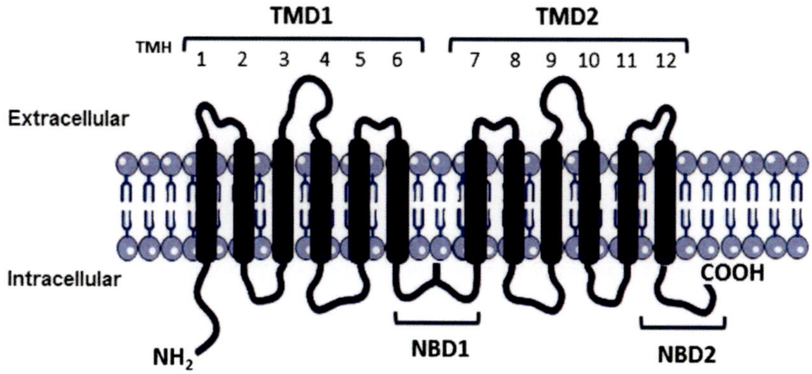

FIG. 2 Topological model of ABC transporter, P-gp. *Adapted from Silva R, Vilas-Boas V, Carmo H, Dinis-Oliveira RJ, Carvalho F, de Lourdes Bastos M, Remião F. Modulation of P-glycoprotein efflux pump: induction and activation as a therapeutic strategy. Pharmacol Therapeut 2015;149:1–123, with permission [136].*

human diseases, such as various Mendelian disorders, cystic fibrosis, progressive familial intrahepatic cholestasis, adrenoleukodystrophy, retinal degeneration, and hypercholesterolemia [134].

Overactivity and overexpression of these transporters result in therapeutic failures of diverse set of drugs. Numerous ABC transporters from various subfamily are involved in drug efflux, with at least 15 transporters being shown to be able to efflux anticancer drugs in vitro so far [138]. However, three main ABC transporters, namely, P-glycoprotein (P-gp; MDR, ABCB1), multidrug resistance-associated protein 1 (MRP1, ABCC1), and ABCG2 (breast cancer resistance protein; BCRP, mitoxantrone resistance protein; MXR), are responsible for export of most drugs [132].

ABCB1, commonly known as P-glycoprotein or MDR1, was discovered in 1976 and is the most extensively studied and characterized efflux transporter [139]. It is found on the apical surface of the cells and is widely expressed in the liver, kidney, adrenal glands, placenta, testes, intestine, and endothelial cells of the blood-brain barrier. It plays an essential role in protecting the body by exporting numerous xenobiotics and toxins as its normal physiological function. However, its substantial role in drug efflux results in drug resistance and consequent therapeutic failures [136,137,140,141]. The role of P-gp in anticancer drug resistance was first unveiled by Kartner et al. in 1983, and numerous studies following revealed its contribution in cancer chemoresistance [140]. P-gp is highly overexpressed in many cancers and associated with poor clinical prognosis in breast, liver, kidney, colon, ovarian, leukemia, and adrenocortical cancers [132,133]. It exports many structurally unrelated chemotherapeutic agents, such as vinca alkaloids (vinblastine, vincristine, vindesine, vinorelbine), anthracyclines (doxorubicin, daunorubicin), epipodophyllotoxines, and taxanes (paclitaxel, docetaxel) [142].

Furthermore, chemotherapy-induced expression of efflux transporters was reported for several cancers, a phenomenon termed as acquired resistance. For example, in a clinical study, Abolhoda el al. observed a 3–15-fold (median, 6.8) increase in P-gp RNA levels in tumors of unresectable sarcoma pulmonary metastases patients after only 20–50 min of lung chemoperfusion with doxorubicin [143]. Another mechanism of acquired drug resistance is reported by Levchenko et al., with authors observing the intercellular transfer of functional P-gp from P-gp-positive to P-gp-negative cells in neuroblastoma in vivo and in multiple cell lines in vitro. This type of P-gp transfer can protect sensitive cells from chemotherapeutic agents and ultimately lead to the formation of permanently resistant cells, thus increasing the possibility of chemoresistance [144]. Induction of expression was also observed during malignant progression. For example, P-gp was shown to express in glioblastoma tumors where it is not expressed normally [145]. In another study, P-gp was shown to be responsible for resistance toward caspase-dependent cell apoptosis induced by various stimuli such as Fas ligand, tumor necrosis factor (TNF), UV radiation, and serum starvation. Thus, P-gp overexpressing tumor cells have less apoptotic death, which results in higher survival rate for these cells [146].

ABCC1, commonly known as *multidrug resistance-associated protein 1 (MRP1)*, is the first member of ABCC transporter family. It was identified in 1992, when researchers observed drug resistance independent of P-gp pathway in a human small-cell lung cancer cell line not expressing P-gp [147]. Since then, MRP1 has been linked with chemoresistance of various chemotherapeutic drugs in several cancer cell lines. It was found to localize at the basolateral membrane of polarized epithelial cells, where it exports substrates into the blood vessels. It is widely expressed, in high intensity, in many organs such as kidney, lung, stomach, spleen, colon, bladder, bone marrow, adrenal gland, placenta, and testes [148]. MRP1 was shown to be upregulated in numerous cancers such as lung, breast, prostate, leukemia, melanoma, ovarian cancer, and gastrointestinal carcinoma [149]. Consequently, it has been assumed that it has a significant role in drug resistance and it was linked with poor drug response and overall survival in several cancers. MRP1 transports a wide variety of substrates, including some that overlap with P-gp, including folate-based antimetabolites (methotrexate), vinca alkaloids (vincristine, vinblastine, etoposide), anthracyclines (doxorubicin), antiandrogens (flutamide), and glutathione (GSH) and glucuronide conjugates of these drugs, as well as endogenous substrates such as folates, glutathione, and glucuronide conjugates of steroids, leukotrienes, and prostaglandins. Despite significant role in MRP1 in cancer drug chemoresistance, only few specific inhibitors for MRP1 have been developed so far.

ABCG2 (Breast cancer resistance protein/BCRP/MXR) is a 95-kDa membrane protein, first identified in 1990 and found to linked with doxorubicin resistance in MCF-7/AdrVp(100) cells, therefore named as Breast Cancer Resistance Protein (BCRP) [150]. It is localized at the apical membrane of epithelial cells, with physiological expression reported to various organs such as liver, brain, prostate, intestine, and placenta [151]. It is reported to play important role in the final stages of erythropoiesis by exporting protoporphyrin IX in hematopoietic cells [152]. It is also highly expressed in stem cells, where it plays important protective function [153]. ABCG2 was reported to be a high-capacity urate exporter and mediates urate homeostasis in the human body through both renal and extrarenal urate excretion [154]. Upregulation of ABCG2 has been correlated with poor chemotherapeutic response of several anticancer, including mitoxantrone, topotecan, and methotrexate in leukemia and breast cancers [155–157].

P-gp/MDR1 inhibitors

Numerous P-gp inhibitors have been developed so far and are currently categorized into four generations based on their potency, selectivity, and drug interaction profiles (Table 2)

TABLE 2 P-gp/MDR1 inhibitors.

Generation	Drugs
First generation	Verapamil, flupenthixol, chlorpromazine, cyclosporin A, reserpine, quinine, quinidine, tamoxifen and toremifen
Second generation	Dexverapamil, valspodar, biricodar, timcodar, dofequidar, dexniguldipine, quinine homodimer Q2
Third generation	Elacridar, tariquidar, zosuquidar, laniquidar, DP7, PGP-4008, CBT-1, annamycin and mitotane
Fourth generation	Natural products (e.g., quercetin, tangeretin, pervilleine F, cnidiadin, cannabidiol) Peptidomimetics (e.g., reversin 121, reversin 205) Surfactants and lipids (e.g., pluronic P85, tween-20, triton X-100, cremophor EL, poly (ethylene glycol)-300) Dual ligands [aminated thioxanthones such as 1-[2-(1H-benzimidazol-2yl)ethanamine]-4-propoxy-9H-thioxanthen-9-one]

[136,158]. These inhibitors act through several mechanisms by either inhibiting the substrate binding-site(s) through competitive, noncompetitive inhibition and allosteric modification [159], or prohibiting the ATP hydrolysis [160], or altering the integrity of cell membrane lipids [161].

First-generation inhibitors are various P-gp substrates, already clinically established molecules for different therapeutic targets such as calcium channel blockers (verapamil), antipsychotic drug (flupenthixol and chlorpromazine), immunosuppressants (cyclosporin A), antihypertensives (reserpine), antiarrhythmics (quinine and quinidine), and antiestrogens (tamoxifen and toremifen) [159]. They were investigated for their ability to inhibit P-gp in an effort to reverse chemoresistance. These compounds mainly act by competing with other P-gp substrates for efflux by P-gp through competitive inhibition. For example, verapamil and cyclosporin A were shown to inhibit P-gp competitively and reverse chemoresistance of vincristine, vinblastine, and daunorubicin in vivo and in vitro [162–164]. Unfortunately, they usually have low affinity for P-gp resulting in toxicity (as higher doses are required to achieve effective therapeutic concentration), which ultimately limits their clinical applications [136,158].

The second-generation P-gp inhibitors were developed through structural modifications of first-generation inhibitors in order to improve the potency and selectivity toward P-gp and to reduce their primary therapeutic activity (Table 2) [158]. Consequently, these compounds lack pharmacological activity of their corresponding parent compounds and usually have higher P-gp affinity and generate less cytotoxicity. For example, Valsopodar, analog of immunosuppressive agent cyclosporin A, developed by Novartis, lacks immunosuppressive activity and showed 5–20-fold higher potency against P-gp compared to cyclosporin A [165–167]. This is one of the best-characterized specific P-gp inhibitors and was extensively studied in many clinical trials in combination various chemotherapeutic agents such as doxorubicin, paclitaxel, etoposide, to overcome P-gp-mediated chemoresistance [168]. Few other notable second-generation inhibitors developed to date are dexverapamil (*R*-enantiomer of verapamil), biricodar, timcodar, and dofequidar (Chart 3) [169,170].

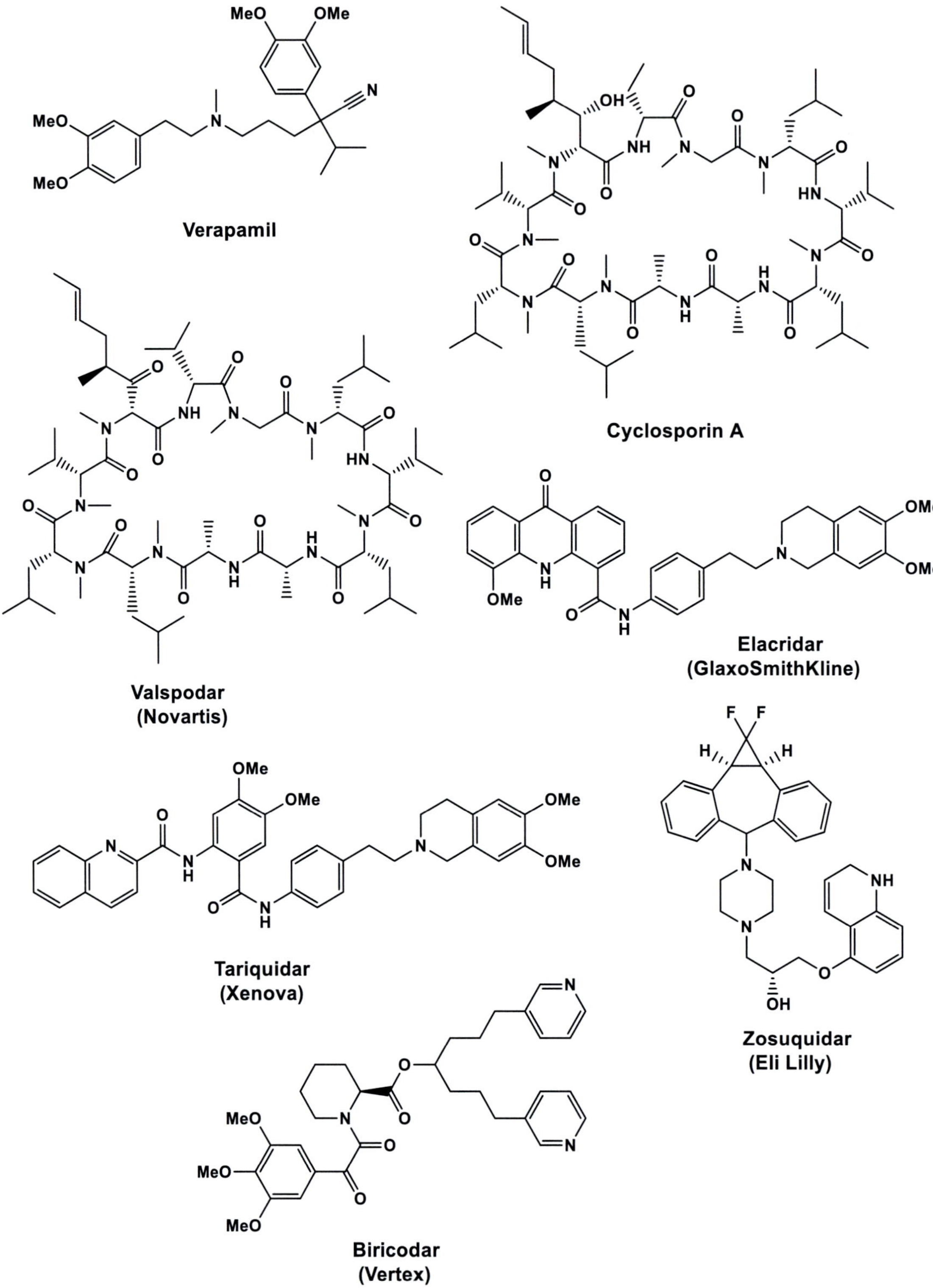

CHART 3 Representative P-gp inhibitors.

Unfortunately, second-generation P-gp inhibitors showed unwanted side effects and toxicity despite substantial improvement in potency and selectivity toward P-gp. These undesirable effects arise from unpredictable pharmacokinetic interactions with multiple other ABC transporters and metabolism by major drug metabolizing enzyme, CYP3A4. For instance, valspodar and biricodar were reported to interact with other ABC transporters MRP1 and with Cyp3A4. This interaction with other transporter and metabolizing enzyme further increases the possibility of adverse effects by reducing the cells capacity to efflux toxins and xenobiotics in the liver, kidney, and other drug-excreting organs [170].

The third-generation P-gp inhibitors were developed through different medicinal chemistry design approaches to further improve the potency, selectivity, and efficacy toward P-gp, with the aim to improve the treatment of multidrug-resistant tumors. These inhibitors possess excellent potency and selectivity for P-gp and are the most effective P-gp inhibitors developed to date, with up to 10-fold increase in potency obtained with some inhibitors relative to previous two generations, resulting in entry of several inhibitors into clinical trials. Some of the most promising and well-studied inhibitors are presented in Table 2. From these inhibitors, Elacridar, an acridone carboxamide derivative, found to inhibit efflux of doxorubicin, enhance uptake, and restore the antiproliferative action of doxorubicin in two animal models of MDR, namely, the P388/Dox leukemia and the C26 colon carcinoma. It acts by binding to the allosteric site of P-gp, with median effective dose of around 20 nM on different human and mice MDR cell lines. It is about 100–200 times more potent than verapamil and 2-fold more potent than other MDR inhibitors such as amiodarone or cyclosporin [171]. Based on the same chemical scaffold, Xenova developed tariquidar, which inhibits P-gp through a noncompetitive inhibition mechanism with superior potency and duration of action compared to previous generation inhibitors. It was shown to reverse resistance to doxorubicin, vinblastine, and paclitaxel in advanced breast cancers. Unfortunately, both elacridar and tariquidar were reported to bind to breast cancer resistance protein (BCRP), and, therefore, to encounter complex pharmacokinetic interactions. However, other third-generation agents such as zosuquidar and laniquidar were reported to overcome this limitation and to display excellent specificity and nanomolar potency toward P-gp in vivo and in vitro. Particularly, zosuquidar was reported to have significantly lower affinity toward metabolizing enzymes CYP3A4 and appeared to lack MRP1 or MRP2 modulation [136,158].

Even after all these improvements in specific P-gp modulations, third-generation P-gp still showed unwanted toxic effects, which again limited their clinical uses. For example, tariquidar was shown to exhibit higher incidence of adverse effects and poor responses in clinical trials, when coadministered either with paclitaxel and carboplatin or vinorelbine in non-small-cell lung cancer patients, which ultimately lead to the discontinuation of the clinical trials. Similar occurrences of adverse pharmacokinetic interactions were also observed in several clinical trials with zosuquidar in combination with doxorubicin, vinorelbine, docetaxel, and paclitaxel [136,158].

Finally, fourth-generation P-gp inhibitors were developed in order to further improve the P-gp modulations. These inhibitors were generated through different strategies and consist of natural products and their analogs, surfactants and lipids, peptidomimetics and dual functional agents. Several classes of natural products showed potent P-gp inhibition, but unfortunately were not specific and were proved to interact with drug metabolizing enzymes, which ultimately generated complex pharmacokinetic interactions and significant toxicities of co-administered chemotherapeutic drugs. In contrast, the use of surfactants is an interesting idea, as many surfactants such as Tween-20, Span-80, sodium dodecyl sulfate are currently approved as pharmaceutical excipients. These agents act by inserting themselves into lipid membranes and cause modifications in P-gp secondary and tertiary structure, resulting in the loss of P-gp functionality [136,158].

ABCC1 (MRP1) and ABCG2 (BCRP/MXR) inhibitors

In contrast with P-gp, a much lower number of studies dealt with MRP1 and ABCG2 inhibition. Some lead compounds include the leukotriene LTD4 receptor antagonist MK571 and fumitremorgin C (FTC), a fungal toxin [172,173]. Some P-gp inhibitors such as biricodar, dofequidar (MS-209), elacridar (GF20918), and CBT-1 were also proved to be MRP1 and/or ABCG2 inhibitors (48, 55–61 and Shukla) [132, 170, 174–180]. The efficacy and safety of these pan-transporter inhibitors is currently under evaluation.

Conclusions and perspectives

The modulation of the activity of drug transporters through selective inhibitors remains a very hot field of research, with recent years seeing progress across inhibition of all classes of transporters. The constant progress translated in many cases into an increased potency, selectivity, and safety profile of the inhibitors. However, the field is at the beginning, with inhibition studies evolving in parallel with physiological, biochemical, and pharmacological studies. More insights are needed to ensure the exploitation of the full potential of this class of compounds.

Acknowledgments

The financial support of NIH (grant R03EB026189) is gratefully acknowledged. U.K.M. acknowledges the financial support of Temple Graduate School through a Dissertation Completion Grant.

References

[1] Nigam SK. What do drug transporters really do? Nat Rev Drug Discov 2015;14(1):29–44.
[2] Du D, Wang-Kan X, Neuberger A, van Veen HW, Pos KM, Piddock LJV, Luisi BF. Multidrug efflux pumps: structure, function and regulation. Nat Rev Microbiol 2018;16(9):523–39.
[3] Liu X. Overview: role of drug transporters in drug disposition and its clinical significance. Adv Exp Med Biol 2019;1141:1–12.
[4] Holohan C, Van Schaeybroeck S, Longley DB, Johnston PG. Cancer drug resistance: an evolving paradigm. Nat Rev Cancer 2013;13(10):714–26.
[5] Shabana AM, Ilies MA. Drug delivery to hypoxic tumors targeting carbonic anhydrase IX. In: Targeted nanosystems for therapeutic applications: new concepts, dynamic properties, efficiency, and toxicity. American Chemical Society; 2019. p. 223–52.
[6] Shabana AM, Mondal UK, Alam MR, Spoon T, Ross CA, Madesh M, Supuran CT, Ilies MA. pH-sensitive multiligand gold nanoplatform targeting carbonic anhydrase IX enhances the delivery of doxorubicin to hypoxic tumor spheroids and overcomes the hypoxia-induced chemoresistance. ACS Appl Mater Interfaces 2018;10(21):17792–808.

[7] Halestrap AP. The SLC16 gene family – structure, role and regulation in health and disease. Mol Asp Med 2013;34(2):337–49.
[8] Halestrap AP, Meredith D. The SLC16 gene family—from monocarboxylate transporters (MCTs) to aromatic amino acid transporters and beyond. Pflügers Arch 2004;447(5):619–28.
[9] Pérez-Escuredo J, Van Hée VF, Sboarina M, Falces J, Payen VL, Pellerin L, Sonveaux P. Monocarboxylate transporters in the brain and in cancer. Biochim Biophys Acta – Mol Cell Res 2016;1863(10):2481–97.
[10] Halestrap AP, Wilson MC. The monocarboxylate transporter family—role and regulation. IUBMB Life 2012;64(2):109–19.
[11] Payen VL, Mina E, Van Hée VF, Porporato PE, Sonveaux P. Monocarboxylate transporters in cancer. Mol Metab 2020;33:48–66.
[12] Halestrap AP, Prince NT. The proton-linked monocarboxylate transporter (MCT) family: structure, function and regulation. Biochem J 1999;343(2):281–99.
[13] Halestrap AP. The monocarboxylate transporter family—structure and functional characterization. IUBMB Life 2012;64(1):1–9.
[14] Dimmer K-S, Friedrich B, Lang F, Deitmer JW, Bröer S. The low-affinity monocarboxylate transporter MCT4 is adapted to the export of lactate in highly glycolytic cells. Biochem J 2000;350(1):219–27.
[15] Bröer S, Bröer A, Schneider H-P, Stegen C, Halestrap AP, Deitmer JW. Characterization of the high-affinity monocarboxylate transporter MCT2 in *Xenopus laevis* oocytes. Biochem J 1999;341(3):529–35.
[16] Bröer S, Schneider H-P, Bröer A, Rahman B, Hamprecht B, Deitmer JW. Characterization of the monocarboxylate transporter 1 expressed in *Xenopus laevis* oocytes by changes in cytosolic pH. Biochem J 1998;333(1):167–74.
[17] Murakami Y, Kohyama N, Kobayashi Y, Ohbayashi M, Ohtani H, Sawada Y, Yamamoto T. Functional characterization of human monocarboxylate transporter 6 (SLC16A5). Drug Metab Dispos 2005;33(12):1845–51.
[18] Hugo SE, Cruz-Garcia L, Karanth S, Anderson RM, Stainier DYR, Schlegel A. A monocarboxylate transporter required for hepatocyte secretion of ketone bodies during fasting. Genes Dev 2012;26(3):282–93.
[19] Visser WE, Friesema ECH, Visser TJ. Minireview: thyroid hormone transporters: the knowns and the unknowns. Mol Endocrinol 2011;25(1):1–14.
[20] Suhre K, Shin S-Y, Petersen A-K, Mohney RP, Meredith D, Wägele B, Altmaier E, Deloukas P, Erdmann J, Grundberg E, Hammond CJ, de Angelis MH, Kastenmüller G, Köttgen A, Kronenberg F, Mangino M, Meisinger C, Meitinger T, Mewes H-W, Milburn MV, Prehn C, Raffler J, Ried JS, Römisch-Margl W, Samani NJ, Small KS, Erich Wichmann H, Zhai G, Illig T, Spector TD, Adamski J, Soranzo N, Gieger C, CardioGram. Human metabolic individuality in biomedical and pharmaceutical research. Nature 2011;477(7362):54–60.
[21] Pinheiro C, Longatto-Filho A, Azevedo-Silva J, Casal M, Schmitt FC, Baltazar F. Role of monocarboxylate transporters in human cancers: state of the art. J Bioenerg Biomembr 2012;44(1):127–39.
[22] Manoharan C, Manoharan C, Wilson MC, Manoharan C, Wilson MC, Sessions RB, Halestrap AP. The role of charged residues in the transmembrane helices of monocarboxylate transporter 1 and its ancillary protein basigin in determining plasma membrane expression and catalytic activity. Mol Membr Biol 2006;23(6): 486–98.
[23] Wilson MC, Meredith D, Bunnun C, Sessions RB, Halestrap AP. Studies on the DIDS-binding site of monocarboxylate transporter 1 suggest a homology model of the open conformation and a plausible translocation cycle. J Biol Chem 2009;284(30):20011–21.
[24] Van Rymenant E, Abrankó L, Tumova S, Grootaert C, Van Camp J, Williamson G, Kerimi A. Chronic exposure to short-chain fatty acids modulates transport and metabolism of microbiome-derived phenolics in human intestinal cells. J Nutr Biochem 2017;39:156–68.
[25] Tamai I, Takanaga H, Ogihara T, Higashida H, Maeda H, Sai Y, Tsuji A. Participation of a proton-cotransporter, MCT1, in the intestinal transport of monocarboxylic acids. Biochem Biophys Res Commun 1995;214(2):482–9.
[26] Ritzhaupt A, Wood IS, Ellis A, Hosie KB, Shirazi-Beechey SP. Identification and characterization of a monocarboxylate transporter (MCT1) in pig and human colon: its potential to transport L-lactate as well as butyrate. J Physiol 1998;513(3):719–32.
[27] Majumdar S, Gunda S, Pal D, Mitra AK. Identification and characterization of a monocarboxylic acid transport system on the human retinal pigment epithelium cell line – ARPE-19. Invest Ophthalmol Vis Sci 2003;44(13): 3464.

[28] Orsenigo MN, Tosco M, Bazzini C, Laforenza U, Faelli A. A monocarboxylate transporter MCT1 Islocated at the basolateral pole of rat jejunum. Exp Physiol 1999;84(6):1033–42.

[29] Kirat D, Inoue H, Iwano H, Hirayama K, Yokota H, Taniyama H, Kato S. Monocarboxylate transporter 1 gene expression in the ovine gastrointestinal tract. Vet J 2006;171(3):462–7.

[30] Iwanaga T, Takebe K, Kato I, Karaki S-I, Kuwahara A. Cellular expression of monocarboxylate transporters (MCT) in the digestive tract of the mouse, rat, and humans, with special reference to slc5a8. Biomed Res 2006;27(5):243–54.

[31] Kuchiiwa T, Nio-Kobayashi J, Takahashi-Iwanaga H, Yajima T, Iwanaga T. Cellular expression of monocarboxylate transporters in the female reproductive organ of mice: implications for the genital lactate shuttle. Histochem Cell Biol 2011;135(4):351–60.

[32] Garcia CK, Goldstein JL, Pathak RK, Anderson RGW, Brown MS. Molecular characterization of a membrane transporter for lactate, pyruvate, and other monocarboxylates: implications for the Cori cycle. Cell 1994;76(5): 865–73.

[33] Koho NM, Väihkönen LK, Pösö AR. Lactate transport in red blood cells by monocarboxylate transporters. Equine Vet J 2002;34(S34):555–9.

[34] Merezhinskaya N, Ogunwuyi SA, Mullick FG, Fishbein WN. Presence and localization of three lactic acid transporters (MCT1, -2, and -4) in separated human granulocytes, lymphocytes, and monocytes. J Histochem Cytochem 2004;52(11):1483–93.

[35] Stanley WC, Gertz EW, Wisneski JA, Neese RA, Morris DL, Brooks GA. Lactate extraction during net lactate release in legs of humans during exercise. J Appl Physiol 1986;60(4):1116–20.

[36] McCullagh KJ, Poole RC, Halestrap AP, O'Brien M, Bonen A. Role of the lactate transporter (MCT1) in skeletal muscles. Am J Physiol Endocrinol Metab 1996;271(1):E143–50.

[37] Brooks GA. Intra- and extra-cellular lactate shuttles. Med Sci Sports Exerc 2000;32(4):790–9.

[38] Mac M, Nałęcz KA. Expression of monocarboxylic acid transporters (MCT) in brain cells: implication for branched chain α-ketoacids transport in neurons. Neurochem Int 2003;43(4):305–9.

[39] Bröer S, Rahman B, Pellegri G, Pellerin L, Martin J-L, Verleysdonk S, Hamprecht B, Magistretti PJ. Comparison of lactate transport in astroglial cells and monocarboxylate transporter 1 (MCT 1) expressing *Xenopus laevis* oocytes: expression of two different monocarboxylate transporters in astroglial cells and neurons. J Biol Chem 1997;272(48):30096–102.

[40] Moreira TJTP, Pierre K, Maekawa F, Repond C, Cebere A, Liljequist S, Pellerin L. Enhanced cerebral expression of MCT1 and MCT2 in a rat ischemia model occurs in activated microglial cells. J Cereb Blood Flow Metab 2009;29(7):1273–83.

[41] Pierre K, Pellerin L, Debernardi R, Riederer BM, Magistretti PJ. Cell-specific localization of monocarboxylate transporters, MCT1 and MCT2, in the adult mouse brain revealed by double immunohistochemical labeling and confocal microscopy. Neuroscience 2000;100(3):617–27.

[42] Takebe K, Nio-Kobayashi J, Takahashi-Iwanaga H, Yajima T, Iwanaga T. Cellular expression of a monocarboxylate transporter (MCT1) in the mammary gland and sebaceous gland of mice. Histochem Cell Biol 2009;131(3):401–9.

[43] Garcia CK, Brown MS, Pathak RK, Goldstein JL. cDNA cloning of MCT2, a second monocarboxylate transporter expressed in different cells than MCT1. J Biol Chem 1995;270(4):1843–9.

[44] Fisel P, Schaeffeler E, Schwab M. Clinical and functional relevance of the monocarboxylate transporter family in disease pathophysiology and drug therapy. Clin Transl Sci 2018;11(4):352–64.

[45] Becker HM, Mohebbi N, Perna A, Ganapathy V, Capasso G, Wagner CA. Localization of members of MCT monocarboxylate transporter family Slc16 in the kidney and regulation during metabolic acidosis. Am J Physiol Renal Physiol 2010;299(1):F141–54.

[46] Pellerin L, Halestrap AP, Pierre K. Cellular and subcellular distribution of monocarboxylate transporters in cultured brain cells and in the adult brain. J Neurosci Res 2005;79(1–2):55–64.

[47] Fishbein WN, Merezhinskaya N, Foellmer JW. Relative distribution of three major lactate transporters in frozen human tissues and their localization in unfixed skeletal muscle. Muscle Nerve 2002;26(1):101–12.

[48] Philp NJ, Yoon H, Lombardi L. Mouse MCT3 gene is expressed preferentially in retinal pigment and choroid plexus epithelia. Am J Physiol Cell Physiol 2001;280(5):C1319–26.

[49] Yoon H, Fanelli A, Grollman EF, Philp NJ. Identification of a unique monocarboxylate transporter (MCT3) in retinal pigment epithelium. Biochem Biophys Res Commun 1997;234(1):90–4.

[50] Philp NJ, Wang D, Yoon H, Hjelmeland LM. Polarized expression of monocarboxylate transporters in human retinal pigment epithelium and ARPE-19 cells. Invest Ophthalmol Vis Sci 2003;44(4):1716–21.

[51] Pilegaard H, Terzis G, Halestrap A, Juel C. Distribution of the lactate/H^+ transporter isoforms MCT1 and MCT4 in human skeletal muscle. Am J Physiol Endocrinol Metab 1999;276(5):E843–8.

[52] Merezhinskaya N, Ogunwuyi SA, Fishbein WN. Expression of monocarboxylate transporter 4 in human platelets, leukocytes, and tissues assessed by antibodies raised against terminal versus pre-terminal peptides. Mol Genet Metab 2006;87(2):152–61.

[53] Domènech-Estévez E, Baloui H, Repond C, Rosafio K, Médard J-J, Tricaud N, Pellerin L, Chrast R. Distribution of monocarboxylate transporters in the peripheral nervous system suggests putative roles in lactate shuttling and myelination. J Neurosci 2015;35(10):4151–6.

[54] Settle P, Mynett K, Speake P, Champion E, Doughty IM, Sibley CP, D'Souza SW, Glazier J. Polarized lactate transporter activity and expression in the syncytiotrophoblast of the term human placenta. Placenta 2004;25(6):496–504.

[55] Juel C, Halestrap AP. Lactate transport in skeletal muscle – role and regulation of the monocarboxylate transporter. J Physiol 1999;517(Pt 3):633–42.

[56] Colen CB, Seraji-Bozorgzad N, Marples B, Galloway MP, Sloan AE, Mathupala SP. Metabolic remodeling of malignant gliomas for enhanced sensitization during radiotherapy: an in vitro study. Neurosurgery 2006;59(6):1313–24.

[57] Colen CB, Shen Y, Ghoddoussi F, Yu P, Francis TB, Koch BJ, Monterey MD, Galloway MP, Sloan AE, Mathupala SP. Metabolic targeting of lactate efflux by malignant glioma inhibits invasiveness and induces necrosis: an in vivo study. Neoplasia 2011;13(7):620–32.

[58] Filipa M-S, Vera M-G, Sílvia P, André FV, Joana P, Fernando CS, Fátima B, Céline P. Differential sensitivities to lactate transport inhibitors of breast cancer cell lines. Endocr Relat Cancer 2014;21(1):27–38.

[59] Perez-Escuredo J, Van Hee VF, Sboarina M, Falces J, Payen VL, Pellerin L, Sonveaux P. Monocarboxylate transporters in the brain and in cancer. Biochim Biophys Acta 2016;1863(10):2481–97.

[60] Park SJ, Smith CP, Wilbur RR, Cain CP, Kallu SR, Valasapalli S, Sahoo A, Guda MR, Tsung AJ, Velpula KK. An overview of MCT1 and MCT4 in GBM: small molecule transporters with large implications. Am J Cancer Res 2018;8(10):1967–76.

[61] Ekberg H, Qi Z, Pahlman C, Veress B, Bundick RV, Craggs RI, Holness E, Edwards S, Murray CM, Ferguson D, Kerry PJ, Wilson E, Donald DK. The specific monocarboxylate transporter-1 (MCT-1) inhibitor, AR-C117977, induces donor-specific suppression, reducing acute and chronic allograft rejection in the rat. Transplantation 2007;84(9):1191–9.

[62] Le Floch R, Chiche J, Marchiq I, Naiken T, Ilc K, Murray CM, Critchlow SE, Roux D, Simon M-P, Pouysségur J. CD147 subunit of lactate/H^+ symporters MCT1 and hypoxia-inducible MCT4 is critical for energetics and growth of glycolytic tumors. Proc Natl Acad Sci U S A 2011;108(40):16663–8.

[63] Ovens MJ, Davies AJ, Wilson MC, Murray CM, Halestrap AP. AR-C155858 is a potent inhibitor of monocarboxylate transporters MCT1 and MCT2 that binds to an intracellular site involving transmembrane helices 7-10. Biochem J 2010;425(3):523–30.

[64] Curtis NJ, Mooney L, Hopcroft L, Michopoulos F, Whalley N, Zhong H, Murray C, Logie A, Revill M, Byth KF, Benjamin AD, Firth MA, Green S, Smith PD, Critchlow SE. Pre-clinical pharmacology of AZD3965, a selective inhibitor of MCT1: DLBCL, NHL and Burkitt's lymphoma anti-tumor activity. Oncotarget 2017;8(41):69219–36.

[65] ClinicalTrials.gov. A phase I trial of AZD3965 in patients with advanced cancer, Available from https://clinicaltrials.gov/ct2/show/study/NCT01791595; 2013.

[66] Lee JY, Lee I, Chang WJ, Ahn SM, Lim SH, Kim HS, Yoo KH, Jung KS, Song H-N, Cho JH, Kim SY, Kim K-M, Lee S, Kim ST, Park SH, Lee J, Park JO, Park YS, Lim HY, Kang WK. MCT4 as a potential therapeutic target for metastatic gastric cancer with peritoneal carcinomatosis. Oncotarget 2016;7(28):43492–503.

[67] Polański R, Hodgkinson CL, Fusi A, Nonaka D, Priest L, Kelly P, Trapani F, Bishop PW, White A, Critchlow SE, Smith PD, Blackhall F, Dive C, Morrow CJ. Activity of the monocarboxylate transporter 1 inhibitor AZD3965 in small cell lung cancer. Clin Cancer Res 2014;20(4):926–37.

[68] Quanz M, Bender E, Kopitz C, Grünewald S, Schlicker A, Schwede W, Eheim A, Toschi L, Neuhaus R, Richter C, Toedling J, Merz C, Lesche R, Kamburov A, Siebeneicher H, Bauser M, Hägebarth A. Preclinical efficacy of the novel monocarboxylate transporter 1 inhibitor BAY-8002 and associated markers of resistance. Mol Cancer Ther 2018;17(11):2285–96.

[69] Voss DM, Spina R, Carter DL, Lim KS, Jeffery CJ, Bar EE. Disruption of the monocarboxylate transporter-4-basigin interaction inhibits the hypoxic response, proliferation, and tumor progression. Sci Rep 2017;7(1) 4292.

[70] Fuster DG, Alexander RT. Traditional and emerging roles for the SLC9 Na+/H+ exchangers. Pflugers Arch - Eur J Physiol 2014;466(1):61–76.

[71] Orlowski J, Grinstein S. Diversity of the mammalian sodium/proton exchanger SLC9 gene family. Pflugers Arch 2004;447(5):549–65.

[72] Brett CL, Donowitz M, Rao R. Evolutionary origins of eukaryotic sodium/proton exchangers. Am J Phys Cell Physiol 2005;288(2):C223–39.

[73] Donowitz M, Ming Tse C, Fuster D. SLC9/NHE gene family, a plasma membrane and organellar family of Na^+/H^+ exchangers. Mol Asp Med 2013;34(2–3):236–51.

[74] Kurashima K, Szabó EZ, Lukacs G, Orlowski J, Grinstein S. Endosomal recycling of the Na+/H+exchanger NHE3 isoform is regulated by the phosphatidylinositol 3-kinase pathway. J Biol Chem 1998;273(33):20828–36.

[75] Warmerdam PA, Nabben NM, van de Graaf SA, van de Winkel JG, Capel PJ. The human low affinity immunoglobulin G fc receptor IIC gene is a result of an unequal crossover event. J Biol Chem 1993;268(10):7346–9.

[76] Counillon L, Pouyssegur J, Reithmeier RAF. The Na+/H+ exchanger NHE-1 possesses N- and O-linked glycosylation restricted to the first N-terminal extracellular domain. Biochemistry 1994;33(34):10463–9.

[77] Tse C-M, Levine SA, Yun CHC, Khurana S, Donowitz M. The plasma membrane Na+/H+ exchanger 2 is an O-linked but not an N-linked sialoglycoprotein: use of a polyclonal antibody to identify and characterize glycosylation. Biochemistry 1994;33(44):12954–61.

[78] Shrode LD, Gan BS, D'Souza SJA, Orlowski J, Grinstein S. Topological analysis of NHE1, the ubiquitous Na+/H+exchanger using chymotryptic cleavage. Am J Phys Cell Physiol 1998;275(2):C431–9.

[79] Cabado AG, Yu FH, Kapus A, Lukacs G, Grinstein S, Orlowski J. Distinct structural domains confer cAMP sensitivity and ATP dependence to the Na/H exchanger NHE3 isoform. J Biol Chem 1996;271(7):3590–9.

[80] Levine SA, Nath SK, Yun CHC, Yip JW, Montrose M, Donowitz M, Tse CM. Separate C-terminal domains of the epithelial specific brush border Na+/H+ exchanger isoform NHE3 are involved in stimulation and inhibition by protein kinases/growth factors. J Biol Chem 1995;270(23):13716–25.

[81] Wakabayashi S, Fafournoux P, Sardet C, Pouysségur J. The Na+/H+ antiporter cytoplasmic domain mediates growth factor signals and controls "H(+)-sensing" Proc Natl Acad Sci 1992;89(6):2424–8.

[82] Wakabayashi S, Pang T, Su X, Shigekawa M. A novel topology model of the human Na+/H+ exchanger isoform 1. J Biol Chem 2000;275(11):7942–9.

[83] Gazmuri RJ, Radhakrishnan J, Ayoub IM. Sodium-hydrogen exchanger isoform-1 inhibition: a promising pharmacological intervention for resuscitation from cardiac arrest. Molecules 2019;24(9):1765.

[84] Boedtkjer E, Bunch L, F. Pedersen S. Physiology, pharmacology and pathophysiology of the pH regulatory transport proteins NHE1 and NBCn1: similarities, differences, and implications for cancer therapy. Curr Pharm Des 2012;18(10):1345–71.

[85] Stock C, Ludwig FT, Schwab A. Is the multifunctional Na+/H+ exchanger isoform 1 a potential therapeutic target in cancer? Curr Med Chem 2012;19(5):647–60.

[86] Simonin A, Fuster D. Nedd4-1 and beta-arrestin-1 are key regulators of Na+/H+ exchanger 1 ubiquitylation, endocytosis, and function. J Biol Chem 2010;285(49):38293–303.

[87] Padan E, Landau M. Sodium-proton (Na+/H+) antiporters: properties and roles in health and disease. In: Sigel A, Sigel H, Sigel RKO, editors. The alkali metal ions: their role for life. Cham: Springer International Publishing; 2016. p. 391–458.

[88] Reshkin SJ, Bellizzi A, Caldeira S, Albarani V, Malanchi I, Poignee M, Alunni-Fabbroni M, Casavola V, Tommasino M. Na+/H+ exchanger-dependent intracellular alkalinization is an early event in malignant transformation and plays an essential role in the development of subsequent transformation-associated phenotypes. FASEB J 2000;14(14):2185–97.

[89] Harguindey S, Orive G, Luis Pedraz J, Paradiso A, Reshkin SJ. The role of pH dynamics and the Na+/H+ antiporter in the etiopathogenesis and treatment of cancer. Two faces of the same coin—one single nature. Biochim Biophys Acta Rev Cancer 2005;1756(1):1–24.

[90] Harguindey S, Arranz JL, Polo Orozco JD, Rauch C, Fais S, Cardone RA, Reshkin SJ. Cariporide and other new and powerful NHE1 inhibitors as potentially selective anticancer drugs – an integral molecular/biochemical/metabolic/clinical approach after one hundred years of cancer research. J Transl Med 2013;11(1):282.

[91] Di Sario A, Bendia E, Omenetti A, De Minicis S, Marzioni M, Kleemann HW, Candelaresi C, Saccomanno S, Alpini G, Benedetti A. Selective inhibition of ion transport mechanisms regulating intracellular pH reduces proliferation and induces apoptosis in cholangiocarcinoma cells. Dig Liver Dis 2007;39(1):60–9.
[92] Grinstein S, Rothstein A, Cohen S. Mechanism of osmotic activation of Na+/H+ exchange in rat thymic lymphocytes. J Gen Physiol 1985;85(5):765–87.
[93] Lee CH, Cragoe Jr EJ, Edwards AM. Control of hepatocyte DNA synthesis by intracellular pH and its role in the action of tumor promoters. J Cell Physiol 2003;195(1):61–9.
[94] McLean LA, Roscoe J, Jørgensen NK, Gorin FA, Cala PM. Malignant gliomas display altered pH regulation by NHE1 compared with nontransformed astrocytes. Am J Phys Cell Physiol 2000;278(4):C676–88.
[95] Huang Z, Huang Y. The change of intracellular pH is involved in the cisplatin-resistance of human lung adenocarcinoma A549/DDP cells. Cancer Investig 2005;23(1):26–32.
[96] Miraglia E, Viarisio D, Riganti C, Costamagna C, Ghigo D, Bosia A. Na+/H+ exchanger activity is increased in doxorubicin-resistant human colon cancer cells and its modulation modifies the sensitivity of the cells to doxorubicin. Int J Cancer 2005;115(6):924–9.
[97] Boedtkjer E, Moreira JMA, Mele M, Vahl P, Wielenga VT, Christiansen PM, Jensen VED, Pedersen SF, Aalkjaer C. Contribution of Na+,HCO3−-cotransport to cellular pH control in human breast cancer: a role for the breast cancer susceptibility locus NBCn1 (SLC4A7). Int J Cancer 2013;132(6):1288–99.
[98] Yang X, Wang D, Dong W, Song Z, Dou K. Expression and modulation of Na+/H+ exchanger 1 gene in hepatocellular carcinoma: a potential therapeutic target. J Gastroenterol Hepatol 2011;26(2):364–70.
[99] Chiang Y, Chou C-Y, Hsu K-F, Huang Y-F, Shen M-R. EGF upregulates Na+/H+ exchanger NHE1 by post-translational regulation that is important for cervical cancer cell invasiveness. J Cell Physiol 2008;214(3):810–9.
[100] Guan B, Hoque A, Xu X. Amiloride and guggulsterone suppression of esophageal cancer cell growth in vitro and in nude mouse xenografts. Front Biol 2014;9(1):75–81.
[101] Cong D, Zhu W, Shi Y, Pointer KB, Clark PA, Shen H, Kuo JS, Hu S, Sun D. Upregulation of NHE1 protein expression enables glioblastoma cells to escape TMZ-mediated toxicity via increased H+ extrusion, cell migration and survival. Carcinogenesis 2014;35(9):2014–24.
[102] Harguindey S, Orive G, Cacabelos R, Hevia EM, de Otazu RD, Arranz JL, Anitua E. An integral approach to the etiopathogenesis of human neurodegenerative diseases (HNDDs) and cancer. Possible therapeutic consequences within the frame of the trophic factor withdrawal syndrome (TFWS). Neuropsychiatr Dis Treat 2008;4(6):1073–84.
[103] Rich I, Worthington-White D, Garden O, Musk P. Apoptosis of leukemic cells accompanies reduction in intracellular pH after targeted inhibition of the Na+/H+ exchanger. Blood 2000;95:1427–34.
[104] Che X-F, Zheng C-L, Akiyama S-I, Tomoda A. 2-Aminophenoxazine-3-one and 2-amino-4,4α-dihydro-4α,7-dimethyl-3H-phenoxazine-3-one cause cellular apoptosis by reducing higher intracellular pH in cancer cells. Proc Jpn Acad B Phys Biol Sci 2011;87(4):199–213.
[105] Granja S, Tavares-Valente D, Queirós O, Baltazar F. Value of pH regulators in the diagnosis, prognosis and treatment of cancer. Semin Cancer Biol 2017;43:17–34.
[106] Mazzo DJ, Hydrochloride A. Florey K, editor. Analytical profiles of drug substances. Academic Press; 1986. p. 1–34.
[107] Sparks RL, Pool TB, Smith NKR, Cameron IL. Effects of amiloride on tumor growth and intracellular element content of tumor cells *in vivo*. Cancer Res 1983;43(1):73–7.
[108] Matthews H, Ranson M, Kelso MJ. Anti-tumour/metastasis effects of the potassium-sparing diuretic amiloride: an orally active anti-cancer drug waiting for its call-of-duty? Int J Cancer 2011;129(9):2051–61.
[109] Tatsuta M, Iishi H, Baba M, Yano H, Iseki K, Uehara H, Nakaizumi A. Inhibition by amiloride of experimental carcinogenesis induced by azaserine in rat pancreas. Cancer Lett 1996;106(1):23–8.
[110] Tatsuta M, Iishi H, Baba M, Yano H, Sakai N, Uehara H, Nakaizumi A. Chemoprevention by amiloride against experimental hepatocarcinogenesis induced by N-nitrosomorpholine in Sprague-Dawley rats. Cancer Lett 1997;119(1):109–13.
[111] Jankun J, Keck RW, Skrzypczak-Jankun E, Swiercz R. Inhibitors of urokinase reduce size of prostate cancer xenografts in severe combined immunodeficient mice. Cancer Res 1997;57(4):559–63.
[112] Alliegro MC, Alliegro MA, Cragoe EJ, Glaser BM. Amiloride inhibition of angiogenesis in vitro. J Exp Zool 1993;267(3):245–52.

[113] Iishi H, Tatsuta M, Baba M, Yano H, Uehara H, Nakaizumi A. Suppression by amiloride of bombesin-enhanced peritoneal metastasis of intestinal adenocarcinomas induced by azoxymethane. Int J Cancer 1995;63(5):716–9.
[114] Harley W, Floyd C, Dunn T, Zhang X-D, Chen T-Y, Hegde M, Palandoken H, Nantz MH, Leon L, Carraway KL, Lyeth B, Gorin FA. Dual inhibition of sodium-mediated proton and calcium efflux triggers non-apoptotic cell death in malignant gliomas. Brain Res 2010;1363:159–69.
[115] Yang X, Wang D, Dong W, Song Z, Dou K. Inhibition of Na+/H+ exchanger 1 by 5-(N-ethyl-N-isopropyl) amiloride reduces hypoxia-induced hepatocellular carcinoma invasion and motility. Cancer Lett 2010; 295(2):198–204.
[116] Wong P, Kleemann HW, Tannock IF. Cytostatic potential of novel agents that inhibit the regulation of intracellular pH. Br J Cancer 2002;87(2):238–45.
[117] Tatsuta M, Iishi H, Baba M, Uehara H, Nakaizumi A, Taniguchi H. Inhibition by amiloride of gastric carcinogenesis induced by N-methyl-N'-nitro-N-nitrosoguanidine in Wistar rats. Br J Cancer 1993;67(5):1011–4.
[118] Tatsuta M, Iishi H, Baba M, Uehara H, Nakaizumi A. Chemoprevention by amiloride of experimental carcinogenesis in rat colon induced by azoxymethane. Carcinogenesis 1995;16(4):941–2.
[119] Evans D, Stakleff K. Suppression of the invasive capacity of human breast cancer cells by inhibition of urokinase plasminogen activator via amiloride and B428. Am Surg 2000;66:460–4.
[120] Evans DM, Sloan Stakleff KD. Control of pulmonary metastases of rat mammary cancer by inhibition of uPA and COX-2, singly and in combination. Clin Exp Metastasis 2004;21(4):339–46.
[121] Vigne P, Frelin C, Cragoe EJ, Lazdunski M. Structure-activity relationships of amiloride and certain of its analogues in relation to the blockade of the Na+/H+ exchange system. Mol Pharmacol 1984;25(1):131–6.
[122] Kleyman TR, Cragoe EJ. Amiloride and its analogs as tools in the study of ion transport. J Membr Biol 1988;105(1):1–21.
[123] Cragoe EJ, Woltersdorf OW, Bicking JB, Kwong SF, Jones JH. Pyrazine diuretics. II. N-amidino-3-amino-5-substituted 6-halopyrazinecarboxamides. J Med Chem 1967;10(1):66–75.
[124] Chen Y-X, O'Brien ER. Ethyl isopropyl amiloride inhibits smooth muscle cell proliferation and migration by inducing apoptosis and antagonizing urokinase plasminogen activator activity. Can J Physiol Pharmacol 2003;81(7):730–9.
[125] Garcı´a-Cañero R, Trilla C, Pérez de Diego J, Dı´az-Gil JJ, Cobo JM. Na+:H+ exchange inhibition induces intracellular acidosis and differentially impairs cell growth and viability of human and rat hepatocarcinoma cells. Toxicol Lett 1999;106(2):215–28.
[126] Reshkin SJ, Bellizzi A, Cardone RA, Tommasino M, Casavola V, Paradiso A. Paclitaxel induces apoptosis via protein kinase A- and p38 mitogen-activated protein-dependent inhibition of the Na^+/H^+ exchanger (NHE) NHE isoform 1 in human breast cancer cells. Clin Cancer Res 2003;9(6):2366–73.
[127] Poon AC, Inkol JM, Luu AK, Mutsaers AJ. Effects of the potassium-sparing diuretic amiloride on chemotherapy response in canine osteosarcoma cells. J Vet Intern Med 2019;33(2):800–11.
[128] Jin W, Li Q, Wang J, Chang G, Lin Y, Li H, Wang L, Gao W, Pang T. Na+/H+ exchanger 1 inhibition contributes to K562 leukaemic cell differentiation. Cell Biol Int 2012;36(8):739–45.
[129] Lindner D, Raghavan D. Intra-tumoural extra-cellular pH: a useful parameter of response to chemotherapy in syngeneic tumour lines. Br J Cancer 2009;100(8):1287–91.
[130] Torigoe T, Izumi H, Ise T, Murakami T, Uramoto H, Ishiguchi H, Yoshida Y, Tanabe M, Nomoto M, Kohno K. Vacuolar H+-ATPase: functional mechanisms and potential as a target for cancer chemotherapy. Anti-Cancer Drugs 2002;13(3):237–43.
[131] Atwal KS, O'Neil SV, Ahmad S, Doweyko L, Kirby M, Dorso CR, Chandrasena G, Chen B-C, Zhao R, Zahler R. Synthesis and biological activity of 5-aryl-4-(4-(5-methyl-1H-imidazol-4-yl)piperidin-1-yl)pyrimidine analogs as potent, highly selective, and orally bioavailable NHE-1 inhibitors. Bioorg Med Chem Lett 2006;16(18):4796–9.
[132] Shukla S, Ohnuma S, V. Ambudkar S. Improving cancer chemotherapy with modulators of ABC drug transporters. Curr Drug Targets 2011;12(5):621–30.
[133] Gottesman MM, Ambudkar SV. Overview: ABC transporters and human disease. J Bioenerg Biomembr 2001; 33(6):453–8.
[134] Dean M, Annilo T. Evolution of the ATP-binding cassette (ABC) transporter superfamily in vertebrates. Annu Rev Genomics Hum Genet 2005;6(1):123–42.
[135] Linton KJ. Structure and function of ABC transporters. Phys Ther 2007;22(2):122–30.
[136] Silva R, Vilas-Boas V, Carmo H, Dinis-Oliveira RJ, Carvalho F, de Lourdes Bastos M, Remião F. Modulation of P-glycoprotein efflux pump: induction and activation as a therapeutic strategy. Pharmacol Ther 2015;149:1–123.

[137] Modok S, Mellor HR, Callaghan R. Modulation of multidrug resistance efflux pump activity to overcome chemoresistance in cancer. Curr Opin Pharmacol 2006;6(4):350–4.

[138] Schinkel AH, Jonker JW. Mammalian drug efflux transporters of the ATP binding cassette (ABC) family: an overview. Adv Drug Deliv Rev 2003;55(1):3–29.

[139] Juliano RL, Ling V. A surface glycoprotein modulating drug permeability in Chinese hamster ovary cell mutants. Biochim Biophys Acta Biomembr 1976;455(1):152–62.

[140] Kartner N, Shales M, Riordan JR, Ling V. Daunorubicin-resistant Chinese hamster ovary cells expressing multidrug resistance and a cell-surface P-glycoprotein. Cancer Res 1983;43(9):4413–9.

[141] El-Awady R, Saleh E, Hashim A, Soliman N, Dallah A, Elrasheed A, Elakraa G. The role of eukaryotic and prokaryotic ABC transporter family in failure of chemotherapy. Front Pharmacol 2017;7:535.

[142] Gottesman MM, Fojo T, Bates SE. Multidrug resistance in cancer: role of ATP–dependent transporters. Nat Rev Cancer 2002;2(1):48–58.

[143] Abolhoda A, Wilson AE, Ross H, Danenberg PV, Burt M, Scotto KW. Rapid activation of *MDR1* gene expression in human metastatic sarcoma after in vivo exposure to doxorubicin. Clin Cancer Res 1999;5(11):3352–6.

[144] Levchenko A, Mehta BM, Niu X, Kang G, Villafania L, Way D, Polycarpe D, Sadelain M, Larson SM. Intercellular transfer of P-glycoprotein mediates acquired multidrug resistance in tumor cells. Proc Natl Acad Sci U S A 2005;102(6):1933–8.

[145] Spiegl-Kreinecker S, Buchroithner J, Elbling L, Steiner E, Wurm G, Bodenteich A, Fischer J, Micksche M, Berger W. Expression and functional activity of the ABC-transporter proteins P-glycoprotein and multidrug-resistance protein 1 in human brain tumor cells and astrocytes. J Neuro-Oncol 2002;57(1):27–36.

[146] Ruefli A, Smyth M, Johnstone R. HMBA induces activation of a caspase-independent cell death pathway to overcome P-glycoprotein-mediated multidrug resistance. Blood 2000;95:2378–85.

[147] Cole S, Bhardwaj G, Gerlach J, Mackie J, Grant C, Almquist K, Stewart A, Kurz E, Duncan A, Deeley R. Overexpression of a transporter gene in a multidrug-resistant human lung cancer cell line. Science 1992; 258(5088):1650–4.

[148] Eckford PDW, Sharom FJ. ABC efflux pump-based resistance to chemotherapy drugs. Chem Rev 2009; 109(7):2989–3011.

[149] Hipfner DR, Deeley RG, Cole SPC. Structural, mechanistic and clinical aspects of MRP1. Biochim Biophys Acta Biomembr 1999;1461(2):359–76.

[150] Chen YN, Mickley LA, Schwartz AM, Acton EM, Hwang JL, Fojo AT. Characterization of adriamycin-resistant human breast cancer cells which display overexpression of a novel resistance-related membrane protein. J Biol Chem 1990;265(17):10073–80.

[151] Kirtane AR, Kalscheuer SM, Panyam J. Exploiting nanotechnology to overcome tumor drug resistance: challenges and opportunities. Adv Drug Deliv Rev 2013;65(13–14):1731–47.

[152] Scharenberg CW, Harkey MA, Torok-Storb B. The ABCG2 transporter is an efficient Hoechst 33342 efflux pump and is preferentially expressed by immature human hematopoietic progenitors. Blood 2002;99(2):507–12.

[153] Zhou S, Schuetz JD, Bunting KD, Colapietro A-M, Sampath J, Morris JJ, Lagutina I, Grosveld GC, Osawa M, Nakauchi H, Sorrentino BP. The ABC transporter Bcrp1/ABCG2 is expressed in a wide variety of stem cells and is a molecular determinant of the side-population phenotype. Nat Med 2001;7(9):1028–34.

[154] Nakayama A, Matsuo H, Takada T, Ichida K, Nakamura T, Ikebuchi Y, Ito K, Hosoya T, Kanai Y, Suzuki H, Shinomiya N. ABCG2 is a high-capacity urate transporter and its genetic impairment increases serum uric acid levels in humans. Nucleosides Nucleotides Nucleic Acids 2011;30(12):1091–7.

[155] Sarkadi B, Homolya L, Szakács G, Váradi A. Human multidrug resistance ABCB and ABCG transporters: participation in a chemoimmunity defense system. Physiol Rev 2006;86(4):1179–236.

[156] Robey RW, Polgar O, Deeken J, To KW, Bates SE. ABCG2: determining its relevance in clinical drug resistance. Cancer Metastasis Rev 2007;26(1):39–57.

[157] Noguchi K, Katayama K, Mitsuhashi J, Sugimoto Y. Functions of the breast cancer resistance protein (BCRP/ABCG2) in chemotherapy. Adv Drug Deliv Rev 2009;61(1):26–33.

[158] Palmeira A, Sousa E, H. Vasconcelos M, M. Pinto M. Three decades of P-gp inhibitors: skimming through several generations and scaffolds. Curr Med Chem 2012;19(13):1946–2025.

[159] Varma MVS, Ashokraj Y, Dey CS, Panchagnula R. P-glycoprotein inhibitors and their screening: a perspective from bioavailability enhancement. Pharmacol Res 2003;48(4):347–59.

[160] Shapiro AB, Ling V. Effect of quercetin on hoechst 33342 transport by purified and reconstituted p-glycoprotein. Biochem Pharmacol 1997;53(4):587–96.

[161] Drori S, Eytan GD, Assaraf YG. Potentiation of anticancer-drug cytotoxicity by multidrug-resistance chemosensitizers involves alterationsin membrane fluidity leading to increased membrane permeability. Eur J Biochem 1995;228(3):1020–9.

[162] Tsuruo T, Iida H, Tsukagoshi S, Sakurai Y. Overcoming of vincristine resistance in P388 Leukemia *in vivo* and *in vitro* through enhanced cytotoxicity of vincristine and vinblastine by verapamil. Cancer Res 1981; 41(5):1967–72.

[163] Slater LM, Sweet P, Stupecky M, Gupta S. Cyclosporin a reverses vincristine and daunorubicin resistance in acute lymphatic leukemia in vitro. J Clin Invest 1986;77(4):1405–8.

[164] Miller TP, Grogan TM, Dalton WS, Spier CM, Scheper RJ, Salmon SE. P-glycoprotein expression in malignant lymphoma and reversal of clinical drug resistance with chemotherapy plus high-dose verapamil. J Clin Oncol 1991;9(1):17–24.

[165] te Boekhorst PAW, van Kapel J, Schoester M, Sonneveld P. Reversal of typical multidrug resistance by cyclosporin and its non-immunosuppressive analogue SDZ PSC 833 in Chinese hamster ovary cells expressing themdr1 phenotype. Cancer Chemother Pharmacol 1992;30(3):238–42.

[166] Kusunoki N, Takara K, Tanigawara Y, Yamauchi A, Ueda K, Komada F, Ku Y, Kuroda Y, Saitoh Y, Okumura K. Inhibitory effects of a cyclosporin derivative, SDZ PSC 833, on transport of doxorubicin and vinblastine via human P-glycoprotein. Jpn J Cancer Res 1998;89(11):1220–8.

[167] Twentyman PR, Bleehen NM. Resistance modification by PSC-833, a novel non-immunosuppressive cyclosporin A. Eur J Cancer Clin Oncol 1991;27(12):1639–42.

[168] Chauncey TR, Rankin C, Anderson JE, Chen IM, Kopecky KJ, Godwin JE, Kalaycio ME, Moore DF, Shurafa MS, Petersdorf SH, Kraut EH, Leith CP, Head DR, Luthardt FW, Willman CL, Appelbaum FR. A phase I study of induction chemotherapy for older patients with newly diagnosed acute myeloid leukemia (AML) using mitoxantrone, etoposide, and the MDR modulator PSC 833: a southwest oncology group study 9617. Leuk Res 2000;24(7):567–74.

[169] Cree I, Knight L, Di Nicolantonio F, Sharma S, Gulliford T. Chemosensitization of solid tumors by modulation of resistance mechanisms. Curr Opin Investig Drugs 2002;2000(3):634–40.

[170] Minderman H, O'Loughlin KL, Pendyala L, Baer MR. VX-710 (biricodar) increases drug retention and enhances chemosensitivity in resistant cells overexpressing P-glycoprotein, multidrug resistance protein, and breast cancer resistance protein. Clin Cancer Res 2004;10(5):1826–34.

[171] Hyafil F, Vergely C, Du Vignaud P, Grand-Perret T. In vitro and in vivo reversal of multidrug resistance by GF120918, an acridonecarboxamide derivative. Cancer Res 1993;53(19):4595–602.

[172] Gekeler V, Ise W, Sanders KH, Ulrich WR, Beck J. The leukotriene LTD4 receptor antagonist MK571 specifically modulates MRP associated multidrug resistance. Biochem Biophys Res Commun 1995;208(1):345–52.

[173] Rabindran SK, Ross DD, Doyle LA, Yang W, Greenberger LM. Fumitremorgin C reverses multidrug resistance in cells transfected with the breast cancer resistance protein. Cancer Res 2000;60(1):47–50.

[174] Naito M, Matsuba Y, Sato S, Hirata H, Tsuruo T. MS-209, a quinoline-type reversal agent, potentiates antitumor efficacy of docetaxel in multidrug-resistant solid tumor xenograft models. Clin Cancer Res 2002;8(2):582–8.

[175] Germann UA, Shlyakhter D, Mason VS, Zelle RE, Duffy JP, Galullo V, Armistead DM, Saunders JO, Boger J, Harding MW. Cellular and biochemical characterization of VX-710 as a chemosensitizer: reversal of P-glycoprotein-mediated multidrug resistance in vitro. Anti-Cancer Drugs 1997;8(2):125–40.

[176] Seiden MV, Swenerton KD, Matulonis U, Campos S, Rose P, Batist G, Ette E, Garg V, Fuller A, Harding MW, Charpentier D. A phase II study of the MDR inhibitor biricodar (INCEL, VX-710) and paclitaxel in women with advanced ovarian cancer refractory to paclitaxel therapy. Gynecol Oncol 2002;86(3):302–10.

[177] Toppmeyer D, Seidman AD, Pollak M, Russell C, Tkaczuk K, Verma S, Overmoyer B, Garg V, Ette E, Harding MW, Demetri GD. Safety and efficacy of the multidrug resistance inhibitor Incel (biricodar; VX-710) in combination with paclitaxel for advanced breast cancer refractory to paclitaxel. Clin Cancer Res 2002;8(3):670–8.

[178] Gandhi L, Harding MW, Neubauer M, Langer CJ, Moore M, Ross HJ, Johnson BE, Lynch TJ. A phase II study of the safety and efficacy of the multidrug resistance inhibitor VX-710 combined with doxorubicin and vincristine in patients with recurrent small cell lung cancer. Cancer 2007;109(5):924–32.

[179] Robey RW, Shukla S, Finley EM, Oldham RK, Barnett D, Ambudkar SV, Fojo T, Bates SE. Inhibition of P-glycoprotein (ABCB1)- and multidrug resistance-associated protein 1 (ABCC1)-mediated transport by the orally administered inhibitor, CBT-1((R)). Biochem Pharmacol 2008;75(6):1302–12.

[180] Robey RW, To KK, Polgar O, Dohse M, Fetsch P, Dean M, Bates SE. ABCG2: a perspective. Adv Drug Deliv Rev 2009;61(1):3–13.

CHAPTER

8

Innovative pH-interfering agents in association with anticancer agents

Daniela Secci and Paolo Guglielmi

Department of Drug Chemistry and Technology, Sapienza University of Rome, Rome, Italy

Abstract

Although the common strategies to efficiently interfere with cellular pH lie in the direct inhibition of the enzymes/transporters/pumps constituting the cooperative system that controls proton concentration, other approaches can be exploited to obtain similar results. In this respect, it is worthwhile to take into account the effects exhibited by drugs that are not usually considered as pH-interfering agents like HDAC inhibitors and H_2S-releasing compounds.

Abbreviations

AEs	anion exchangers
CAs	carbonic anhydrases
CNCn	α-cyano-4-hydroxycinnamate
CoA	coenzyme A
HAT	histone acetyl transferase
HDACs	histone deacetylases
HEPES	4-(2-hydroxyethyl)piperazine-1-ethanesulfonic acid
HNSCC	head and neck squamous cell carcinoma
IHC	immunohistochemistry
LAMP	lysosomal-associated membrane protein
LMP	lysosomal membrane permeabilization
MCTs	monocarboxylate transporters
NAD	nicotinamide adenine dinucleotide
pH_e	extracellular pH
pH_i	intracellular pH
TSA	Trichostatin A
ZBG	zinc-binding group

pH-Interfering Agents as Chemosensitizers in Cancer Therapy
https://doi.org/10.1016/B978-0-12-820701-7.00019-1

Conflict of interest

No potential conflicts of interest were disclosed.

Introduction

In the efforts aimed at developing survival and "escaping" abilities, cancer cells take advantage of a series of adjustments in order to be, among the other, energetically independent from the oxygen, whose concentration dramatically falls down in the innermost part of the tumor tissues due to the lack and/or aberrant growth of vascular structures [1]. The consequence of this circumstance is the well-known Warburg effect, involving the transition of the cells from an O_2-dependent metabolism (glycolysis followed by Krebs cycle and oxidative phosphorylation), to an O_2-independent one, in which the cells feed their energetic retrieval mainly through the glycolytic step (Scheme 1). However, in order to regenerate the NAD^+ consumed during glycolysis, that normally is restored in the oxidative phosphorylation phase, the hypoxic cells channel the pyruvate molecules into the anaerobic respiration, leading to the production of lactate (Scheme 1) [2].

Since massive production of this organic acid, and more generally the boosted glycolysis, would alter internal pH, a series of countermeasures reducing this change need to be adopted by the cells. In this regard, a series of proteins that are involved in pH_i balance, including monocarboxylate transporters (MCTs), proton pump vacuolar ATPase, carbonic anhydrases (CAs), anion exchangers (AEs), and sodium/proton exchangers, play an important role [3].

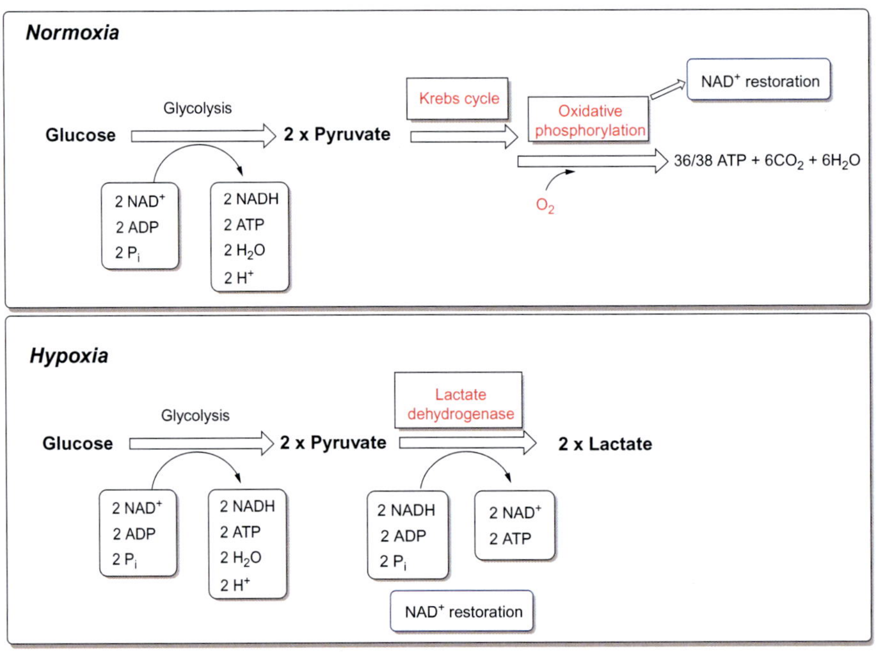

SCHEME 1 Difference between normal metabolism (normoxia) and anaerobic metabolism (hypoxia).

All these proteins cover a precise role in the phenotype acquired by the cancer cells and can be considered as their blessing and curse, because if, on one hand, they confer resistance and in most cases aggressive and metastatic behavior, on the other, most of them represent a target that can be pointed to fight cancer. Hence, one of the approaches that has been largely exploited, to obtain anticancer as well as diagnostic tools, is the interference with the complex machinery regulating the intracellular pH (pH_i) [4–6]. This can be realized through the inhibition of targets (enzymes/transporters/pumps) such as: (i) carbonic anhydrase enzymes, (ii) V-ATPase proton pump, (iii) HCO_3^- transporters, (iv) Na^+/H^+ exchanger and (v) monocarboxylate transporters. Although this strategy is the most immediate one to interfere with cellular pH, other insights, involving the proton balance and not directly connected with the inhibition of these systems, have been discovered and deepened. These approaches take advantage of the interference with metabolic/physiological pathways, which, in turn, are linked to the pH_i control, affecting the survival capability of the cancer cell.

pH-interfering agents with uncommon mechanism in association with anticancer drugs

In the light of the above, one of the opportunities relies on the evidence of the connection between pH_i and the activity of histone deacetylases (HDACs). Specific acetylation of lysine residues belonging to histone proteins regulates almost all DNA-template processes (e.g., transcription, replication, repair, recombination, and conversion to heterochromatin), affecting the chromatin structure and so, the ability to interact with bromo-domain containing proteins [7, 8]. The extent of histone acetylation can differ, spanning from the local/site-specific levels to the cellular one. By using immunohistochemistry (IHC) technique, studies on primary cancer cells tissues were performed, showing that lower acetylation levels account for more aggressive cancer and poorer clinical outcomes [9–14]. The level of histone acetylation is maintained through the balance between acetylation, performed by histone acetyl transferase (HAT) through the consumption of acetyl-CoA (the substrate used to transfer the acetyl group to the lysine residues) and deacetylation, performed by HDAC, which generates negatively charged acetate anion. Since acetyl-CoA and acetate anion also participate in many metabolic processes, McBrian *et al.* [15] proposed a hypothetic relationship between histone acetylation levels and metabolic/physiological paths. In this regard the researchers investigated most of the aspects influencing the acetylation/deacetylation cycle of histones. Interestingly, reducing the extracellular pH (pH_e) the authors observed the decreasing of pH_i, albeit the latter was altered in a lesser extent than the former. Although the cellular response to pH_e changes has already been demonstrated [16, 17], the pH lowering led to a significant reduction of histone acetylation regardless the buffer media used (bicarbonate or HEPES), accounting for a pH-dependent effect instead of buffer-type mediated one. The reduction of histone acetylation was associated with the activity of class I and II HDACs. In fact, treatment of the cells with HDAC inhibitors Trichostatin A (TSA) and sodium butyrate (Fig. 1) prior to pH_e decreasing, counteracted the effects of pH changes on global levels of histone acetylation. Taking into account the acetylation levels regulation in response to pH, the opposite, that is the influence of histone acetylation levels on pH_i, was also considered. In fact, albeit the lysine deacetylation has not effects on pH_i *per se*, the acetate anion produced can be

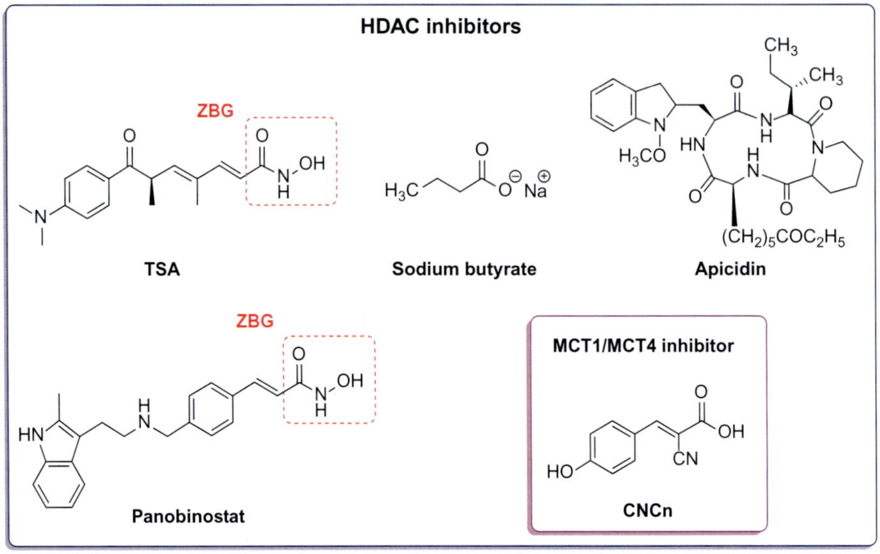

FIG. 1 Structure of HDAC and MCTs inhibitors.

cotransported along with proton through H^+-coupled MCTs, playing a buffer role via the lowering of the proton concentration. Using histones bearing ^{3}H-labeled acetyl moiety, it was noticed that the pH reduction stimulated the increase of the tritium released. These results led to the conclusion that the global deacetylation of histones induced by pH drop is associated with the enhanced excretion of acetate anion from the cells.

These remarks were further corroborated through the treatment of the cells with the MCT1/MCT4 inhibitor α-cyano-4-hydroxycinnamate (CNCn, Fig. 1) that diminished the amount of excreted tritium after the placement of the cells at low pH_e. Similarly, treatment with TSA caused comparable results suggesting that the deacetylation of the histone at low pH takes place in a HDAC-dependent manner, followed by the excretion of the produced acetate anion through the H^+-coupled MCTs (Fig. 2). In view of this, the acetate anion coming from the HDAC activity on histones is exploited to feed MCTs and extrude protons; therefore, the lack of the acetate anion could block this cotransport leading to intracellular pH imbalance.

Indeed, the treatment of the cells incubated at normal pH_e with increasing concentrations of HDAC inhibitor TSA elicited a dose-dependent lowering of pH_i in both HeLa and MDA-MB-231 cells. Similar results were also obtained with other HDAC inhibitors such as Panobinostat and the cyclic peptide antibiotic Apicidin. The capacity of HDAC inhibitors to interfere with pH-buffering ability of cells could be exploited in combination with other drugs. In this respect, Eriksson and colleagues evaluated the capability of TSA to increase the effectiveness of cisplatin treatment, against head and neck squamous cell carcinoma (HNSCC) cancer cells (cell line UT-SCC-77) [18]. However, in this work TSA was examined for its ability to facilitate lysosomal cell death provoked by cisplatin. Cisplatin (cis-diamminedichloroplatinum(II), Fig. 3) is a potent chemotherapeutic agent that exerts its antitumor effect through the formation of intra- or interstrand crosslinks with DNA; however, the

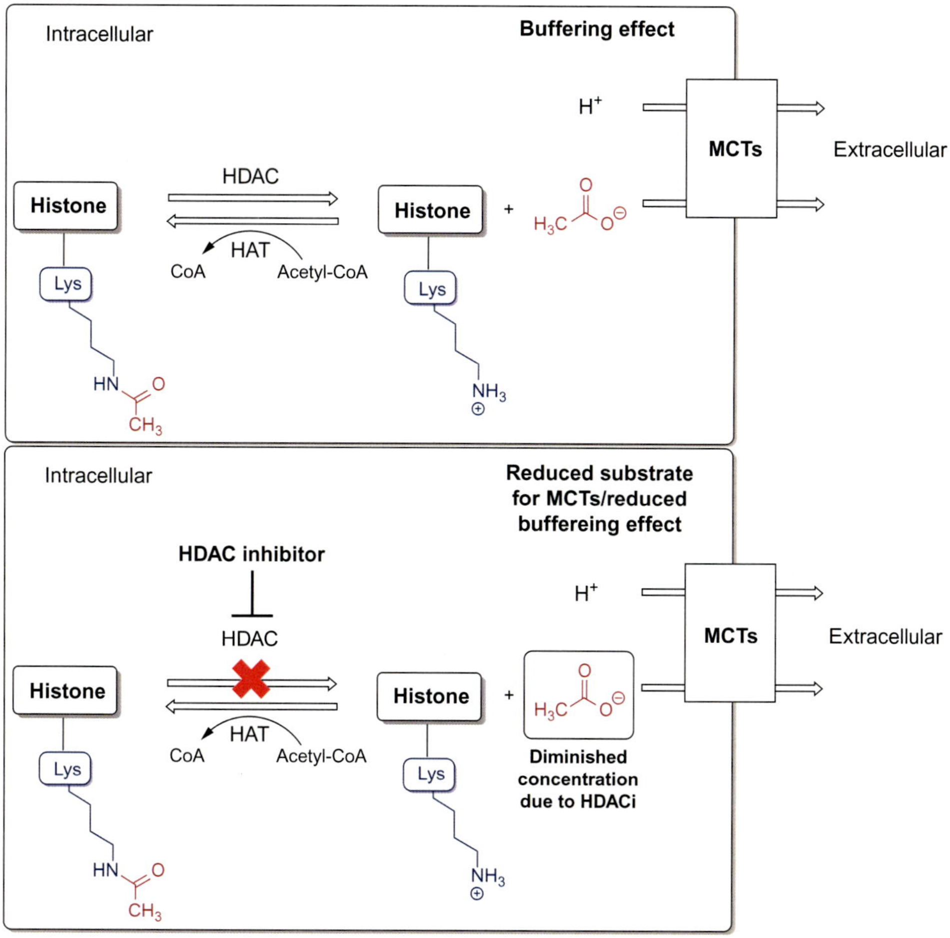

FIG. 2 Scheme of the connection between HDAC activity and buffering effect mediated by MCTs.

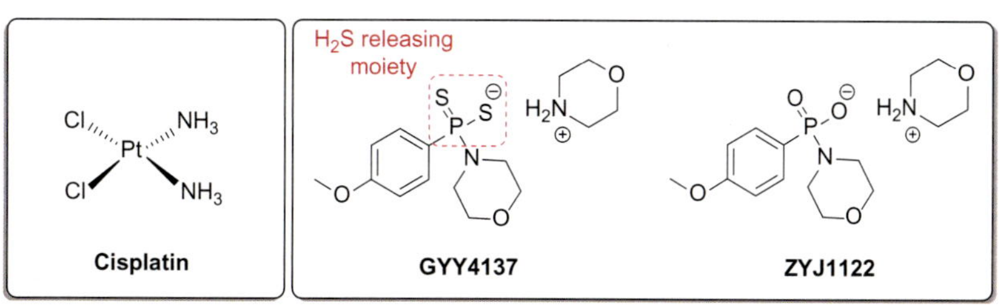

FIG. 3 Structures of cisplatin, GYY4137, and ZYJ1122.

majority of the intracellular drug interacts with other sites, accounting for additional toxic mechanisms [18]. In this regard, the cell death caused by lysosomal membrane permeabilization (LMP) followed by the release of lysosomal enzymes (proteases such as cathepsins B, D, and L), has gained growing interest in recent years. Cisplatin exposure favored the increase of LMP in HNSCC cells, leading to apoptotic cell death; interestingly, the sensitivity to cisplatin was linearly related to the lysosomal pH [19].

A study aimed at investigating the role of the pan-HDAC inhibitor Trichostatin A during cisplatin induced cell death, evaluated the possible impact on lysosomal death pathway [18]. A diminished value of lysosomal pH was detected for cells exposed to TSA, causing an improvement of treatment efficacy due to the increased sensitization of cells to cisplatin. The pH dependency of cisplatin toxicity was confirmed through ammonium chloride treatment of cells. In fact, the addition of NH_4Cl reduced the lysosomal pH gradient, counteracting the effect elicited by TSA and leading to apoptosis levels similar to that observed for the cells treated with cisplatin alone. The augmented efficacy of cisplatin could be related to the improved activity that the lysosomal proteases exert at the acidic pH induced by TSA treatment. As a consequence the destabilization of lysosomal-associated membrane protein (LAMP)-2 is observed, leading in turn the increase of LMP and the release of lysosomal peptidases in the cytosol. In our opinion, since TSA is also able to reduce pH_i as demonstrated by McBrian and colleagues [15], is putative that the proteases released in the cytosol after the increase of LMP, could exert better their activity being favored at acidic pH, accounting for the increased apoptosis induced by cisplatin. However, this is only speculative and further studies should be performed.

Other options to interfere with the pH balance of cells come from compounds able to release hydrogen sulfide. Hydrogen sulfide (H_2S) is a gasotransmitter that covers a series of physiological and pathophysiological functions, being involved in vasodilatation, neuromodulation, and inflammatory signaling [20]. Different evidence underlined the implication of this molecule in regulating cell growth and survival, although counterposing findings have been obtained about its pro- or anticell survival activity. Nevertheless, it is even more evident as the conundrum of whether H_2S is beneficial or not for cell survival is directly related to the manner of its "supplying" to cells. For example, Lee and coworkers proved as the prolonged exposition of cancer cells to low levels of H_2S led to cancer cell death *in vitro* by promoting caspase activity and causing apoptosis [21]. On the contrary, sodium hydrogen sulfide (NaHS), that is able to release large amount of H_2S, did not elicit the same inhibitory effect on the growth of the cancer cell lines. The same authors struggled to evaluate how the H_2S-donor GYY4137 (Fig. 3) exerts its antitumor activity against a broad range of cancer cells types; GYY4137 is endowed with high selectivity for tumor cells being devoid of noticeable effect toward normal lung fibroblasts [22]. In order to demonstrate that the antitumor effect is related to H_2S delivery, a structural analogue of GYY4137, that does not release H_2S, was used as negative control (Fig. 3, ZYJ1122). The release of H_2S prompted anticancer activity thanks to the combined effect of increased metabolic acid production (due to the increased glycolysis) and defective pH regulation, resulting in intracellular acidification. In fact, it seems that the GYY4137 is able, through H_2S release, to alter the glycolysis rate, increasing lactate production. Simultaneously, this effect is coupled with the impairing of activity of pH regulators (AEs and NHEs), leading to reduction of pH_i due to acid accumulation and so, to the cell death.

Conclusion

Novel approaches affecting pH balance of the cells have been shown and proposed as useful strategies for cancer treatment. The use of HDAC inhibitors or H_2S releasers can induce sensitization of the cancer cells, increasing the effectiveness of classical antitumor drugs like cisplatin. Further studies need to be performed in order to improve the knowledge about these agents that act differently from the classical pH-interfering drugs.

References

[1] Ferreira LMR. Cancer metabolism: the Warburg effect today. Exp Mol Pathol 2010;89:372–80. https://doi.org/10.1016/j.yexmp.2010.08.006.

[2] Gatenby RA, Gillies RJ. Why do cancers have high aerobic glycolysis? Nat Rev Cancer 2004;4:891–9. https://doi.org/10.1038/nrc1478.

[3] Semenza GL. Evaluation of HIF-1 inhibitors as anticancer agents. Drug Discov Today 2007;12:853–9. https://doi.org/10.1016/j.drudis.2007.08.006.

[4] Potter CPS, Harris AL. Diagnostic, prognostic and therapeutic implications of carbonic anhydrases in cancer. Br J Cancer 2003;89:2–7. https://doi.org/10.1038/sj.bjc.6600936.

[5] Webb BA, Chimenti M, Jacobson MP, Barber DL. Dysregulated pH: a perfect storm for cancer progression. Nat Rev Cancer 2011;11:671–7. https://doi.org/10.1038/nrc3110.

[6] Neri D, Supuran CT. Interfering with pH regulation in tumours as a therapeutic strategy. Nat Rev Drug Discov 2011;10:767–77. https://doi.org/10.1038/nrd3554.

[7] Shogren-Knaak M, Ishii H, Sun JM, Pazin MJ, Davie JR, Peterson CL. Histone H4-K16 acetylation controls chromatin structure and protein interactions. Science 2006;311:844–7. https://doi.org/10.1126/science.1124000.

[8] Eberharter A, Becker PB. Histone acetylation: a switch between repressive and permissive chromatin. Second in review on chromatin dynamics. EMBO Rep 2002;3:224–9. https://doi.org/10.1093/embo-reports/kvf053.

[9] Seligson DB, Horvath S, Shi T, Yu H, Tze S, Grunstein M, et al. Global histone modification patterns predict risk of prostate cancer recurrence. Nature 2005;435:1262–6. https://doi.org/10.1038/nature03672.

[10] Fraga MF, Ballestar E, Villar-Garea A, Boix-Chornet M, Espada J, Schotta G, et al. Loss of acetylation at Lys16 and trimethylation at Lys20 of histone H4 is a common hallmark of human cancer. Nat Genet 2005;37:391–400. https://doi.org/10.1038/ng1531.

[11] Iwabata H, Yoshida M, Komatsu Y. Proteomic analysis of organ-specific post-translational lysine-acetylation and -methylation in mice by use of anti-acetyllysine and -methyllysine mouse monoclonal antibodies. Proteomics 2005;5:4653–64. https://doi.org/10.1002/pmic.200500042.

[12] Elsheikh SE, Green AR, Rakha EA, Powe DG, Ahmed RA, Collins HM, et al. Global histone modifications in breast cancer correlate with tumor phenotypes, prognostic factors, and patient outcome. Cancer Res 2009;69:3802–9. https://doi.org/10.1158/0008-5472.CAN-08-3907.

[13] Seligson DB, Horvath S, McBrian MA, Mah V, Yu H, Tze S, et al. Global levels of histone modifications predict prognosis in different cancers. Am J Pathol 2009;174:1619–28. https://doi.org/10.2353/ajpath.2009.080874.

[14] Manuyakorn A, Paulus R, Farrell J, Dawson NA, Tze S, Cheung-Lau G, et al. Cellular histone modification patterns predict prognosis and treatment response in resectable pancreatic adenocarcinoma: results from RTOG 9704. J Clin Oncol 2010;28:1358–65. https://doi.org/10.1200/JCO.2009.24.5639.

[15] McBrian MA, Behbahan IS, Ferrari R, Su T, Huang TW, Li K, et al. Histone acetylation regulates intracellular pH. Mol Cell 2013;49:310–21. https://doi.org/10.1016/j.molcel.2012.10.025.

[16] Fellenz MP, Gerweck LE. Influence of extracellular pH on intracellular pH and cell energy status: relationship to Hyperthermic sensitivity. Radiat Res 1988;116:305. https://doi.org/10.2307/3577466.

[17] Wong P, Kleemann HW, Tannock IF. Cytostatic potential of novel agents that inhibit the regulation of intracellular pH. Br J Cancer 2002;87:238–45. https://doi.org/10.1038/sj.bjc.6600424.

[18] Eriksson I, Joosten M, Roberg K, Öllinger K. The histone deacetylase inhibitor trichostatin A reduces lysosomal pH and enhances cisplatin-induced apoptosis. Exp Cell Res 2013;319:12–20. https://doi.org/10.1016/j.yexcr.2012.10.004.

[19] Nilsson C, Roberg K, Grafström RC, Öllinger K. Intrinsic differences in cisplatin sensitivity of head and neck cancer cell lines: correlation to lysosomal pH. Head Neck 2010;32:1185–94. https://doi.org/10.1002/hed.21317.
[20] Wang R. Physiological implications of hydrogen sulfide: a whiff exploration that blossomed. Physiol Rev 2012;92:791–896. https://doi.org/10.1152/physrev.00017.2011.
[21] Lee ZW, Zhou J, Chen CS, Zhao Y, Tan CH, Li L, et al. The slow-releasing hydrogen sulfide donor, GYY4137, exhibits novel anti-cancer effects in vitro and in vivo. PLoS One 2011;6. https://doi.org/10.1371/journal.pone.0021077.
[22] Lee ZW, Teo XY, Tay EYW, Tan CH, Hagen T, Moore PK, et al. Utilizing hydrogen sulfide as a novel anti-cancer agent by targeting cancer glycolysis and pH imbalance. Br J Pharmacol 2014;171:4322–36. https://doi.org/10.1111/bph.12773.

CHAPTER

9

Patent survey on chemosensitizers (2015–2019)

Ahmed M. Shabana[a,b] *and Suleyman Akocak*[c]

[a]Institute for NanoBioTechnology, Johns Hopkins University, Baltimore, MD, United States [b]Department of Pharmaceutics and Industrial Pharmacy, Faculty of Pharmacy, Cairo University, Cairo, Egypt [c]Department of Pharmaceutical Chemistry, Faculty of Pharmacy, Adiyaman University, Adiyaman, Turkey

Abstract

Chemotherapy represents the gold standard method for cancer treatment during the past decades. However, many commercially available cytotoxic drugs suffer from chemoresistance, which results in lowering the therapeutic efficacy of drugs, cancer recurrence, progression, and metastasis with a subsequent decline in the overall survival rate of cancer patients. This discrepancy raises the demand for developing novel approaches to enhance the sensitivity of resistant tumor cells to chemotherapeutic drugs, a process known by chemosensitization. In this chapter, we discuss the various drug-resistant mechanisms developed within the tumor cells along with their implications for chemotherapy. Chemoresistance is particularly driven by a group of regulatory proteins and biological pathways, which either promote drug efflux, induce DNA damage repair mechanisms, or inhibit apoptosis. We also present herein the recent inventions describing the successful design of different classes of small molecule chemosensitizers, and their significant role in counteracting the cancer cells' inherent drug resistance mechanisms, rendering them capable of either eliciting a cytotoxic effect on their own or synergistically acting with existing chemotherapeutic drugs toward maximizing the therapeutic outcome, enhancing tumor regression while minimizing the unwanted toxic side effects associated with chemotherapy administration.

Abbreviations

ABC	ATP-binding cassette
ABCB4	ATP-binding cassette subfamily B member 4
AML	acute myeloid leukemia
BCRP	breast cancer-resistant protein
CDK	cyclin-dependent kinase
CLL	chronic lymphocytic leukemia
GLI1	glioma-associated oncogene

https://doi.org/10.1016/B978-0-12-820701-7.00003-8

GSH	glutathione
hPNKP	human polynucleotide kinase phosphatase
MAPK	mitogen-activated protein kinase
MRP1	multidrug resistance protein-1
MXR	mitoxantrone resistance protein
NF-kB	nuclear factor kB
P-gp	P-glycoprotein
SLL	small lymphocytic lymphoma
TMPK	thymidylate kinase
TR-FRET	time-resolved fluorescence energy transfer

Conflict of interest

No potential conflicts of interest were disclosed.

Introduction

Cancer is a recognized as a major life-threatening disease, accounting for more than one million death cases worldwide each year [1]. It arises as a consequence of aberrant genetic mutations that trigger the transformation of normal, healthy cells into cancerous phenotypes [2]. Over the past century, surgery and radiotherapy were the dominant key treatments for different kinds of cancer specifically solid tumors [3]. However, the cure rates in cancer patients did not significantly improve, particularly due to the development of uncontrolled micrometastases in advanced cancer stages [4]. This fact raised the demands for alternative strategies toward cancer therapy with chemotherapy being implemented as an adjunct therapy following surgery and/or radiotherapy to solve the issue of micrometastases. Despite the major success of utilizing chemotherapy in controlling tumor growth, many cancer cells develop either inherent (primary) or acquired (secondary) resistance to chemotherapeutic drugs through various mechanisms, as shown in Fig. 1, resulting in a significant decline in the overall survival rate of cancer patients [6].

One major mechanism of cancer cell resistance to chemotherapy is the abundance of efflux transporters that prevent the accumulation of drugs intracellularly below the toxic levels. The efflux of chemotherapeutic drugs is mediated by several members of the ATP-binding cassette (ABC) superfamily of membrane transporters, particularly the elevated expression of P-glycoprotein (P-gp), multidrug resistance protein-1 (MRP1) and breast cancer resistant protein (BCRP), or mitoxantrone resistance protein (MXR) [7]. Another prominent mechanism of chemoresistance is the ability of many cancer cells to respond to DNA damages induced by the chemotherapeutic drugs using repair pathways such as nucleotide excision repair and homologous recombination repair [8–10].

In this context, the coadministration of compounds with chemotherapeutic drugs that inhibit cancer cell drug resistance and their self-defensive mechanisms would sensitize tumors toward chemotherapy and promote the efficacy of the cytotoxic drugs, a process known as chemosensitization [11, 12]. Chemosensitizers provide the advantage of lowering the dosages of chemotherapeutic drugs, which, will in turn, minimize the toxicity to nontarget healthy cells and lower the overall adverse effects associated with cytotoxic drug administration [13].

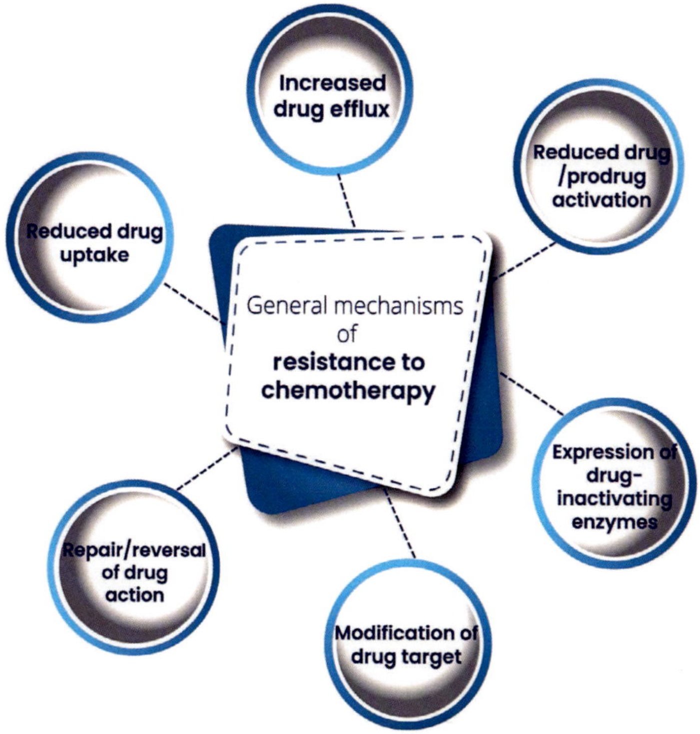

FIG. 1 General mechanisms of cancer cells resistance to chemotherapeutic drugs [5].

To date, there has been a plethora of chemosensitizing agents reported in the literature affecting pharmacodynamic processes in tumor cells. Chemosensitizers can be classified according to the drug resistance mechanism that they inhibit in cancer cells. Based on this, we can classify chemosensitizers into three main categories, (i) Chemosensitizers of resistance mechanisms responsible for overexpression of the actual drug target such as inhibitors of drug efflux transporters, (ii) Chemosensitizers of drug-target site mechanisms such as inhibitors of specific target proteins and inhibitors of DNA repair mechanisms and (iii) Chemosensitizers of resistance mechanisms responsible for downregulation of the actual drug target such as inhibitors of negative regulators in autophagy, apoptosis, and other modes of programmed cell death.

Chemosensitizers inhibiting the activity of drug efflux transporters

Role of drug efflux transporters in chemoresistance

Drug efflux transporters represent the most common mechanism of resistance to chemotherapy. They are recognized as membrane-bound transporters, which belong to the class of ATP-binding cassette (ABC) superfamily and function as a hydrophobic vacuum cleaner driving lipid, multiple drugs, natural products, and peptides from the membrane bilayer to outside the cells in an ATP-dependent process [14]. They are particularly responsible for pumping anticancer drugs and other xenobiotics outside the cells, therefore minimizing their intracellular accumulation and lowering their therapeutic efficacy [15, 16]. Literature has cited more than 100 ABC transporters that are distributed from prokaryotes to humans. The P-glycoprotein (P-gp), multidrug resistance protein-1 (MRP1), and breast cancer–resistant protein (BCRP), or mitoxantrone resistance protein (MXR) are foreseen as the dominant efflux transporters responsible for developing chemoresistance [17, 18]. Their overexpression in many types of cancer has been associated with the poor therapeutic outcome of several commercially available chemotherapeutic drugs such as anthracyclines, taxanes, and vinca alkaloids [19, 20]. In this context, the administration of chemosensitizers that function as inhibitors of these efflux transporters will potentiate the therapeutic activity of many chemotherapeutic drugs suffering from inadequate intracellular accumulation while minimizing the unwanted toxic side effects to healthy tissues. To date, three generations of chemosensitizers have been reported in the literature as inhibitors of P-gp efflux chemoresistance based on their pharmacodynamics characteristics as shown in Table 1. The first generation encompasses calcium channel blockers, antimalarial drugs such as quinidine and immunosuppressants such as cyclosporine A [21–23], while the second generation includes drugs such as valspodar and dexverapamil [24, 25]. On the other hand, the third generation of inhibitors consists of drugs such as tariquidar and zosuquidar [26, 27]. However, most of these inhibitors impair the process of drug clearance through interfering with the activity of cytochrome P450 [28, 29].

TABLE 1 Classes of P-glycoprotein inhibitors [5].

Class	Examples
Natural products	Flavonoids from Chinese plants Neochamaejasmin B Cytarabine and halaven
Monoclonal antibodies	MRK-16 MRK-17
First generation	Verapamil, Felodipine, Tamoxifen, Diltiazem Flupenthixol, Chlorpromazine, CSA, Quinine, and Quinidine
Second generation	Dexverapamil, Gallopamil Biricodar citrate Valspodar
Third generation	Tariquidar and Laniquidar Elacridar Mitotane and Annamycin
Others	Gatifloxacin Celecoxib

Chemosensitizing compounds as inhibitors of drug efflux transporters

Extensive research has been applied to come up with novel compounds that would selectively target and inhibit the activity of efflux transporters with the aim of enhancing the sensitivity of cancer cells to cytotoxic drugs. A research collaboration between McGill University and Hong Kong Polytechnic University has resulted in filing a patent for a series of novel triazole-bridged flavonoid dimer compounds as well as their precursor alkyne- and azide-containing flavonoids discussing their potential ability in reversing the activity of P-gp-, MRP1-, and BCRP-mediated drug resistance. They claim that these novel compounds can efficiently lower the EC_{50} values of paclitaxel, doxorubicin, vincristine, and topotecan in different resistant cancer cell lines. Moreover, these flavonoids demonstrated a significant anticancer activity with EC_{50} lower than 10nM in BCRP-overexpressed cell lines, which reflects the selectivity of that these bivalent triazoles toward inhibiting BCRP transporter than the P-gp and MRP1 transporters [30].

In another invention, researchers have patented a novel class of P-gp inhibitors based on a tripeptide structure represented by the sequence XAB, wherein X, A, and B are 7 amino acids, wherein X can be any amino acid and A and B are basic amino acids. These basic tripeptides are substrates for P-gp and can therefore competitively inhibit the efflux of drugs from the cell resulting in reversing the chemoresistance. These peptides show this activity in the low micromolar range when added to cells, and even higher activity when import into the cytoplasm is enhanced, which can be achieved by conjugation to a cell-penetrating peptide or by formulation into a carrier such as liposomes or such as (albumin) nanoparticles. Moreover, these tripeptides possess excellent solubility, smaller in size than other P-gp inhibitors, and do not have any effect on cytochrome P-450. All these features render this class of tripeptides an ideal chemosensitizer the treatment, delay, prevention, cure, or stabilization of cancer [31].

Another patented study by researchers at Macau Univ of Science and Technology where they discuss the coadministration of a citrus methoxyflavone and a chemotherapeutic drug in order to overcome multidrug resistance. The citrus methoxyflavone is called tangeretin that act by inhibiting the activity of P-gp efflux transporters, thus improving intracellular accumulation of the chemotherapeutic drug. Researchers claim that the coadministration of tangeretin with paclitaxel, docetaxel, doxorubicin, or daunorubicin showed significant improvement in the anticancer activity, particularly in the paclitaxel-resistant cancer (non-small-cell lung cancer) or paclitaxel-resistant ovarian cancer [32] (Fig. 2).

Another invention by Gerson Silva Paiva reported an antitumor drug cocktail directed toward all types of solid tumors and comprising a mixture of commercially available tumor acid selective anticancer drug such as 9-acridinecarboxylic acid, in combination with an

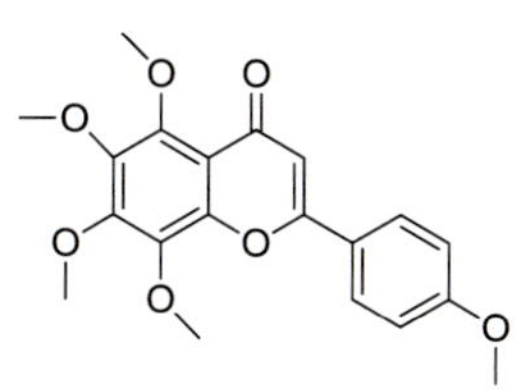

FIG. 2 Chemical structure of tangeretin used as a chemosensitizer to improve the efficacy of chemotherapeutic drugs through inhibiting the activity of P-glycoprotein (P-gp) [32].

antioxidant such as Coenzyme Q10, and a P-glycoprotein inhibitor such as metformin, ketoconazole, quinidine, verapamil, or cyclosporin. This drug cocktail is intended to selectively destroy solid tumor cells while avoiding unwanted toxic effects toward rapidly replicating healthy cells. Selectivity toward tumor acidity is achieved using an anticancer compound (acridine derivative), possessing a negatively charged carboxylic group in its chemical structure, which is attracted by electrostatic forces within the tumor microenvironment due to the characteristic low pH and high acidity of solid tumors, owing to the high production of lactic and glycolic acids. Therefore, the solid tumor acidity favors the neutralization of the negatively charged carboxylic group of acridine resulting in the free uptake of the neutral acridine drug to the tumor cell DNA, since only neutral molecules can freely cross the cell membrane, leading to apoptosis or tumor cell self-destruction. Moreover, the presence of an antioxidizing agent, such as Coenzyme Q10, in the antitumor drug cocktail acts to reduce the toxicity of naturally occurring free radicals generated by the antitumor agent, thereby helps in protecting healthy cells from the harmful effect of these free radicals during treatment. Furthermore, the cocktail according to the invention also comprise a P-glycoprotein (P-gp) inhibitor such as metformin, ketoconazole, quinidine, verapamil, or cyclosporine, whose function is to inhibit chemoresistance due to P-gp drug efflux transporters normally present in large quantities in tumor cells and responsible for pumping the chemotherapy outside the tumor cells, rendering the concentration of these chemotherapeutic drugs below their therapeutic window, which often results in lower efficacy of chemotherapy [33].

An invention disclosure by John G. Wise et al. at Southern Methodist University describes the treatment of chemotherapy-resistant cancers, primary cancers, and cancer stem cells using inhibitors of the multidrug resistance transporter P-glycoprotein (P-gp). It provides a method of sensitization and resensitization of a cancer cell to a chemotherapeutic drug by identifying a subject having cancer cells in need of sensitization and resensitization to one or more chemotherapeutics, administering to the subject a pharmaceutically effective amount of a P-gp inhibitor compound that can interact with the exporter protein, with drug-toxin pumping structures of a P-gp or with ATP-binding domains of a P-gp. This strategy can be applied for the treatment of several types of cancer, including lymphoma, leukemia, liver cancer carcinoma of prostate, intestine cancer, myeloma tumor, lymphoma, breast carcinoma, and ovarian cancer [34].

In a study conducted by Quan et al. [35], the authors designed a series of novel quinoline compound derivatives with a lead compound named YS-7a demonstrating a significant inhibition in the activity of P-gp as compared to first-generation P-gp inhibitor verapamil. It is suggested that YS-7a exerts its chemosensitization effect through suppressing the transport activity of P-gp without affecting its level of expression. Interestingly YS-7a did not induce cytochrome P40 expression, thus preventing any drug-drug interactions. Moreover, the coadministration of YS-7a with vincristine showed a better inhibitory effect than verapamil in vivo without a negative effect on mouse weight. These findings render YS-7a a novel therapeutic agent with a significant potential in reversing multidrug resistance through the inhibition of P-gp transport function in vitro and in vivo.

A study conducted by Wen et al. [36] revealed the advantage of coadministering curcumin with doxorubicin in order to reverse multidrug resistance in doxorubicin-resistant breast cancer cell lines. The authors found out that curcumin improves the intracellular accumulation of doxorubicin in a manner that was inversely dependent on the activity of ATP-binding cassette subfamily B member 4 (ABCB4) without altering the protein expression. This resulted in significant decrease in doxorubicin efflux in cancer cells overexpressing ABCB4 and outlining the potential role of curcumin as a chemosensitizer in breast cancer [36].

Role of drug delivery systems in chemosensitization and reduction of drug efflux transporters activity

Many of the developed chemosensitizers, which reduce the activity of drug efflux transporters, are classified as small molecules incapable of achieving significant accumulation at the tumor site to promote the therapeutic efficacy of chemotherapeutic drugs following systemic administration [37]. For this purpose, a variety of nano drug delivery systems have been fabricated that possess the ability of overcoming delivery barriers and selectively codelivering chemosensitizers and chemotherapeutic drugs to the tumor site, thus promoting optimal accumulation while minimizing the toxic side effects to healthy tissues [38]. Moreover, the encapsulation of chemotherapeutic drugs into nano drug delivery systems decorated with ligands targeting overexpressed receptors in cancer cells promotes endocytosis and therefore bypasses the drug efflux transporters activity resulting in a significant increase in the therapeutic efficacy of many cancerous drugs [39, 40].

In this context, an invention disclosed by Modi [41] described the design of an efficient oral drug delivery nanosystem for the codelivery of taxanes and cyclosporine as shown in Fig. 3. Cyclosporine functions as both an absorption enhancing agent and a P-gp inhibitor in order

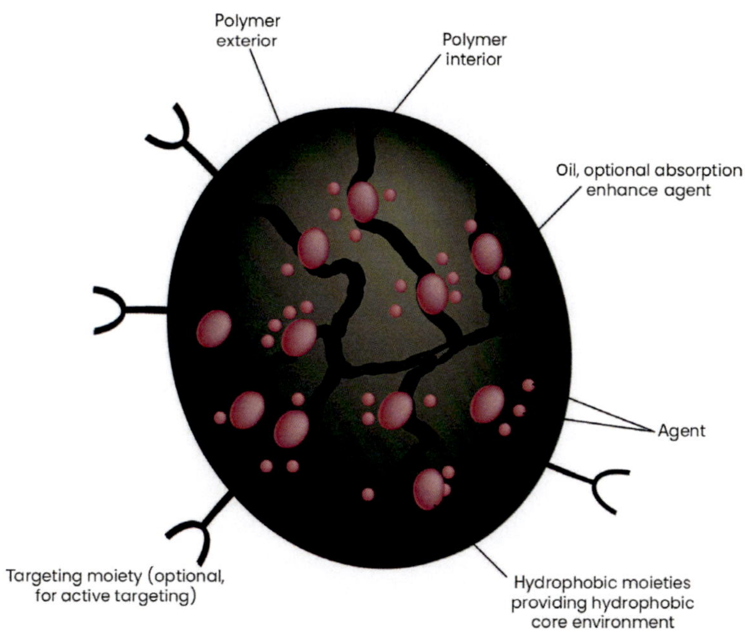

FIG. 3 Design and composition of a nano-drug delivery system used for the delivery of chemotherapeutic drugs and P-gp inhibitors to enhance the oral bioavailability and therapeutic efficacy of taxane drugs [41].

to enhance the bioavailability of the taxane chemotherapeutic drugs. The drug delivery vehicle is made up of polymer or lipids, where the chemotherapeutic drug is physically entrapped in the hydrophobic core, thus allowing for high loading efficiency of poorly water-soluble drugs. The surface of the drug delivery vehicle is decorated with targeting moieties, which are directed toward receptors overexpressed in cancerous tissues as compared to normal healthy ones. This formulation is claimed to improve the therapeutic efficacy of chemotherapeutic drugs delivered orally, which suffers from chemoresistance due to overexpression of P-gp.

On the other hand, a Chinese patent discussed the design of a polymeric drug delivery system for the treatment of drug-resistant tumors. The polymeric micelles are made of a pH-sensitive amphipathilic diblock or triblock or polymer of poly-(2-alkyl-2-oxazoline) or PAOz. The hydrophobic segment is selected from polylactic acid, polycaprolactone, or poly- butyrolactone, while the alkyl chain in the hydrophilic segment of PAOz preferably contains 1–4 carbon atom, such as methyl, ethyl, *n*-propyl, *n*-butyl. These micelles are used to encapsulate and deliver a combination of hydrophobic cytotoxic drugs such as taxanes and a P-gp inhibitor such as curcumin. The cytotoxicity of the drug-loaded polymeric micelles was evaluated in vitro using MCF-7/ADR breast cancer cells. Results showed that the IC_{50} for drug-loaded micelles was $1.75 \pm 0.19\,\mu g/mL$ as compared to $18.10 \pm 1.22\,\mu g/mL$ for taxol (a commercially available form of paclitaxel) [42].

In a research study, Guo et al. reported the design of a nanoplatform prechemosensitizer (DOX@PCDA-PEG-Biotin NPs) made via the self-assembly of amphiphilic poly(curcumin-dithiodipropionic acid)-b-poly(ethylene glycol)-biotin (PCDA-PEG-Biotin) copolymer [43]. This nanoparticulate system possessed the advantages of high multidrug-loading capacity of the chemosensitizer curcumin as well as the chemotherapeutic drug, doxorubicin. It also triggered the intracellular release of the multidrug cargo in response to glutathione (GSH). Moreover, the surface decoration of the nanoparticles with biotin facilitated the interaction with overexpressed biotin receptors on the surface of MCF-7/ADR drug-resistant breast cancer cells, thus promoting endocytosis and bypassing the activity of drug efflux transporters. The delivery of curcumin significantly downregulated the P-gp expression and inhibited the ATP activity of MCF-7/ADR cells, thus minimizing the efflux of doxorubicin and facilitating its intracellular accumulation. This resulted in a significant enhancement in the cytotoxic activity of doxorubicin against MCF-7/ADR cells as compared to the delivery of the chemotherapeutic drug alone.

In another study, Shabana et al. reported the design of a pH-sensitive carbonic anhydrase IX (CA IX)-targeted gold nanoplatform to enhance therapeutic efficacy doxorubicin to hypoxic tumor spheroids and to overcome hypoxia-induced chemoresistance [44]. Hypoxia is known to be a characteristic pathophysiological feature of many solid tumors, which accounts for chemoresistance through the upregulation of drug efflux transporters particularly P-gp. This results in a significant decrease in the effectiveness of many cytotoxic drugs, including doxorubicin, a substrate for P-gp. The designed gold nanoplatform was decorated with carbonic anhydrase inhibitor (CAI) ligands, which target the overexpressed CA IX in hypoxia and promoted the intracellular uptake of the gold nanoplatform, thus bypassing the P-gp activity. The designed gold nanoplatform demonstrated significant cytotoxic activity under hypoxic conditions in HT-29 colorectal cancer cells. They also delivered doxorubicin inside the cell in a controlled fashion. Moreover, they enhanced the delivery of doxorubicin in HT-29 tumor spheroids and showed better tissue penetrability than free drug.

Chemosensitizers inhibiting DNA repair mechanisms

Many chemotherapeutic drugs such as doxorubicin- and platinum-based drugs act by inducing DNA damage in rapidly dividing cancer cells, which results in apoptosis [45, 46]. However, cancer cells possess characteristic biological features to overcome this damage by activating distinct DNA repair mechanisms [47]. Moreover, cancer stem cells, various oncogenes, transcription factors, and hypoxic tumor microenvironment play a central role in repairing DNA damage caused by cytotoxic drugs [48–50]. This defensive response renders cancer cells resistant to chemotherapy resulting in recurrence of the disease and development of a more aggressive form of the tumors [47]. The various regulatory proteins, which are responsible for repairing DNA damages in cancer cells, are shown in Fig. 4. One may notice in Fig. 6 that the regulatory proteins can be modified either directly or indirectly during the process of DNA damage, which will result in aggressive repair of DNA damage such that the increased activity of DNA repair counteracts the damage induced by cytotoxic drugs resulting in chemoresistance. A key regulatory protein of the homologous recombination DNA repair is Breast cancer type 1 susceptibility (BRCA1) protein [52, 53]. Mutations of BRCA1 commonly result in ovarian and breast cancer [54, 55]. A study has shown that the BRCA1-activated cancer cells trigger cell cycle arrest when exposed to Irofulven compared to BRCA1-deficient cancer cells. This cell cycle arrest allows the cancer cells to repair the Irofulven-induced DNA damage and causes chemoresistance [56]. On the other hand, it has been shown that glioma-associated oncogene (GLI1) is upregulated in several types of cancer [57–59]. GLI1 is particularly involved in DNA damage response and epithelial to mesenchymal transition [59]. Therefore, its upregulation has been linked to chemoresistance due

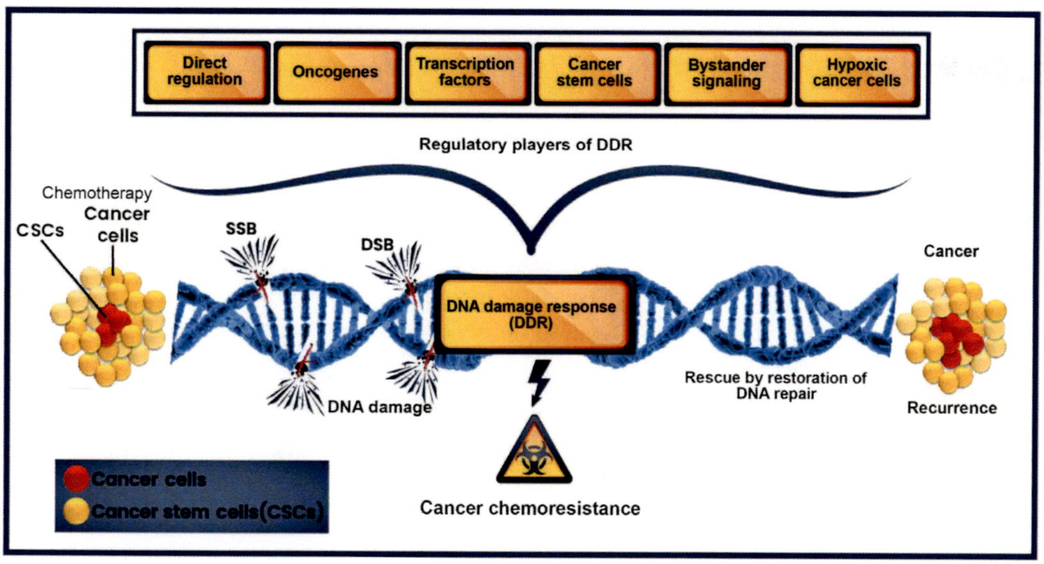

FIG. 4 Regulatory players of the DNA damage response network and their role in cancer chemoresistance by using various factors/signaling mechanisms, including oncogenes, transcriptional factors, cancer stem cells, bystander signaling, and hypoxia [47].

to the activation of cell cycle checkpoint and a study has shown the inhibition of GLI1 in the cancer cells enhanced the sensitivity of the cancer cells to camptothecin (CPT)-induced DNA damage [60]. Besides the regulatory proteins, it has been found that cancer stem cells or tumor initiating cells possess significant DNA damage repair capabilities, which render them resistant to the cytotoxic drugs [61]. The cancer stem cells maintain a quiescent phase and have the ability of self-renewal, differentiation, pluripotency, and resistance to conventional chemo- and radiotherapies [47]. As an example, studies have shown that $CD133^+$ colon cancer stem cells demonstrated upregulation of phosphoinositide 3-kinases (P13Ks) that triggers the activation of the ATR/CHK1-mediated cell cycle arrest upon DNA damage causing resistance to interstrand crosslinking agents [62]. Similarly, pancreatic cancer stem cells showed overexpression of DNA repair proteins, particularly BRCA1, rendering them capable of repairing the gemcitabine-induced DNA damage as compared to their control pancreatic cancer cells [63]. On the other hand, hypoxia is a characteristic pathophysiological feature of many solid tumors, which upregulates a cascade of proteins favoring tumor cell survival, progression, and metastasis [64, 65]. It also contributes significantly to chemoresistance, thus accounting for the lower efficacy of cytotoxic drugs [20]. Hypoxia induces chemoresistance through complex mechanisms, including the downregulation of p53 tumor suppressor gene, upregulation of drug efflux transporters, particularly P-gp, and induction of various DNA damage repair mechanisms [66].

For this purpose, the administration of compounds that could effectively target and inhibit the DNA repair mechanisms would potentiate the therapeutic activity of cytotoxic drugs, overcome the cancer cells induced chemoresistance, prevent disease recurrence, and lower the severity of tumor progression. Several patents have been filed reporting the design of many small molecules that efficiently inhibit essential pathways and prominent mechanisms responsible for repairing DNA single- and double-strand breaks. In this context, a patent filed by a group of researchers in National Taiwan University [67], the inventors presents the design and synthetic scheme of a novel class of inhibitors of thymidylate kinase (TMPK), a key enzyme responsible for the phosphorylation of thymidine monophosphate (dTMP) to thymidine diphosphate (dTDP), which is then converted to thymidine triphosphate (dTTP) by nucleotide diphosphate kinases (NDPKs) for DNA synthesis. The double-strand breaks (DSBs) in proliferating cells are mainly repaired by homologous recombination (HR) in which the repair of a single DSB needs more than 10,000 of deoxyribonucleotide triphosphate (dNTPs) new incorporation. Since dTDP formation, specifically requires TMPK function, the authors demonstrate that blocking TMPK decreases the efficiency of DSBs repair and sensitizes tumor cells to genotoxic insults. Therefore, the TMPK inhibitor can be used for chemosensitization and inhibiting the growth of cancer. The TMPK inhibitor treatment has low general genotoxicity as compared to 5-fluorouracil, the chemotherapeutic drug responsible for blocking dTMP formation [67].

In another invention, researchers from Canada have patented the design of novel imidopiperidine compounds that function as inhibitors of human polynucleotide kinase phosphatase (hPNKP) [68]. This enzyme is responsible for the phosphorylation of 5′-hydroxyl termini and dephosphorylation of 3′-phosphate termini in human cells. This process is an integral component of the nonhomologous end-joining (NHEJ) pathway. Besides the role of (hPNKP) in the repair of strand breaks produced directly by cytotoxic agents, hPNKP has been also involved in the repair of strand breaks produced by the enzymatic processes, including strand breaks introduced by the 135-AP lyase activity of DNA glycosylases. In this context, this class of imidopiperidine compounds would allow chemosensitization and/or radiosensitization method to treat a cell in vitro and/or in vivo through blocking down key processes involved in DNA damage repair [68].

An invention disclosure by Delmar Pharmaceuticals Inc. [69] reported the chemosensitizing activity of substituted hexitol derivatives such as dianhydrogalactitol for the treatment of glioblastoma, non-small-cell lung carcinoma as well as ovarian cancer through induction of double-strand breaks in the DNA of cancer cells. The activity of dianhydrogalactitol is attributed to its ability to create N^7 methylation and interstrand DNA crosslinks. The inventors claim that this compound can function as a chemosensitizer to potentiate the therapeutic efficacy of several anti-cancer drugs, including topoisomerase inhibitors, antitubulin agents, cisplatin or platinum analogs, and 5-fluorouracil [69].

Another invention disclosure by Janssen Pharmaceuticals [70] reported the design, synthesis, and chemosensitizing activity of quinoxaline and pyridopyrazine derivatives that function as phosphoinositide-3-kinases (PI3Kβ) inhibitors. These compounds are particularly active in tumors with Phosphatase and Tensin Homolog (PTEN) deficiency. It is known that PI3Ks are activated in a plethora of solid and malignant tumors via mutation or deletion of the tumor suppressor PTEN. This activation catalyzes the phosphorylation of phosphatidyl 4,5-diphosphate leading to the generation of phosphatidyl-3,4,5-triphosphate (PIP3), which is responsible for the upstream of numerous biologically relevant processes such as metabolism, differentiation, proliferation, longevity, and apoptosis. Therefore, the inhibition of PI3Ks using the quinoxaline and pyridopyrazine derivatives would antagonize the activity of the PI3Ks through the dephosphorylation PIP3 resulting in tumor shrinkage. The claimed compounds can also be foreseen as chemosensitizers to be used in combination with anticancer drugs to establish chemosensitivity in tumor cells resistant to commercially available chemotherapy and to potentiate the efficacy of these drugs in drug-sensitive tumors [70].

A patent invented by Godavari Biorefineries Ltd. [71], Mumbai, India, described a series of small-molecule cancer stem cells' inhibitors based on thieno[2,3-*d*]pyrimidine scaffold. The mechanism of action of these compounds is thought to be inhibition of the activity of PI3Ks according to their chemical structure. These compounds demonstrated an enhanced anti-CSC activity as compared to cisplatin in MDA-MB-231, MCF-7, and T47D breast and PC3, DU145, and LNCap prostate cancer cell lines when evaluated systematically in MTT, soft agar, 3D sphere assays [71].

Chemosensitizers as modulators of apoptotic pathways

Many of the commercially available chemotherapeutic drugs elicit their cytotoxic activity by triggering apoptosis through inducing the release of cytochrome *c* from mitochondria [72, 73]. Apoptosis is normally modulated by a complex set of genes that interact to control the progression of the cell cycle [74]. The alteration in the expression of these regulatory genes results in the development of chemoresistance within many solid tumors [75, 76]. Among the apoptotic regulatory genes is the p53 tumor suppressor gene, which is foreseen as a key player in preserving genetic stability [77]. The p53 accumulation in the cell triggers cell cycle arrest and apoptosis [78]. Therefore, downregulation or mutation of the p53 gene enables cell proliferation and develops resistance to chemotherapeutic drugs [79]. Studies have found that p53 tumor suppressor gene is mutated in most malignant tumors and its lack of expression is correlated with tumor cell growth, progression, and apoptosis [80]. On the other hand, B-cell lymphoma-2 (Bcl-2) family proteins are another set of genes regulating apoptosis [81]. This protein family comprises both proapoptotic Bax and Bak and antiapoptotic B-cell lymphoma-2 (Bcl-2), B-cell lymphoma-extra large (Bcl-x_L), and Myeloid Cell Leukemia − 1 (MCL-1) [82]. The relative expression level of these proteins dictates the sensitivity or resistance of cells to apoptosis such that the overexpression of antiapoptotic proteins or downregulation of proapoptotic proteins prevents the cytochrome *c* release from the mitochondria, which is considered to be an early stage of apoptosis preceding the activation of caspase and endonuclease [82, 83]. During apoptosis, Bax and Bak proteins are translocated from the cytoplasm to the outer mitochondrial membrane where they oligomerize to form pores to stimulate the release of cytochrome *c* [83]. These effects are antagonized by the overexpression of the antiapoptotic Bcl-2, Bcl-x_L, and MCL-1, which prevents oligomerization and cytochrome *c* release, therefore blocking apoptosis. This process has been linked with the development of chemoresistance toward a variety of chemotherapeutic drugs. Another regulatory group of apoptosis is the Inhibitor of Apoptosis (IAP) Family Proteins [84]. They are known to be upregulated in different types of cancer and responsible for cancer cell growth, tumor progression as well as development of chemoresistance [85]. These proteins act by downregulating the expression of caspases. The IAP family member XIAP is particularly binding to the effector caspase-9, thus preventing the initiation of downstream apoptotic events, such as activation of pro-caspase-3 [86, 87]. Moreover, IAPs control ubiquitin (Ub)-dependent signaling pathways that regulate activation of nuclear factor kB (NF-kB) and mitogen-activated protein kinase (MAPK) pathways [88]. These pathways trigger the expression of genes responsible for inflammation, immunity, cell survival, and migration [89]. In this context, a variety of strategies have been developed to boost the sensitivity of cancer cells to chemotherapy through targeting and controlling the expression of the previously mentioned apoptosis regulatory proteins [90].

A patent application filed by Nikolovska-Coleska et al. [91] discussed the design of a new class of selective MCL-1 inhibitors based on small-molecule pyrazolopyridine structure. MCL-1 is classified as a potent antiapoptotic protein and a crucial survival factor for different types of cancer, including breast, pancreatic, colon, lung, ovarian, prostate, melanoma, multiple myeloma, and acute myeloid leukemia. The overexpression of MCL-1 has been associated with tumor initiation, progression, and resistance to commercially available chemotherapeutic drugs. Recent studies demonstrated that using a genetic approach to downregulation of Mcl-l provided a significant proof-of-concept that selective, small-molecule Mcl-l inhibitors may have potential as a new treatment for human cancers by overcoming the apoptosis resistance of cancer cells to current therapeutic agents. The new class of small-molecule inhibitors was discovered through high-throughput screening (HTS) accompanied by molecular docking and nuclear magnetic resonance (NMR) in order to verify the binding and inhibition activity. The general structure of the pyrazolopyridine is highlighted in Fig. 5. Their most potent and lead compound in their study demonstrated a K_i (MCL-1) = 130 nM, K_i (BCL-2) = 1.95 μM, and K_i (BCL-xL) > 5 μM. Moreover, they confirmed through biological assay that a treatment combination of animals with a therapeutically effective amount of the lead compound of the present invention and a course of an anticancer agent resulted in a greater anticancer activity and clinical benefit in such animals compared to those

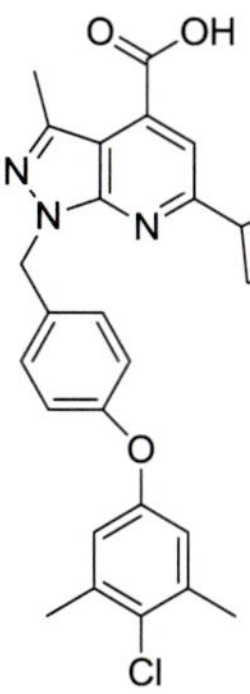

FIG. 5 Chemical structure of lead compound of the new class of MCL-1 inhibitors based on pyrazolopyridine scaffold [91].

treated with the compound or anticancer drugs/radiation alone. Furthermore, the lead compound inhibited the activity of MCL-1 at the mitochondrial level and controlled the growth of the MCL-1-dependent mice transgenic lymphoma cell line Eμ-myc.

Another patent filed by Amgen Inc. [92], Sean P. Brown et al. published the preparation of a class of small-molecule MCL-1 inhibitors based on a tetrahydronaphthalene structure. In their study, they have identified 83 chemical structural compounds and investigated their MCL-1 binding affinities using an in vitro Time-resolved fluorescence energy transfer (TR-FRET) assay. Moreover, they also examined their IC_{50}'s to inhibit MCL-1/BAK protein-protein interactions using a split luciferase assay in human embryonic kidney 293 (HEK293) cells; and IC_{50}'s to induce cytotoxic in human multiple myeloma cells (OPM2) in a 10% FBS environment. Their results revealed that their lead compound demonstrated a nanomolar potency IC_{50} (MCL-1)= 0.25nM in TR-FRET assay as well as IC_{50} (MCL-1)=31.2nM in the luciferase assay, and IC_{50} (OPM2)=259.9nM. Furthermore, they tested the activity of their compounds in vivo using an OPM2 xenograft mouse model. The compounds demonstrated a significant antitumor activity over a period of 10-day treatment. Interestingly, their lead compound exhibited a superior decline in tumor growth at a dose of 25mg/kg compared to 1mg/kg of bortezomib [92].

Venetoclax (chemical structure shown in Fig. 6) is an FDA-approved selective B-cell lymphoma-2 (Bcl-2) inhibitor with a Bcl-2 binding K_i of <0.010nM. It has gained approval

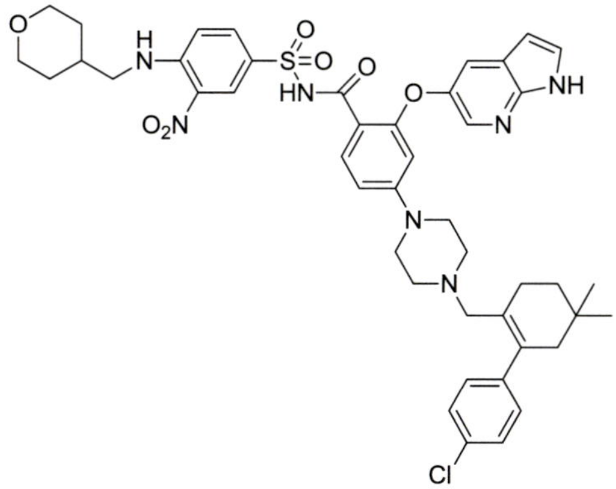

FIG. 6 Chemical structure of Venetoclax, a selective inhibitor of B-cell lymphoma 2 (Bcl-2) [51].

over the years for the treatment of different types of cancer, including chronic lymphocytic leukemia (CLL), small lymphocytic lymphoma (SLL), and acute myeloid leukemia (AML) [93]. A joint US patent filed by Hall Walter and Eliza Institute of Medical Research [94], Genentech Inc. and AbbVie Inc. disclosed the structure of venetoclax and other Bcl-2 inhibitors for the treatment of chronic lymphocytic leukemia, lymphoblastic leukemia, follicular lymphoma, myeloid leukemia, myeloma, breast cancer, and a lymphoid malignancy of T-cell or B-cell origin [94].

Another patent invented by Alan R. Eastman [95] described a synergistic combination of venetoclax and a vascular disrupting agent, named BNC105P, for the treatment of resistant chronic lymphocytic leukemia. This combination treatment comprises a combination of 1 μM of venetoclax and 1 μM of BNC105P and demonstrated 80% apoptosis in vitro as compared to the administration of venetoclax alone (10% apoptosis) or BNC105P alone (60% apoptosis) [95].

Another invention published by Peter Strack and Robert Booher [51] described a synergistic combination of venetoclax and a cyclin-dependent kinase (CDK) inhibitor, dinaciclib for the treatment of small-cell lung cancer (SCLC), and diffuse large B-cell lymphoma (DLBCL). This drug combination was tested in vitro and showed a 64.9% decrease in cell viability compared to single-agent activity after 8-h treatment period. Moreover, this drug combination demonstrated a synergistic effect in SCLC xenografts eliciting a 79% tumor growth inhibition effect as compared to single drugs, which showed no significant effect in the same xenograft model. When tested in DLBCL xenograft models, the drug combination showed enhanced antitumor activity and tumor regression as compared to single agents used alone [51].

Conclusions and perspectives

The inventions presented herein showcase the significant contribution of chemosensitizers in cancer therapy owing to their critical role in inhibiting the activity of drug efflux transporters, counteracting the DNA damage repair defensive mechanisms, and modulating the expression of proteins that trigger apoptosis. Chemosensitizers have the potential to elicit their anticancer activity on their own or they can be used in combination with commercially available chemotherapeutic drugs to induce a synergistic cytotoxic effect and they can be delivered via nanoplatforms specifically targeting cancer cells to provide better tumor accumulation and minimize unwanted toxic effects to normal healthy cells. Novel design strategies and smart delivery technologies are systematically required to boost the therapeutic efficacy of chemosensitizers for the purpose of achieving optimal selectivity toward their therapeutic targets, overcoming various drug resistance mechanisms within cancer cells and preventing tumor recurrence, progression, and metastasis.

References

[1] Siegel RL, Miller KD, Jemal A. Cancer statistics, 2019. CA Cancer J Clin 2019;69:7–34.
[2] Lazebnik Y. What are the hallmarks of cancer? Nat Rev Cancer 2010;10:232–3.
[3] Arruebo M, Vilaboa N, Saez-Gutierrez B, Lambea J, Tres A, Valladares M, Gonzalez-Fernandez A. Assessment of the evolution of cancer treatment therapies. Cancers (Basel) 2011;3:3279–330.

[4] DeVita VT, Chu JE. A history of cancer chemotherapy. Cancer Res 2008;68:8643–53.
[5] El-Awady R, Saleh E, Hashim A, Soliman N, Dallah A, Elrasheed A, Elakraa G. The role of eukaryotic and prokaryotic ABC transporter family in failure of chemotherapy. Front Pharmacol 2016;7:535.
[6] Hasan S, Taha R, Omri HE. Current opinions on chemoresistance: an overview. Bioinformation 2018;14:80–5.
[7] Sauna ZE, Ambudkar SV. Characterization of the catalytic cycle of ATP hydrolysis by human P-glycoprotein. The two ATP hydrolysis events in a single catalytic cycle are kinetically similar but affect different functional outcomes. J Biol Chem 2001;276:11653–61.
[8] Wilcox JE. DNA repair by ERCC1 in non-small-cell lung cancer. N Engl J Med 2006;355:2590.
[9] Thacker J. Homologous recombination repair. In: Schwab M, editor. Encyclopedia of cancer. Berlin, Heidelberg: Springer; 2011. p. 1725–9.
[10] Bridges BA. SOS repair. In: Brenner S, Miller JH, editors. Encyclopedia of genetics. New York: Academic Press; 2001. p. 1853–5.
[11] Gupta SC, Kannappan R, Reuter S, Kim JH, Aggarwal BB. Chemosensitization of tumors by resveratrol. Ann N Y Acad Sci 2011;1215:150–60.
[12] Wcislo G. Resveratrol inhibitory effects against a malignant tumor: a molecular introductory review. In: Watson RR, Preedy VR, Zibadi S, editors. Polyphenols in human health and disease. San Diego: Academic Press; 2014. p. 1269–81 [Chapter 95].
[13] Campbell B, Chan K, Kim JH. Chemosensitization as a means to augment commercial antifungal agents. Front Microbiol 2012;3.
[14] Sharom FJ, Liu R, Qu Q, Romsicki Y. Exploring the structure and function of the P-glycoprotein multidrug transporter using fluorescence spectroscopic tools. Semin Cell Dev Biol 2001;12:257–65.
[15] Choi CH. ABC transporters as multidrug resistance mechanisms and the development of chemosensitizers for their reversal. Cancer Cell Int 2005;5:30.
[16] Nanayakkara AK, Follit CA, Chen G, Williams NS, Vogel PD, Wise JG. Targeted inhibitors of P-glycoprotein increase chemotherapeutic-induced mortality of multidrug resistant tumor cells. Sci Rep 2018;8:967.
[17] Cole SP, Bhardwaj G, Gerlach JH, Mackie JE, Grant CE, Almquist KC, Stewart AJ, Kurz EU, Duncan AM, Deeley RG. Overexpression of a transporter gene in a multidrug-resistant human lung cancer cell line. Science 1992;258:1650–4.
[18] Doyle LA, Yang W, Abruzzo LV, Krogmann T, Gao Y, Rishi AK, Ross DD. A multidrug resistance transporter from human MCF-7 breast cancer cells. Proc Natl Acad Sci U S A 1998;95:15665–70.
[19] Sharom FJ. ABC multidrug transporters: structure, function and role in chemoresistance. Pharmacogenomics 2008;9:105–27.
[20] Shabana AM, Ilies MA. Drug delivery to hypoxic tumors targeting carbonic anhydrase IX. In: Targeted nanosystems for therapeutic applications: new concepts, dynamic properties, efficiency, and toxicity. American Chemical Society; 2019. p. 223–52.
[21] Tsuruo T, Iida H, Nojiri M, Tsukagoshi S, Sakurai Y. Circumvention of vincristine and adriamycin resistance in vitro and in vivo by calcium influx blockers. Cancer Res 1983;43:2905–10.
[22] Tsuruo T, Iida H, Kitatani Y, Yokota K, Tsukagoshi S, Sakurai Y. Effects of quinidine and related compounds on cytotoxicity and cellular accumulation of vincristine and adriamycin in drug resistant tumor cells. Cancer Res 1984;44:4303–7.
[23] Slater LM, Sweet P, Stupecky M, Gupta S. Cyclosporin a reverses vincristine and daunorubicin resistance in acute lymphatic leukemia in vitro. J Clin Invest 1986;77:1405–8.
[24] Gruol DJ, Zee MC, Trotter J, Bourgeois S. Reversal of multidrug resistance by RU 486. Cancer Res 1994;54:3088–91.
[25] Pirker R, FitzGerald DJ, Raschack M, Frank Z, Willingham MC, Pastan I. Enhancement of the activity of immunotoxins by analogues of verapamil. Cancer Res 1989;49:4791–5.
[26] Krishna R, Mayer LD. Multidrug resistance (MDR) in cancer. Mechanisms, reversal using modulators of MDR and the role of MDR modulators in influencing the pharmacokinetics of anticancer drugs. Eur J Pharm Sci 2000;11:265–83.
[27] Thomas H, Coley HM. Overcoming multidrug resistance in cancer: an update on the clinical strategy of inhibiting P-glycoprotein. Cancer Control 2003;10:159–65.
[28] Pusztai L, Wagner P, Ibrahim N, Rivera E, Theriault R, Booser D, Symmans FW, Wong F, Blumenschein G, Fleming DR, Rouzier R, Boniface G, Hortobagyi GN. Phase II study of tariquidar, a selective P-glycoprotein inhibitor, in patients with chemotherapy-resistant, advanced breast carcinoma. Cancer 2005;104:682–91.

[29] Kathawala RJ, Gupta P, Ashby CR, Chen ZS. The modulation of ABC transporter-mediated multidrug resistance in cancer: a review of the past decade. Drug Resist Updat 2015;18:1–17.

[30] LMC C, THC C, Chan KF, ILKW W, Law MC. Alkyne-, azide- and triazole-containing flavonoids as modulators for multidrug resistance in cancer. WO2013127361; 2013.

[31] Brock R, Favretto ME. Novel inhibitors of p-glycoprotein. WO2017013172A1; 2017.

[32] Ma EZ, Feng SL, Yao XJ, Yuan ZW, Liu L, Xie Y. Use of tangeretin in cancer treatment. US9808439B2; 2017.

[33] Paiva GS. Anti-tumor cocktail comprising an anti-tumor agent selective for tumor acidity, an antioxidant, and p-glycoprotein inhibitor. WO2017173509A1; 2017.

[34] Wise JG, Vogel PD, Lippert AR, Nanyakkara AK, Aljowni MA. Inhibitors of multidrug resistance transporter p-glycoprotein. US20180185366A1; 2018.

[35] Quan X, Du H, Xu J, Hou X, Gong X, Wu Y, Zhou Y, Jiang J, Lu L, Yuan S, Yang X, Shi L, Sun L. Novel quinoline compound derivatives of NSC23925 as potent reversal agents against P-glycoprotein-mediated multidrug resistance. Front Chem 2019;7.

[36] Wen C, Fu L, Huang J, Dai Y, Wang B, Xu G, Wu L, Zhou H. Curcumin reverses doxorubicin resistance via inhibition the efflux function of ABCB4 in doxorubicinresistant breast cancer cells. Mol Med Rep 2019;19:5162–8.

[37] Caster JM, Sethi M, Kowalczyk S, Wang E, Tian X, Nabeel Hyder S, Wagner KT, Zhang YA, Kapadia C, Man Au K, Wang AZ. Nanoparticle delivery of chemosensitizers improve chemotherapy efficacy without incurring additional toxicity. Nanoscale 2015;7:2805–11.

[38] Brigger I, Dubernet C, Couvreur P. Nanoparticles in cancer therapy and diagnosis. Adv Drug Deliv Rev 2002;54:631–51.

[39] Allen TM. Ligand-targeted therapeutics in anticancer therapy. Nat Rev Cancer 2002;2:750–63.

[40] Ding J, Xiao C, Li Y, Cheng Y, Wang N, He C, Zhuang X, Zhu X, Chen X. Efficacious hepatoma-targeted nanomedicine self-assembled from galactopeptide and doxorubicin driven by two-stage physical interactions. J Control Release 2013;169:193–203.

[41] Modi P. Oral formulations of chemotherapeutic agents. US20150231069A1; 2015.

[42] The polymer micelle composition of antitumor drug resistance and transfer. CN109303767A; 2019.

[43] Guo S, Lv L, Shen Y, Hu Z, He Q, Chen X. A nanoparticulate pre-chemosensitizer for efficacious chemotherapy of multidrug resistant breast cancer. Sci Rep 2016;6:21459.

[44] Shabana AM, Mondal UK, Alam MR, Spoon T, Ross CA, Madesh M, Supuran CT, Ilies MA. pH-sensitive multiligand gold nanoplatform targeting carbonic anhydrase IX enhances the delivery of doxorubicin to hypoxic tumor spheroids and overcomes the hypoxia-induced chemoresistance. ACS Appl Mater Interfaces 2018;10:17792–808.

[45] Basu A, Krishnamurthy S. Cellular responses to cisplatin-induced DNA damage. J Nucleic Acids 2010.

[46] Thorn CF, Oshiro C, Marsh S, Hernandez-Boussard T, McLeod H, Klein TE, Altman RB. Doxorubicin pathways: pharmacodynamics and adverse effects. Pharmacogenet Genomics 2011;21:440–6.

[47] Sakthivel KM, Hariharan S. Regulatory players of DNA damage repair mechanisms: role in cancer chemoresistance. Biomed Pharmacother 2017;93:1238–45.

[48] Zona S, Bella L, Burton MJ, Nestal de Moraes G, EWF L. FOXM1: an emerging master regulator of DNA damage response and genotoxic agent resistance. Biochim Biophys Acta Gene Regul Mech 2014;1839:1316–22.

[49] Jackson SP, Bartek J. The DNA-damage response in human biology and disease. Nature 2009;461:1071–8.

[50] Chang HY, Chang TC, Huang WY, Lee CT, Yen CJ, Tsai YS, Tzai TS, Chen SH, Chow NH. RON nuclear translocation under hypoxia potentiates chemoresistance to DNA double-strand break–inducing anticancer drugs. Mol Cancer Ther 2016;15:276–86.

[51] Strack P, Booher R. Method for treating cancer. WO2015130585A1; 2015.

[52] Tripathi K, Mani C, Clark DW, Palle K. Rad18 is required for functional interactions between FANCD2, BRCA2, and Rad51 to repair DNA topoisomerase 1-poisons induced lesions and promote fork recovery. Oncotarget 2016;7:12537–53.

[53] Scata KA, El-Deiry WS. p53, BRCA1 and breast cancer chemoresistance. Adv Exp Med Biol 2007;608:70–86.

[54] Takada M, Nagai S, Haruta M, Sugino RP, Tozuka K, Takei H, Ohkubo F, Inoue K, Kurosumi M, Miyazaki M, Sato-Otsubo A, Sato A, Ogawa S, Kaneko Y. BRCA1 alterations with additional defects in DNA damage response genes may confer chemoresistance to BRCA-like breast cancers treated with neoadjuvant chemotherapy. Genes Chromosom Cancer 2017;56:405–20.

[55] Swisher EM, Sakai W, Karlan BY, Wurz K, Urban N, Taniguchi T. Secondary BRCA1 mutations in BRCA1-mutated ovarian carcinomas with platinum resistance. Cancer Res 2008;68:2581–6.

[56] Wiltshire T, Senft J, Wang Y, Konat GW, Wenger SL, Reed E, Wang W. BRCA1 contributes to cell cycle arrest and chemoresistance in response to the anticancer agent irofulven. Mol Pharmacol 2007;71:1051–60.
[57] Haaf A, Bektas N, Serenyi S, Losen I, Arweiler AC, Hartmann A, Knuchel R, Dahl E. Expression of the glioma-associated oncogene homolog (GLI) 1 in human breast cancer is associated with unfavourable overall survival. BMC Cancer 2009;9:298.
[58] Ding YL, Zhou Y, Xiang L, Ji ZP, Luo ZH. Expression of glioma-associated oncogene homolog 1 is associated with invasion and postoperative liver metastasis in colon cancer. Int J Med Sci 2012;9:334–8.
[59] Palle K, Mani C, Tripathi K, Athar M. Aberrant GLI1 activation in DNA damage response, carcinogenesis and chemoresistance. Cancers (Basel) 2015;7:2330–51.
[60] Tripathi K, Mani C, Barnett R, Nalluri S, Bachaboina L, Rocconi RP, Athar M, Owen LB, Palle K. Gli1 protein regulates the S-phase checkpoint in tumor cells via bid protein, and its inhibition sensitizes to DNA topoisomerase 1 inhibitors. J Biol Chem 2014;289:31513–25.
[61] Zhao J. Cancer stem cells and chemoresistance: the smartest survives the raid. Pharmacol Ther 2016;160:145–58.
[62] Gallmeier E, Hermann PC, Mueller MT, Machado JG, Ziesch A, De Toni EN, Palagyi A, Eisen C, Ellwart JW, Rivera J, Rubio-Viqueira B, Hidalgo M, Bunz F, Goke B, Heeschen C. Inhibition of ataxia telangiectasia- and Rad3-related function abrogates the in vitro and in vivo tumorigenicity of human colon cancer cells through depletion of the CD133(+) tumor-initiating cell fraction. Stem Cells 2011;29:418–29.
[63] Mathews LA, Cabarcas SM, Hurt EM, Zhang X, Jaffee EM, Farrar WL. Increased expression of DNA repair genes in invasive human pancreatic cancer cells. Pancreas 2011;40:730–9.
[64] Supuran CT. Carbonic anhydrases: novel therapeutic applications for inhibitors and activators. Nat Rev Drug Discov 2008;7:168–81.
[65] Zamanova S, Shabana AM, Mondal UK, Ilies MA. Carbonic anhydrases as disease markers. Expert Opin Ther Pat 2019;29:509–33.
[66] Tredan O, Galmarini CM, Patel K, Tannock IF. Drug resistance and the solid tumor microenvironment. J Natl Cancer Inst 2007;99:1441–54.
[67] Chang ZF, Fang JM. Targeting human thymidylate kinase induces DNA repair toxicity in malignant tumor cells. US9662328B2; 2017.
[68] Hall D, Weinfeld M, Bernard S, Verdelet T, Morgan T, Lamontagne V, Shire Z, Lavasanifar A. Imidopiperidine compounds as inhibitors of human polynucleotide kinase phosphatase. CA3058927A1; 2018.
[69] Bacha JA, Zhai B, Steino A, Daugaard M. Use of dianhydrogalactitol and derivatives thereof in the treatment of glioblastoma, lung cancer, and ovarian cancer. US20190091195A1; 2019.
[70] Angibaud PR, Querolle OAG, Berthelot DJC, Meyer C, Willot MPV, Meerpoel L. Quinoxaline and pyridopyrazine derivatives as pi3kbeta inhibitors. WO2017060406A1; 2017.
[71] Srivastava S, Athavale M, Shukre K, More G. Compounds for eradicating or inhibiting proliferation of cancer stem cells. WO2015136556A2; 2015.
[72] Kaufmann SH, Earnshaw WC. Induction of apoptosis by cancer chemotherapy. Exp Cell Res 2000;256:42–9.
[73] Ricci MS, Zong WX. Chemotherapeutic approaches for targeting cell death pathways. Oncologist 2006; 11:342–57.
[74] Hickman JA. Apoptosis induced by anticancer drugs. Cancer Metastasis Rev 1992;11:121–39.
[75] Simstein R, Burow M, Parker A, Weldon C, Beckman B. Apoptosis, chemoresistance, and breast cancer: insights from the MCF-7 cell model system. Exp Biol Med (Maywood) 2003;228:995–1003.
[76] Soengas MS, Lowe SW. Apoptosis and melanoma chemoresistance. Oncogene 2003;22:3138–51.
[77] Fridman JS, Lowe SW. Control of apoptosis by p53. Oncogene 2003;22:9030–40.
[78] Amaral JD, Xavier JM, Steer CJ, Rodrigues CM. The role of p53 in apoptosis. Discov Med 2010;9:145–52.
[79] Hientz K, Mohr A, Bhakta-Guha D, Efferth T. The role of p53 in cancer drug resistance and targeted chemotherapy. Oncotarget 2017;8:8921–46.
[80] Knappskog S, Lonning PE. P53 and its molecular basis to chemoresistance in breast cancer. Expert Opin Ther Targets 2012;16:23–30.
[81] Campbell KJ, Tait SWG. Targeting BCL-2 regulated apoptosis in cancer. Open Biol 2018;8.
[82] Delbridge AR, Strasser A. The BCL-2 protein family, BH3-mimetics and cancer therapy. Cell Death Differ 2015;22:1071–80.
[83] Hardwick JM, Soane L. Multiple functions of BCL-2 family proteins. Cold Spring Harb Perspect Biol 2013;5.
[84] Nachmias B, Ashhab Y, Ben-Yehuda D. The inhibitor of apoptosis protein family (IAPs): an emerging therapeutic target in cancer. Semin Cancer Biol 2004;14:231–43.

[85] Silke J, Meier P. Inhibitor of apoptosis (IAP) proteins-modulators of cell death and inflammation. Cold Spring Harb Perspect Biol 2013;5.
[86] Berger AB, Witte MD, Denault JB, Sadaghiani AM, Sexton KM, Salvesen GS, Bogyo M. Identification of early intermediates of caspase activation using selective inhibitors and activity-based probes. Mol Cell 2006;23:509–21.
[87] Deveraux QL, Takahashi R, Salvesen GS, Reed JC. X-linked IAP is a direct inhibitor of cell-death proteases. Nature 1997;388:300–4.
[88] Ikeda F, Deribe YL, Skanland SS, Stieglitz B, Grabbe C, Franz-Wachtel M, van Wijk SJ, Goswami P, Nagy V, Terzic J, Tokunaga F, Androulidaki A, Nakagawa T, Pasparakis M, Iwai K, Sundberg JP, Schaefer L, Rittinger K, Macek B, Dikic I. SHARPIN forms a linear ubiquitin ligase complex regulating NF-kappaB activity and apoptosis. Nature 2011;471:637–41.
[89] Dynek JN, Vucic D. Antagonists of IAP proteins as cancer therapeutics. Cancer Lett 2013;332:206–14.
[90] Shabbits JA, Hu Y, Mayer LD. Tumor chemosensitization strategies based on apoptosis manipulations. Mol Cancer Ther 2003;2:805–13.
[91] Coleska ZN, Abulwerdi F, Showalter H, Miao L, Stuckey J, Mady A. Small molecule inhibitors of mcl-1 and uses thereof. WO2015153959A2; 2015.
[92] Brown SP, Li Y, Lizarzaburu ME, Luca BS, Paras NA, Taygerly J, Vimolratana M, Wang X, Yu M, Zancanella M, Zhu L, Buenrostro AG, Li Z. Tetrahydronaphthalene derivatives that inhibit mcl-1 protein. WO2016033486A1; 2016.
[93] Zigart N, Casar Z. A literature review of the patent publications on venetoclax - a selective Bcl-2 inhibitor: discovering the therapeutic potential of a novel chemotherapeutic agent. Expert Opin Ther Pat 2019;29:487–96.
[94] Bruncko M, Ding H, Doherty GA, Elmore SW, Hasvold LA, Hexamer L, Kunzer AR, Song X, Souers AJ, Sullivan GM, Tao ZF, Wang GT, Wang L, Wang X, Wendt MD, Mantei RA, Hansen TM. Apoptosis-inducing agents for the treatment of cancer and immune and autoimmune diseases. US9174982B2; 2015.
[95] Eastman AR, Bates D. Method for treating chronic lymphocytic leukemia. WO2015149105A1; 2015.

CHAPTER

10

Natural compounds as chemosensitizers: A lesson from plants

Liora Colobatiu, Laura Gavrilas, and Andrei Mocan

Iuliu Hatieganu University of Medicine and Pharmacy, Cluj-Napoca, Romania

Abstract

Cancer is one of the most devastating and haunting diseases, affecting the lives of millions around the globe. Even though chemotherapy represents the upfront method of disease management, most cancer patients treated with chemotherapeutics will experience treatment failure, with metastatic cancer at some point in time, due to chemoresistance - a significant hurdle to the successful treatment of various types of cancer. In this particular context, different strategies, such as the combinatorial therapy (the combination of conventional anticancer drugs and natural products), have emerged as a new and interesting mean to overcome chemoresistance. A wide variety of natural compounds such as curcumin, quercetin, lupeol, ursolic acid, resveratrol, epigallocatechin-3- gallate, α-carotene, β-carotene, silymarin, noscapine, or naringenin have been tested in cancer therapy, being used either as chemotherapeutic agents, either as chemosensitizers, in order to improve the effectiveness of conventional therapy. Curcumin proved to efficiently modulate multiple molecular pathways linked with MDR, posing a synergistic effect when combined with standard chemotherapeutic drugs. Resveratrol could act as a chemosensitizer on 5-fluorouracil-resistant cells, also enhancing apoptosis induced by conventional chemotherapeutic agents, while quercetin was able to attenuate the toxic side effects of doxorubicin in several mammary cells, in addition to increasing the intracellular accumulation of the drug. Moreover, cotreatment with quercetin and either doxorubicin, paclitaxel, or vincristine was found to be successful in eliminating breast cancer stem cells, improving the effectiveness of the chemotherapeutic regimens used in breast cancer treatment.

Even though natural compounds may represent attractive and promising candidates to be used as chemosensitizers, there is just a limited number of clinical trials that have been conducted so far regarding this matter.

Abbreviations

ABC	ATP-binding cassette
ABCB1/MDR1	ATP-binding cassette subfamily B member 1
ABCC1/C2	ATP-binding cassette subfamily C member 1 or 2
ABCG2	ATP-binding cassette subfamily G member 2
ATP	adenosine triphosphate

https://doi.org/10.1016/B978-0-12-820701-7.00009-9

Bcl-2	B-cell lymphoma 2
Bcl-XL	B-cell lymphoma-extra large
BCRP	breast cancer resistance protein
Bmi1	polycomb complex protein BMI-1
Cr/BUN	creatinine/blood urea nitrogen ratio
CRC	colorectal cancer
CYP	cytochrome P450
CYP1A1	cytochrome P450 family 1 subfamily A member 1
DNA	deoxyribonucleic acid
EGCG	epigallocatechin-3-gallate
Ezh2	enhancer of zeste 2 polycomb repressive complex 2 subunit
GOT	glutamic oxaloacetic transaminase ratio
GPT/GOT	glutamic pyruvic transaminase/glutamic oxaloacetic transaminase
GST	glutathione-S-transferase
HER-2	human epidermal growth factor receptor 2
HSP-27	heat shock protein 27
LC3B-II	microtubule-associated protein 1 light-chain 3 beta
MDR	multidrug resistance
miRNA/miR	microribonucleic acid
mRNA	messenger ribonucleic acid
MRP1	multidrug resistance-associated protein 1
NF-kB	nuclear factor kappa B
NSCLC	non-small-cell lung cancer
P53	tumor protein P53
P-gp	p-glycoprotein
Suz12	SUZ12 polycomb repressive complex 2 subunit
TGF-β	transforming growth factor beta
TNF-β	tumor necrosis factor beta
UGT	uridine diphospho-glucuronosyl transferase
Wnt	wingless-related integration site

Conflict of interest

No potential conflicts of interest were disclosed.

Introduction

Cancer is a disease that develops overtime, through the accumulation of mutations and genetic changes in a cell [1]. Several characteristics distinguish cancerous cells from the normal ones, including a sustained proliferative ability, induction of angiogenesis, evasion of growth suppressors, resistance to cell death, replicative immortality, the ability to invade and metastasize, evading immune attack, genomic instability, inflammation, and continuous production of energy, all these being collectively designated as the "hallmarks of cancer" [2].

Cancer is one of the most devastating and haunting diseases, affecting the lives of millions around the globe [3]. Despite the immense efforts involved in cancer research, prevention, detection, and treatment, the number of cancer cases has continued to increase [1, 4].

According to the World Health Organization and the International Agency for Cancer Research, there were 18.1 million new cancer cases and 9.6 million cancer deaths in 2018 [5]. In both sexes combined, lung cancer was the most commonly diagnosed cancer (11.6% of the total cases), as well as the leading cause of cancer death, closely followed by female breast cancer, prostate cancer, and colorectal cancer in terms of incidence. Moreover, malignant neoplasms are ranked the second leading cause of deaths worldwide after cardiovascular diseases [6]. Unfortunately, these worrying estimates are expected to heavily increase in the following years, which constitutes a ringing alarm [5].

Conventional strategies for cancer treatment involve surgery, radiotherapy, immunotherapy, and chemotherapy, depending upon the cancer type, as well as its clinical stage, chemotherapy being the upfront method of disease management in clinically advanced solid malignancies and hematopoietic cancers [1].

Chemotherapy represents one of the main alternatives for cancer treatment, employing a wide group of drugs, including specific molecules capable of inhibiting proliferative signaling pathways, replicative immortality mechanisms, and angiogenesis, as well as inducing the apoptosis of tumor cells [7].

Based on their origins or mechanisms of action, chemotherapeutic agents are broadly classified into different groups such as alkylating agents, cell-cycle inhibitors, antimetabolites, microtubule-damaging agents, antibiotics, targeted antibodies or topoisomerase and small molecule inhibitors, the ultimate goal of chemotherapy being the elimination of tumor cells [1].

Chemoresistance: A therapeutic challenge

Currently, many chemotherapeutic agents used in clinical practice are having a high prevalence of side effects. Moreover, even though cancer cells are initially susceptible to chemotherapy, it is believed that over 90% of the cancer patients treated with chemotherapeutics will experience treatment failure, with metastatic cancer at some point in time [8]. Risk of tumors acquiring resistance to chemotherapy (chemoresistance) represents nowadays a significant hurdle to the successful treatment of various types of cancer [4].

Different possible molecular mechanisms by which a cancer cell develops resistance to a certain drug have been extensively discussed. It is generally believed that resistance in cancer cells could be either "intrinsic," or "acquired" during the course of the treatment [1, 4, 6, 9, 10].

Intrinsic chemoresistance mostly refers to the cells intrinsic ability to inactivate the drug, alter its targets or repair DNA damage, while acquired resistance generally develops after the start of the chemotherapy, over time, or after a second chemotherapy, with tumor relapse [6]. Therefore, cancer cells that initially show sensitivity to the drug, may acquire resistance to that particular drug over time [1].

A specific form of cellular drug resistance in cancer is termed multidrug resistance. MDR is a phenomenon by which cancer cells develop broad resistance to a wide range of structurally and pharmacologically unrelated compounds, ultimately leading to cancer relapse and death [1, 6]. Therefore, cancer cells may not only become resistant to the drug used to treat them, but they may also show cross resistance to other drugs. Due to MDR, cancer cells become nonresponsive to treatment, this ultimately leading to the failure of chemotherapy in 90% of metastatic cancers [11, 12]. Unsuccessful treatments have been attributed to increased rates of drug efflux, enhanced DNA repair systems, drug inactivation, altered drug targets, inhibition of apoptosis, epithelial-mesenchymal transition, autophagy or activation of self-renewal pathways, among other factors [1].

The main chemoresistance mechanism in cancer is represented by the overexpression of multidrug transporters belonging to the ABC family of transporter proteins that effectively efflux diverse chemotherapeutic agents outside the cancer cells, therefore decreasing the intracellular concentration of the drug [6, 13, 14]. Three efflux transporters have been investigated in more detail regarding their role in MDR in cancer: ABCB1 (also known as P-glycoprotein, P-gp, or MDR1), ABCC1 (MDR-associated protein 1 or MRP1), and ABCG2 (also termed breast cancer resistance protein BCRP or mitoxantrone resistance protein-MXR) [6].

The intrinsic ability of a cancer cell to repair DNA damage caused by chemotherapeutics determines its resistance to that particular drug. In such cases, the mutation of p53 is considered to be one of the most important factors associated with intrinsic resistance in most cancers [15]. Moreover, it has been proven that cancer cells are capable of easily reversing the DNA damage caused by drugs, due to a specific enhanced DNA-damage response [1].

Cytochrome P450 family of isozymes, glutathione-*S*-transferase (GST), and uridine diphosphoglucuronosyl transferase (UGT) represent important constituents capable of inactivating chemotherapeutics. Mutations in the gene encoding CYP may change the metabolic capacity of these isozymes, breaking down the drugs [16]. Increased activity of GST leads to the detoxification of many alkylating agents and platinum drugs used in chemotherapy, while epigenetic changes increasing the expression of UGTA1 have been known to provide resistance to irinotecan [17].

Due to the fact that the effectiveness of chemotherapeutics largely depends on their interaction with their corresponding targets, mutations in the target gene, or the high expression level of its target proteins, can also lead to chemoresistance to that particular drug [1].

The epithelial-mesenchymal transition is a phenomenon in which epithelial cells switch from the epithelial to the mesenchymal phenotype, playing a crucial role in embryonic development and in the differentiation of multiple tissues and organs [18]. Different studies have proven that EMT provides a selective growth advantage to cancer cells in the presence of a chemotherapeutic, being associated with resistance against numerous drugs such as oxaliplatin, cisplatin, paclitaxel of 5-fluorouracil [18–20].

A cancer cell is also able to adapt to the stress induced by the presence of a chemotherapeutic agent in the tumor microenvironment, by using the process of autophagy [1]. Several drugs used in the treatment of cancer such as epirubicin, irinotecan, oxaliplatin, or bevacizumab have been known to induce autophagy in order to protect the cancer cells from apoptosis, therefore facilitating drug chemoresistance [21–23].

Even though nowadays there are many different chemotherapeutic agents available for the treatment of cancer, unfortunately, cancer cells have a significant ability to adapt, survive, and become resistant to such therapies, therefore leading to treatment failure. In this particular context, different strategies, such as the combinatorial therapy (the combination of conventional anticancer drugs and natural products), have emerged as a new and interesting mean to overcome chemoresistance in cancer.

Combined approach in cancer treatment: Natural compounds as chemosensitizers

Even though there is a wide variety of conventional anticancer drugs with multiple mechanisms of action, most of them ultimately induce cell death by necrosis, apoptosis, or autophagy by stimulating genotoxic stress [24]. Although the chemotherapeutic agents currently in use mostly target rapidly dividing cells, normal dividing cells are not spared, thus generating a multitude of side effects [25]. These side effects can be short termed (such as nausea, myelosuppression, or alopecia), which will usually resolve within a few months after the completion of the therapy, but others, including infertility, weight gain, cardiac dysfunction, or secondary leukemia, are long term and serious side effects, which may be associated to chemotherapy [25].

As previously mentioned, the development of chemoresistance and especially of MDR, is another significant hindrance in the success of chemotherapy. In order to reverse this effect, important research on molecules that may improve the therapeutic index of antitumoral drugs by making tumor cells more sensitive to chemotherapeutic agents, namely, *chemosensitizers* has been conducted throughout the years. Different chemical compounds have been tested in a number of clinical trials for their MDR modulatory activity [26–28]. All these were classified as first- (verapamil, cyclosporine, quinine, quinidine, reserpine, tamoxifen, and toremipen), second- (dexverapamil, valspodar, biricodar), and third-generation modulators (elacridar, laniquidar, zosuquidar, dofequidar, mitotane, annamycin, tariquidar) [6, 26, 27, 29–33]. First-generation modulators showed major toxicities in clinical trials and did not prove to be effective against tested cancers like lung, ovarian, or multiple myeloma [4, 26]. The majority of the second-generation modulators are considered as analogues of the first generation. Even though they showed improved toxicity, potency, and efficacy compared to first-generation drugs, most of these compounds interact with cytochrome P450 isozymes, decreasing the clearance of other drugs, therefore leading to toxicity and treatment failure [4]. Third-generation modulators are considered to be more specific and potent, with diminutive pharmacokinetic interactions and some of them have initially showed noteworthy success in preclinical and some clinical trials, increasing the sensitivity of anticancer drugs [4, 34]. However, unfortunately, most of them underperformed in clinical trials or showed high toxicity [35]. Due to the lack of nontoxic and specific chemosensitizing MDR modulators, scientists are currently exploring new molecules, such as natural compounds, in order to tackle the important problem of MDR in cancer.

From an ancient time, natural products have been used as healing agents in a variety of ailments, as most plants, fruits, and vegetables are known to contain various bioactive compounds, having numerous therapeutic uses [10]. Natural products are an inexhaustible source of new molecules, with unique structural models and innovative mechanisms of action [7]. Due to these inherent qualities, natural products have always been considered as important sources of new drugs. For example, in the area of cancer, between the 1940s to the end of 2014, 75% of the small molecules approved (131 out of 175) were other than synthetic, with 85 (49%) being actually natural products or directly derived therefrom [36].

The search for novel and more effective anticancer agents has led to the discovery of a series of plant-derived molecules, which are able to target multiple pathways that contribute to carcinogenesis [37]. Most of these phytochemicals or their botanical sources are either consumed routinely as food items, either used as traditional remedies in the treatment of various diseases [1].

A wide variety of natural compounds have been reported for cancer therapy, as such compounds can be used in a versatile manner in cancer management [38, 39]. Firstly, they could be used as chemotherapeutic agents [39]. Secondly, they proved to be efficient in cancer prevention, and could therefore be utilized as chemopreventive agents and, last but not least, natural compounds might be used as chemosensitizing agents, in order to improve the effectiveness of conventional chemotherapy [1, 4, 6, 7, 10, 40].

Due to their high biodiversity, good oral bioavailability and relatively low intrinsic toxicity, natural products (especially edible phytochemicals-nutraceuticals) are attractive partners to be used in combination with chemotherapy in order to enhance the cytotoxic effects of chemotherapeutic agents and reverse MDR [6]. According to some authors, natural products are considered to represent the fourth generation of MDR chemosensitizers [6].

In general, natural compounds used as chemosensitizers act by increasing the resistance time of chemotherapeutics in tumor cells, inducing cell death by upregulation of proapoptotic targets, by promoting DNA damage or by regulating the expression of altered and unaltered drug targets [7]. When associated with conventional chemotherapeutics, plant-based chemosensitizers enhance the cytotoxic effect of the anticancer drugs, promoting a synergistic effect even in cells with acquired resistance [41–43].

Nutraceuticals represent a specific segment of natural compounds, as these are consumed as a part of a normal diet and are considered to be pharmacologically safe [1, 6]. Furthermore, some nutraceuticals were found to target cancerous cells through the regulation of multiple cell signaling pathways and mediators that participate in the manifestation of chemoresistance, or of those which are involved in inflammation or oncogenic transformations such as apoptosis, autophagy, or metastasis [44–49].

However, some challenges regarding the use of nutraceuticals in cancer therapy have been discussed. For example, most nutraceuticals are detectable in the circulation only for a short time, generally from a few minutes to a few hours, this situation being less ideal from the conventional pharmacology point of view [1]. Furthermore, it has been reported that many polyphenols that are possible candidates as anticancer agents are rapidly degraded and metabolized in the human body [1]. In order to overcome the limited bioavailability of these compounds, novel formulations have been developed to ensure a better and sustained systemic release. Novel analogues of nutraceuticals have also been reported, as another approach to increase their efficacy [50, 51].

To date, several classes of chemicals of natural origin have been tested as chemosensitizers, the combination with conventional anticancer drugs emerging as a new mean to overcome chemoresistance in many cancers (Table 1). Among these, curcumin, resveratrol, quercetin, and epigallocatechin-3-gallate are the most cited.

Curcumin

Curcumin is a yellow-pigmented polyphenolic compound isolated from the powdered rhizomes of turmeric (*Curcuma longa*) used for centuries as a traditional medicine throughout Asia. Studies conducted over the past decade have revealed unequivocally the prophylactic or curative value of curcumin based on its anti-inflammatory, antioxidant, anticarcinogenic, hepatoprotective, cardioprotective, and antimicrobial properties [94–96]. Curcumin gained significant interest in cancer chemotherapy as a safe and effective sensitizer for chemoresistant cells counteracting the effects of MDR [55, 56].

TABLE 1 Natural compounds as chemosensitizers.

Chemosensitizer	Associated chemotherapeutic	Source	Reference
Curcumin	Paclitaxel, capecitabine, gemcitabine, 5-fluorouracil, vinblastine, vincristine, cisplatin	*Curcuma longa* (turmeric)	[52–60]
Quercetin	Doxorubicin, glutamic acid	*Ginkgo biloba*	[61–66]
Resveratrol	Gemcitabine, 5-fluorouracil	*Polygonum cuspidatum*, grape skins, red wine	[45, 46, 67–75]
Epigallocatechin-3-gallate	Taxane, capecitabine	Green tea	[40, 76–79]
Ginsenoside Rg3	Paclitaxel	*Panax ginseng*	[80, 81]
Lupeol	5-Fluorouracil	Fruits and vegetables (white cabbage, pepper, cucumber, tomato, mango, strawberry, red grapes etc.)	[82, 83]
Ursolic acid	Gemcitabine, capecitabine, cisplatin, taxol, chloroquine	Plants, herbs, fruits (sage, thyme, lavender, rosemary, peppermint, elder flower, basil, apple, prunes, cranberries etc.)	[10]
α-carotene	Taxol	Orange vegetables (pumpkin, carrots, tangerines)	[84]
β-carotene	5-Fluorouracil	Sweet potatoes, carrots, lettuce, apricots, broccoli, peas	[85]
Noscapine	Gemcitabine, temozolamide, cisplatin	Papaveraceae family	[86, 87]
Naringenin	Doxorubicin, felodipine	*Citrus* fruits	[88, 89]
Icaritin	Doxorubicin	*Herba epimedii*	[90, 91]
Silymarin	Vinblastine, digoxin	Milk thistle	[4]
Piperine	Doxorubicin, mitoxantrone	*Piper nigrum*	[92, 93]

Curcumin can efficiently modulate multiple molecular pathways linked with MDR posing a synergistic effect when combined with standard chemotherapeutic drugs, including 5-fluorouracil, oxaliplatin, irinotecan, doxorubicin, paclitaxel, vincristine, cisplatin, or celecoxib in several cancer cell lines such as colon, breast, bladder, gastric, head and neck, liver, lung, pancreas [55, 58, 97–99].

One of the most studied mechanisms associated with acquired resistance refers to the efflux of the chemotherapeutic agents outside the cells as a result of the overexpression in the ABC membrane transporters. Data from both in vitro and in vivo studies presented the potential of curcumin to inhibit the function of the drug transporters and therefore increase the cellular drug accumulation and reverse MDR [59].

Several studies in the scientific literature investigated the potential of curcumin to sensitize breast cancer cells in vitro and in vivo, contributing to reduced drug toxicity and resistance. In MDA-MB-231 triple-negative breast cancer cell line, a combination of 5-fluorouracil (10 μM) and curcumin (10 μM) had a synergistic cytotoxic and apoptotic potential, along with a chemosensitizing effect of cancer cells, by downregulating NF-kB through a thymidylate-synthase dependent mechanism [98]. In another study conducted by Zhou et al. [100], curcumin (40 μM) significantly sensitized MCF-7 and MDA-MB-231 breast cancer cells to mitomycin C, improving the tumoricidal and apoptotic effect of the drug, showing that the combination therapy using mitomycin C and curcumin resulted in the reduced dosage of the conventional drug to achieve its IC_{50} toxicity. Furthermore, the authors addressed in vivo, using MCF-7 xenografted mice, the well-documented side effects of mitomycin C and investigated whether curcumin can lower the nephrotoxicity of mitomycin by analyzing the level of Cr/BUN and GPT/GOT in mice treated with the drug alone or together with curcumin. They reported that the combined treatment is efficient in decreasing the level of toxicity of mitomycin, by having a protective effect of the renal function [100]. In addition, a recent publication proposed a novel concept that might overcome the limitations related to curcumin bioavailability by using a polycurcumin nanoparticle acting simultaneously as a chemosensitizer, as well as a drug carrier for doxorubicin in the multidrug-resistant MCF-7/ADR xenografted nude mice. This approach was able to reverse doxorubicin resistance by suppressing P-gp expression and ATP-activity, therefore improving the cellular uptake of the drug, as well as minimizing the drug efflux in breast-cancer-resistant cells [101].

Curcumin has been extensively investigated as a supportive agent in the chemotherapy of colorectal cancer, with successful results as a chemosensitizer. In combinatorial studies with different agents (5-fluorouracil, oxaliplatin, vincristine) or regimens (FOLFOX - a chemotherapy regimen for treatment of colorectal cancer, made up of the drugs Folinic acid (leucovorin) "FOL," Fluorouracil (5-FU) "F," and Oxaliplatin "OX"), curcumin enhanced the antitumor effect in vivo and in vitro [55, 60, 102, 103]. Lu et al. concluded that at concentrations above 25 μM curcumin could enhance the sensitivity of vincristine-resistant CRC cells by downregulating P-gp, a well-known MDR-gene from the ABC family [58]. Furthermore, a curcuminoid-rich extract suppressed the expression of P-gp, confirming its potential to decrease the efflux ratio of daunorubicin in drug-resistant Caco-2 cells [57]. In studies conducted by He et al. and Norrato et al., curcumin was also proven effective to reverse the resistance of HCT-8/5-FU and SW-480/5-FU colon cancer cell lines by downregulating MDR1, P-gp and HSP-27, respectively [97, 104].

Furthermore, several studies confirmed that curcumin could also reverse oxaliplatin resistance in human colorectal cancer by multiple mechanisms. The study conducted by Yin et al. indicated that the combination treatment could inhibit the expression of antiapoptotic Bcl-2 while increasing active-caspase 3 [99]. Likewise, curcumin suppressed EMT, which is a critical event on the onset of drug resistance, via regulation of TGF-β/Smad2/3 signaling pathway. The same study confirmed the effectiveness of the combined treatment as well, in vivo [99]. Porras et al. also combined curcumin and oxaliplatin in order to overcome oxaliplatin-acquired resistance in CRC [103]. As expected, they reported that the combination of the two agents had a synergistic effect in several resistant cell lines, resulting in the reversal of their resistant phenotype. This fact was reported to be mediated by the inhibition of the NF-kB signaling cascade. Although curcumin is the most frequently used natural compound in clinical

TABLE 2 Clinical trials based on the combined approach using natural compounds and standard chemotherapeutics in cancer.

Natural compound	Type of cancer	Associated drug	Number of patients	Clinical outcome	Reference
Curcumin	Colorectal cancer	FOLFOX	28	Safe and tolerable adjunct to FOLFOX chemotherapy	[102]
	Breast cancer	Docetaxel	14	Antitumor activity Safe, feasible, and tolerable at 6 g/day	[105]
	Pancreatic cancer	Gemcitabine	17	Low compliance at high dose of 8 g/day	[106]
	Pancreatic cancer	Gemcitabine	21	Safe and feasible at 8 g/day	[107]
	Prostate cancer	Docetaxel Prednisone	30	Good tolerability and patient acceptability at 6 g/day	[108]
Resveratrol	Multiple myeloma	Bortezomib	24	Side effects (nausea and vomiting) leading to renal failure at 5 g/day resveratrol (a micronized oral formulation)	[109]

trials (65 trials registered on *www.Clinicaltrials.gov*), only a few are designed to combine the agent with the conventional treatment (Table 2). The clinical trials, which reported results so far, concluded that curcumin has promising outcomes and is safe and tolerable in multiple malignancies [102, 106–108]. However, even in these circumstances, more clinical trials need to be performed in order to include curcumin in standard cancer care.

Collectively, these approaches, based on combining curcumin with conventional chemotherapeutic agents in order to overcome multidrug resistance, suggest its efficacy unanimously and agree with the fact that curcumin serves as an efficient chemosensitizer in multiple malignancies.

Resveratrol

Resveratrol (3,5,40-trihydroxy-*trans*-stilbene) is a polyphenolic compound being present in more than 70 plant species with the highest concentrations reported in grape skin (50–100 μg/g), red wine (0.98–1.80 μg/g), and peanuts (1.8–7.1 μg/g) [67, 110]. As a dietary component, resveratrol possesses a broad spectrum of biological activities, such as anti-inflammatory, antioxidant, antiviral, antifungal, and antiaging effects [96, 110, 111]. In addition to the properties mentioned above, resveratrol has been extensively studied as an anticarcinogenic agent and lately for its chemosensitizing potential [42, 111]. Resveratrol, when combined with standard chemotherapeutic agents, exerts a sensitizing effect by acting on one or more pathways associated with chemoresistance.

One of the hallmarks of drug resistance is the epithelial-to-mesenchymal transformation. The process of EMT involves molecular reprogramming of the cells that include loss of cell adhesion molecules such as E-cadherin, as well as overexpression of vimentin, N-cadherin, osteopontin, Snail, slug, and other interstitial proteins. These events change the epithelial cell morphology, thus, facilitating tumor cells to obtain higher potential with respect to migration, invasion, antiapoptotic potential, and the degradation of the extracellular matrix [68]. The research group coordinated by Buhrmann conducted extensive studies investigating the mechanisms associated with resveratrol chemosensitizing potential in CRC, especially those linked with EMT [69]. They showed that resveratrol could act as chemosensitizer on 5-fluorouracil-resistant cells by suppressing NF-kB activation and its downstream targets, as well as by inhibiting the process of EMT transformation. Cotreatment with resveratrol and 5-fluorouracil was the most effective approach in downregulating vimentin and slug while upregulating E-cadherin [70]. According to the same authors, resveratrol is an active modulator of the TNF-β signaling pathway in 5-fluorouracil-resistant cells. Furthermore, it can decrease the formation of cancer stem cells and can sensitize cells to 5-fluorouracil in a tumor microenvironment [70, 71].

Acquired resistance represents a significant obstacle to effective therapy of non-small-cell lung cancer. Zhao et al. studied the effect of resveratrol on multidrug-resistant human NSCLC cells and reported that resveratrol significantly decreased cell proliferation in a dose- and time-dependent manner and enhanced apoptosis induced by conventional chemotherapeutic agents such as cisplatin, gefitinib, and paclitaxel by modulating the expression of survivin [72]. Similarly, in another study, Zhu et al. demonstrated that resveratrol synergizes with gefitinib, enhancing the drug sensitivity in several NSCLC-resistant cell lines. The combined treatment was also effective in improving apoptosis, autophagy, cell cycle arrest, and senescence. The underlying mechanisms were mediated by increased expression of cleaved caspase-3, LC3B-II, p53, and p21. Likewise, the cotreatment had increased the cellular accumulation of the drug by upregulating CYP1A1 and ABCG2 [73].

Melphalan, which is a chemotherapeutic agent used in breast cancer therapy, when combined with resveratrol for the treatment of MCF-7 and MDA-MB-231 breast cancer cells, led to a significant decrease in cell viability compared to treatments with either agent alone. Interestingly, the timing of the treatment influenced the extent of the cytotoxic effect, as evidenced by the results obtained when treatment with resveratrol followed melphalan, rather than the other way around. The proposed mechanism of action involved in chemosensitization by resveratrol was associated with increased apoptosis and cell cycle arrest in the S-phase [74]. Similar outcomes were observed by Vinod et al. who demonstrated that pretreatment with resveratrol (24h before drug treatment) followed by docetaxel, blocked HER-2 expression and activation while blocking its downstream signaling pathways, thus highlighting for the first time that resveratrol possesses chemosensitizing potential in HER-2-induced breast cancer resistance [75].

Despite the abundant preclinical evidence showing that resveratrol can prevent tumorigenesis and is useful in combination with standard chemotherapy, there are just a few clinical trials that have been conducted so far. There are 15 clinical trials registered on *www.Clinicatrials.gov*, with only one using a combined approach with a chemotherapeutic drug (Table 2). In contrast to in vitro studies, the above-mentioned clinical trial showed that the combination of resveratrol and bortezomib could lead to severe side effects in patients with relapsed or refractory multiple myeloma, contributing to renal failure [109]. However, this side effect has not been reported in other clinical trials using resveratrol alone in prostate, colorectal, or breast cancer, supporting the idea that the associated nephrotoxicity may have specificity to multiple myeloma patients [112–115].

Epigallocatechin-3-gallate

EGCG is an essential polyphenolic compound in green tea known for a plethora of physiological and pharmacological effects [116]. In oncologic research, it has been shown that EGCG can trigger apoptosis and inhibit both proliferation and migration by modulating multiple cancer-related molecular mechanisms [40, 117]. Also, it has been proposed that EGCG serves as a promising chemosensitizer for conventional cancer treatment in several malignancies [78, 79, 118].

Liang et al. showed that EGCG could sensitize drug-resistant hepatocellular carcinoma cells to doxorubicin by downregulation of MDR1 mRNA with further inhibition of its gene product, suggesting that the combined treatment with the two agents results in an increased intracellular accumulation of doxorubicin due to decreased drug efflux [119]. Besides, their research indicated that the antitumorigenic effect of the combinatorial approach involves autophagy inhibition [118].

The potential of the combined treatment, with EGCG and anticancer drug 5-FU to inhibit cell growth, was assessed in DLD-1, SW480, and COLO201 colorectal cancer cells, with synergistic effects, as reported by Kumazaki et al. [120]. Moreover, in a recent study, EGCG was shown to suppress the formation of cancer stem cells and subsequently sensitize those cells to 5-fluorouracil. The underlying molecular mechanism responsible for this response was associated with the downregulation of Notch1, Bmi1, Suz12, and Ezh2, as well as, with upregulation of key miRNAs such as miR24a, miR145, and miR200c, known to be involved in the inhibition of the self-renewal process [78].

Cotreatment with EGCG, together with a chemotherapeutic agent, has also been proven effective in hormone-sensitive types of cancer. Sun et al. investigated in a recent publication whether EGCG could sensitize MCF7 breast cancer cells to 5-fluorouracil. Their results indicated that EGCG could enhance 5-fluorouracil antitumorigenic activity by regulating the expression of antiapoptotic Bcl-XL [121]. Moreover, the cotreatment with EGCG and cisplatin, as well as with three *trans*-palladiums (TH5, TH6, and TH7), resulted in synergistic inhibition of cell viability for both sensitive and resistant A2780 human ovarian cancer cells. Also, the combination approach has been effective in overcoming drug resistance by increasing intracellular accumulation of platinum and enhancing platinum-DNA binding [122]. Taken together, these results suggest that the combination therapy using EGCG with a conventional chemotherapeutic agent may be a promising therapeutic scheme with improved efficacy in multiple malignancies.

Quercetin

Quercetin is a widespread dietary flavonoid found in many fruits, vegetables, and leaves, such as apples, onions, and green tea. Its use has been associated with multiple beneficial properties, including inhibiting inflammatory response, as well as antiproliferative, proapoptotic, and chemosensitizing activities in several types of cancers [63, 123, 124].

Li et al. demonstrated that quercetin could attenuate the toxic side effects of doxorubicin in several mammary cells, in addition to increasing the intracellular accumulation of the drug by downregulating the level of expression of efflux ABC membrane transporters such as P-gp, BCRP, and MRP1 [64]. Also, the cotreatment with quercetin and either doxorubicin, paclitaxel, or vincristine was found to be successful in both eliminating breast cancer stem cells and significantly downregulating ABCB1 expression. This outcome may improve the effectiveness of the chemotherapeutic regimens used in breast cancer treatment [125].

Similarly, quercetin was effective in the reversal of MDR in BEL hepatocellular carcinoma cells resistant to 5-fluorouracil [65]. According to Chen et al., quercetin suppressed both the function and the expression of efflux transporters ABCB1, ABCC1, and ABCC2 via Wnt/β-catenin signaling pathway [65]. Furthermore, Zhao et al. concluded that cotreatment with quercetin (50 μM) and cisplatin (10 μM) enhanced apoptosis and synergistically inhibited cell proliferation in HepG2 cell, thus improving the outcome of the conventional chemotherapy in hepatocellular carcinoma [66]. Consistent with these results, Yuan et al. confirmed that the sensitizing effect of quercetin applies as well in oral cancer cells resistant to vincristine [126].

Conclusion

Cancer represents a complicated disease, considered difficult to be treated with conventional therapy, a major hurdle of successful cancer chemotherapy being MDR. Different natural compounds have been tested as chemosensitizers in cancer therapy, acting synergistically with conventional chemotherapeutics and enhancing their antitumor activity through various mechanisms. Therefore, natural compounds may represent attractive and promising candidates to be used as chemosensitizers, a combination of these and conventional therapy emerging as a recent and new strategy for the effective management of cancer. However, despite the abundant preclinical evidence proving the efficacy of natural compounds in chemotherapy, there is just a limited number of clinical trials that have been conducted so far regarding this matter, natural compounds not being currently used as chemosensitizers in clinical practice.

Nevertheless, given the fact that there are numerous natural resources yet to be explored at the moment, this particular research field should be rapidly expanding in the coming years, providing novel, efficient, and natural alternatives to manage cancer chemoresistance.

References

[1] Bharti AC, Vishnoi K, Singh SM, Aggarwal BB. Pathways linked to cancer chemoresistance and their targeting by nutraceuticals. In: Role of nutraceuticals in chemoresistance to cancer. 1st ed. vol. 2. Elsevier Inc.; 2018. p. 1–30. https://doi.org/10.1016/B978-0-12-812373-7.00001-2

[2] Hanahan D, Weinberg RA. Hallmarks of cancer: the next generation, Cell 2011;144(5):646–74. Available from: http://www.ncbi.nlm.nih.gov/pubmed/21376230.

[3] Markman JL, Rekechenetskiy A, Holler E, Ljubimova JY. Nanomedicine therapeutic approaches to overcome cancer drug resistance, Adv Drug Deliv Rev 2013;65(13–14):1866–79. Available from: http://www.ncbi.nlm.nih.gov/pubmed/24120656.

[4] Kumar A, Jaitak V. Natural products as multidrug resistance modulators in cancer. Eur J Med Chem 2019;176:268–91.

[5] Bray F, Ferlay J, Soerjomataram I, Siegel RL, Torre LA, Jemal A. Global cancer statistics 2018: GLOBOCAN estimates of incidence and mortality worldwide for 36 cancers in 185 countries, CA Cancer J Clin 2018;68(6):394–424. Available from: http://www.ncbi.nlm.nih.gov/pubmed/30207593.

[6] Hamed AR, Abdel-Azim NS, Shams KA, Hammouda FM. Targeting multidrug resistance in cancer by natural chemosensitizers. Bull Natl Res Cent 2019;43(1).

[7] de Oliveira Júnior RG, Christiane Adrielly AF, da Silva Almeida JRG, Grougnet R, Thiéry V, Picot L. Sensitization of tumor cells to chemotherapy by natural products: a systematic review of preclinical data and molecular mechanisms. Fitoterapia 2018;129:383–400.

[8] Longley D, Johnston P. Molecular mechanisms of drug resistance, J Pathol 2005;205(2):275–92. Available from: http://www.ncbi.nlm.nih.gov/pubmed/15641020.

[9] Kerbel RS, Kobayashi H, Graham CH. Intrinsic or acquired drug resistance and metastasis: are they linked phenotypes?, J Cell Biochem 1994;56(1):37–47. Available from: http://www.ncbi.nlm.nih.gov/pubmed/7806590.

[10] Prasad S, Tyagi AK, Aggarwal BB. Chemosensitization by ursolic acid: a new avenue for cancer therapy. In: Role of nutraceuticals in chemoresistance to cancer. 1st ed. vol. 2. Elsevier Inc.; 2018. p. 99–109. https://doi.org/10.1016/B978-0-12-812373-7.00005-X

[11] Türk D, Hall MD, Chu BF, Ludwig JA, Fales HM, Gottesman MM, et al. Identification of compounds selectively killing multidrug-resistant cancer cells, Cancer Res 2009;69(21):8293–301. Available from: http://www.ncbi.nlm.nih.gov/pubmed/19843850.

[12] Longley DB, Allen WL, Johnston PG. Drug resistance, predictive markers and pharmacogenomics in colorectal cancer, Biochim Biophys Acta Rev Cancer 2006;1766(2):184–96. Available from: http://www.ncbi.nlm.nih.gov/pubmed/16973289.

[13] Yan XJ, Gong LH, Zheng FY, Cheng KJ, Chen ZS, Shi Z. Triterpenoids as reversal agents for anticancer drug resistance treatment, Drug Discov Today 2014;19(4):482–8. Available from: http://www.ncbi.nlm.nih.gov/pubmed/23954181.

[14] Saraswathy M, Gong S. Different strategies to overcome multidrug resistanece in cancer, Biotechnol Adv 2013;31(8):1397–407. Available from: http://www.ncbi.nlm.nih.gov/pubmed/23800690.

[15] Fan S, El-Deiry WS, Bae I, Freeman J, Jondle D, Bhatia K, et al. p53 gene mutations are associated with decreased sensitivity of human lymphoma cells to DNA damaging agents, Cancer Res 1994;54(22):5824–30. Available from: http://www.ncbi.nlm.nih.gov/pubmed/7954409.

[16] Shen H, He MM, Liu H, Wrighton SA, Wang L, Guo B, et al. Comparative metabolic capabilities and inhibitory profiles of CYP2D6.1, CYP2D6.10, and CYP2D6.17, Drug Metab Dispos 2007;35(8):1292–300. Available from: http://www.ncbi.nlm.nih.gov/pubmed/17470523.

[17] Tew KD. Glutathione-associated enzymes in anticancer drug resistance, Cancer Res 1994;54(16):4313–20. Available from: http://www.ncbi.nlm.nih.gov/pubmed/8044778.

[18] Yang Q, Huang J, Wu Q, Cai Y, Zhu L, Lu X, et al. Acquisition of epithelial–mesenchymal transition is associated with Skp2 expression in paclitaxel-resistant breast cancer cells, Br J Cancer 2014;110(8):1958–67. Available from: http://www.ncbi.nlm.nih.gov/pubmed/24642627.

[19] Haslehurst AM, Koti M, Dharsee M, Nuin P, Evans K, Geraci J, et al. EMT transcription factors snail and slug directly contribute to cisplatin resistance in ovarian cancer, BMC Cancer 2012;12(1):91. Available from: http://bmccancer.biomedcentral.com/articles/10.1186/1471-2407-12-91.

[20] Vishnoi K, Mahata S, Tyagi A, Pandey A, Verma G, Jadli M, et al. Human papillomavirus oncoproteins differentially modulate epithelial-mesenchymal transition in 5-FU-resistant cervical cancer cells, Tumor Biol 2016;37(10):13137–54. Available from:http://www.ncbi.nlm.nih.gov/pubmed/27449048.

[21] Paillas S, Causse A, Marzi L, de Medina P, Poirot M, Denis V, et al. MAPK14/p38α confers irinotecan resistance to TP53-defective cells by inducing survival autophagy, Autophagy 2012;8(7):1098–112. Available from: http://www.ncbi.nlm.nih.gov/pubmed/22647487.

[22] Ding ZB, Hui B, Shi YH, Zhou J, Peng YF, Gu CY, et al. Autophagy activation in hepatocellular carcinoma contributes to the tolerance of oxaliplatin via reactive oxygen species modulation, Clin Cancer Res 2011;17(19):6229–38. Available from: http://www.ncbi.nlm.nih.gov/pubmed/21825039.

[23] Sun WL, Chen J, Wang YP, Zheng H. Autophagy protects breast cancer cells from epirubicin-induced apoptosis and facilitates epirubicin-resistance development, Autophagy 2011;7(9):1035–44. Available from: http://www.ncbi.nlm.nih.gov/pubmed/21646864.

[24] Gerl R, Vaux DL. Apoptosis in the development and treatment of cancer, Carcinogenesis 2004;26(2):263–70. Available from: http://www.ncbi.nlm.nih.gov/pubmed/15375012.

[25] Vinod BS, Maliekal TT, Anto RJ. Phytochemicals as chemosensitizers: from molecular mechanism to clinical significance, Antioxid Redox Signal 2013;18(11):1307–48. Available from: http://www.liebertpub.com/doi/10.1089/ars.2012.4573.

[26] Varma MVS, Ashokraj Y, Dey CS, Panchagnula R. P-glycoprotein inhibitors and their screening: a perspective from bioavailability enhancement, Pharmacol Res 2003;48(4):347–59. Available from: http://www.ncbi.nlm.nih.gov/pubmed/12902205.

[27] Kathawala RJ, Gupta P, Ashby CR, Chen Z-S. The modulation of ABC transporter-mediated multidrug resistance in cancer: a review of the past decade, Drug Resist Updat 2015;18:1–17. Available from: http://www.ncbi.nlm.nih.gov/pubmed/25554624.

[28] Li W, Zhang H, Assaraf YG, Zhao K, Xu X, Xie J, et al. Overcoming ABC transporter-mediated multidrug resistance: molecular mechanisms and novel therapeutic drug strategies, Drug Resist Updat 2016;27:14–29. Available from: http://www.ncbi.nlm.nih.gov/pubmed/27449595.

[29] Wilson WH, Jamis-Dow C, Bryant G, Balis FM, Klecker RW, Bates SE, et al. Phase I and pharmacokinetic study of the multidrug resistance modulator dexverapamil with EPOCH chemotherapy, J Clin Oncol 1995;13(8):1985–94. Available from: http://www.ncbi.nlm.nih.gov/pubmed/7636539.

[30] Minderman H, O'Loughlin KL, Pendyala L, Baer MR. VX-710 (biricodar) increases drug retention and enhances chemosensitivity in resistant cells overexpressing P-glycoprotein, multidrug resistance protein, and breast cancer resistance protein, Clin Cancer Res 2004;10(5):1826–34. Available from: http://www.ncbi.nlm.nih.gov/pubmed/15014037.

[31] Gottesman MM, Ludwig J, Xia D, Szakács G. Defeating drug resistance in cancer, Discov Med 2006;6(31):18–23. Available from: http://www.ncbi.nlm.nih.gov/pubmed/17234123.

[32] Kemper EM, Cleypool C, Boogerd W, Beijnen JH, van Tellingen O. The influence of the P-glycoprotein inhibitor zosuquidar trihydrochloride (LY335979) on the brain penetration of paclitaxel in mice, Cancer Chemother Pharmacol 2004;53(2):173–8. Available from: http://link.springer.com/10.1007/s00280-003-0720-y.

[33] Fox E, Bates SE. Tariquidar (XR9576): a P-glycoprotein drug efflux pump inhibitor, Expert Rev Anticancer Ther 2007;7(4):447–59. Available from: http://www.tandfonline.com/doi/full/10.1586/14737140.7.4.447.

[34] Hyafil F, Vergely C, Du Vignaud P, Grand-Perret T. In vitro and in vivo reversal of multidrug resistance by GF120918, an acridonecarboxamide derivative, Cancer Res 1993;53(19):4595–602. Available from: http://www.ncbi.nlm.nih.gov/pubmed/8402633.

[35] Thomas H, Coley HM. Overcoming multidrug resistance in cancer: an update on the clinical strategy of inhibiting P-glycoprotein, Cancer Control 2003;10(2):159–65. Available from: http://www.ncbi.nlm.nih.gov/pubmed/12712010.

[36] Newman DJ, Cragg GM. Natural products as sources of new drugs from 1981 to 2014, J Nat Prod 2016;79(3):629–61. Available from: http://www.ncbi.nlm.nih.gov/pubmed/26852623.

[37] Wang H, Oo Khor T, Shu L, Su ZY, Fuentes F, Lee JH, et al. Plants vs. cancer: a review on natural phytochemicals in preventing and treating cancers and their druggability, Anti Cancer Agents Med Chem 2012;12(10):1281–305. Available from: http://www.ncbi.nlm.nih.gov/pubmed/22583408.

[38] Shukla S, Mehta A. Anticancer potential of medicinal plants and their phytochemicals: a review. Rev Bras Bot 2015;38:199–210.

[39] Khan T, Ali M, Khan A, Nisar P, Jan SA, Afridi S, et al. Anticancer plants: a review of the active phytochemicals, applications in animal models, and regulatory aspects. Biomolecules 2020;10:47.

[40] Kotecha R, Takami A, Espinoza JL. Dietary phytochemicals and cancer chemoprevention: a review of the clinical evidence, Oncotarget 2016;7:52517–29. Available from: http://www.ncbi.nlm.nih.gov/pubmed/27232756.

[41] Hussain SA, Sulaiman AA, Balch C, Chauhan H, Alhadidi QM, Tiwari AK. Natural polyphenols in cancer chemoresistance, Nutr Cancer 2016;68(6):879–91. Available from: http://www.ncbi.nlm.nih.gov/pubmed/27366999.

[42] Gupta SC, Kannappan R, Reuter S, Kim JH, Aggarwal BB. Chemosensitization of tumors by resveratrol. Ann N Y Acad Sci 2011;1215:150–60.

[43] Jacquemin G, Shirley S, Micheau O. Combining naturally occurring polyphenols with TNF-related apoptosis-inducing ligand: a promising approach to kill resistant cancer cells?, Cell Mol Life Sci 2010;67(18):3115–30. Available from http://www.ncbi.nlm.nih.gov/pubmed/20508968.

[44] Bharti AC, Donato N, Aggarwal BB. Curcumin (diferuloylmethane) inhibits constitutive and IL-6-inducible STAT3 phosphorylation in human multiple myeloma cells, J Immunol 2003;171(7):3863–71. Available from: http://www.jimmunol.org/lookup/doi/10.4049/jimmunol.171.7.3863.

[45] Bhardwaj A, Sethi G, Vadhan-Raj S, Bueso-Ramos C, Takada Y, Gaur U, et al. Resveratrol inhibits proliferation, induces apoptosis, and overcomes chemoresistance through down-regulation of STAT3 and nuclear factor-κB-regulated antiapoptotic and cell survival gene products in human multiple myeloma cells, Blood 2007;109 (6):2293–302. Available from: http://www.ncbi.nlm.nih.gov/pubmed/17164350.
[46] Fu Y, Chang H, Peng X, Bai Q, Yi L, Zhou Y, et al. Resveratrol inhibits breast cancer stem-like cells and induces autophagy via suppressing Wnt/β-catenin signaling pathway, PLoS One 2014;9(7)e102535 Available from: http://dx.plos.org/10.1371/journal.pone.0102535.
[47] Pandey A, Vishnoi K, Mahata S, Tripathi SC, Misra SP, Misra V, et al. Berberine and curcumin target survivin and STAT3 in gastric cancer cells and synergize actions of standard chemotherapeutic 5-fluorouracil, Nutr Cancer 2015;67(8):1295–306. Available from: http://www.tandfonline.com/doi/full/10.1080/01635581.2015.1085581.
[48] Sung B, Kunnumakkara AB, Sethi G, Anand P, Guha S, Aggarwal BB. Curcumin circumvents chemoresistance in vitro and potentiates the effect of thalidomide and bortezomib against human multiple myeloma in nude mice model, Mol Cancer Ther 2009;8(4):959–70. Available from: http://www.ncbi.nlm.nih.gov/pubmed/19372569.
[49] Bharti AC, Aggarwal BB. Chemopreventive agents induce suppression of nuclear factor-κB leading to chemosensitization, Ann N Y Acad Sci 2002;973(1):392–5. Available from: http://www.ncbi.nlm.nih.gov/pubmed/12485899.
[50] Sanna V, Siddiqui IA, Sechi M, Mukhtar H. Resveratrol-loaded nanoparticles based on poly (epsilon-caprolactone) and poly(d, l-lactic-*co*-glycolic acid)–poly(ethylene glycol) blend for prostate cancer treatment, Mol Pharm 2013;10(10):3871–81. Available from: https://pubs.acs.org/doi/10.1021/mp400342f.
[51] Mimeault M, Batra SK. Potential applications of curcumin and its novel synthetic analogs and nanotechnology-based formulations in cancer prevention and therapy, Chin Med 2011;6(1):31. Available from: http://www.ncbi.nlm.nih.gov/pubmed/21859497.
[52] Kunnumakkara AB, Diagaradjane P, Anand P, Kuzhuvelil HB, Deorukhkar A, Gelovani J, et al. Curcumin sensitizes human colorectal cancer to capecitabine by modulation of cyclin D1, COX-2, MMP-9, VEGF and CXCR4 expression in an orthotopic mouse model, Int J Cancer 2009;125(9):2187–97. Available from: http://www.ncbi.nlm.nih.gov/pubmed/19623659.
[53] Aggarwal BB, Shishodia S, Takada Y, Banerjee S, Newman RA, Bueso-Ramos CE, et al. Curcumin suppresses the paclitaxel-induced nuclear factor-B pathway in breast cancer cells and inhibits lung metastasis of human breast cancer in nude mice, Clin Cancer Res 2005;11(20):7490–8. Available from: http://www.ncbi.nlm.nih.gov/pubmed/16243823.
[54] Kunnumakkara AB, Guha S, Krishnan S, Diagaradjane P, Gelovani J, Aggarwal BB. Curcumin potentiates antitumor activity of gemcitabine in an orthotopic model of pancreatic cancer through suppression of proliferation, angiogenesis, and inhibition of nuclear factor-κB-regulated gene products, Cancer Res 2007;67 (8):3853–61. Available from: http://www.ncbi.nlm.nih.gov/pubmed/17440100.
[55] Goel A, Aggarwal BB. Curcumin, the golden spice from Indian saffron, is a chemosensitizer and radiosensitizer for tumors and chemoprotector and radioprotector for normal organs. Nutr Cancer 2010;62 (7):919–30.
[56] Bordoloi D, Kunnumakkara AB. The potential of curcumin: a multitargeting agent in cancer cell chemosensitization. In: Role of nutraceuticals in chemoresistance to cancer. 1st ed. vol. 2. Elsevier Inc.; 2018. p. 31–60. https://doi.org/10.1016/B978-0-12-812373-7.00002-4
[57] Ampasavate C, Sotanaphun U, Phattanawasin P, Piyapolrungroj N. Effects of curcuma spp. on P-glycoprotein function. Phytomedicine 2010;17(7):506–12. https://doi.org/10.1016/j.phymed.2009.09.004.
[58] Lu W-D, Qin Y, Yang C, Li L. Effect of curcumin on human colon cancer multidrug resistance in vitro and in vivo. Clinics 2013;68(5):694.
[59] Limtrakul P. Curcumin as chemosensitizer. Adv Exp Med Biol 2007;595:269–300.
[60] Wei Y, Yang P, Cao S, Zhao L. The combination of curcumin and 5-fluorouracil in cancer therapy. Arch Pharm Res 2018;41(1)https://doi.org/10.1007/s12272-017-0979-x.
[61] Shih H, Pickwell GV, Quattrochi LC. Differential effects of flavonoid compounds on tumor promoter-induced activation of the human CYP1A2 enhancer, Arch Biochem Biophys 2000;373(1):287–94. Available from: https://linkinghub.elsevier.com/retrieve/pii/S0003986199915503.

[62] Borska S, Chmielewska M, Wysocka T, Drag-Zalesinska M, Zabel M, Dziegiel P. In vitro effect of quercetin on human gastric carcinoma: targeting cancer cells death and MDR, Food Chem Toxicol 2012;50(9):3375–83. Available from: http://www.ncbi.nlm.nih.gov/pubmed/22750388.

[63] Hashemzaei M, Far AD, Yari A, Heravi RE, Tabrizian K, Taghdisi SM, et al. Anticancer and apoptosis-inducing effects of quercetin in vitro and in vivo. Oncol Rep 2017;38(2):819–28.

[64] Li S, Yuan S, Zhao Q, Wang B, Wang X, Li K. Quercetin enhances chemotherapeutic effect of doxorubicin against human breast cancer cells while reducing toxic side effects of it. Biomed Pharmacother 2018;100:441–7. https://doi.org/10.1016/j.biopha.2018.02.055.

[65] Chen Z, Huang C, Ma T, Jiang L, Tang L, Shi T, et al. Reversal effect of quercetin on multidrug resistance via FZD7/β-catenin pathway in hepatocellular carcinoma cells. Phytomedicine 2018;43:37–45. https://doi.org/10.1016/j.phymed.2018.03.040.

[66] Zhao JL, Zhao J, Jiao HJ. Synergistic growth-suppressive effects of quercetin and cisplatin on HepG2 human hepatocellular carcinoma cells. Appl Biochem Biotechnol 2014;172(2):784–91.

[67] Xiao Q, Zhu W, Feng W, Lee SS, Leung AW, Shen J, et al. A review of resveratrol as a potent chemoprotective and synergistic agent in cancer chemotherapy. Front Pharmacol 2019;9:1–10.

[68] Du B, Shim JS. Targeting epithelial-mesenchymal transition (EMT) to overcome drug resistance in cancer, Molecules 2016;21: Available from: http://www.ncbi.nlm.nih.gov/pubmed/27455225.

[69] Buhrmann C, Yazdi M, Popper B, Kunnumakkara AB, Aggarwal BB, Shakibaei M. Induction of the epithelial-to-mesenchymal transition of human colorectal cancer by human TNF-β (lymphotoxin) and its reversal by resveratrol. Nutrients 2019;11(3).

[70] Buhrmann C, Shayan P, Kraehe P, Popper B, Goel A, Shakibaei M. Resveratrol induces chemosensitization to 5-fluorouracil through up-regulation of intercellular junctions, epithelial-to-mesenchymal transition and apoptosis in colorectal cancer. Biochem Pharmacol 2015;98(1):51–68. https://doi.org/10.1016/j.bcp.2015.08.105.

[71] Buhrmann C, Yazdi M, Popper B, Shayan P, Goel A, Aggarwal BB, et al. Resveratrol chemosensitizes TNF-β-induced survival of 5-FU-treated colorectal cancer cells. Nutrients 2018;10(7).

[72] Zhao W, Bao P, Qi H, You H. Resveratrol down-regulates survivin and induces apoptosis in human multidrug-resistant SPC-A-1/CDDP cells, Oncol Rep 2010;23(1):279–86. Available from: http://www.ncbi.nlm.nih.gov/pubmed/19956893.

[73] Zhu Y, He W, Gao X, Li B, Mei C, Xu R, et al. Resveratrol overcomes gefitinib resistance by increasing the intracellular gefitinib concentration and triggering apoptosis, autophagy and senescence in PC9/G NSCLC cells. Sci Rep 2015;5:1–12. https://doi.org/10.1038/srep17730.

[74] Casanova F, Quarti J, Ferraz Da Costa DC, Ramos CA, Da Silva JL, Fialho E. Resveratrol chemosensitizes breast cancer cells to melphalan by cell cycle arrest. J Cell Biochem 2012;113(8):2586–96.

[75] Vinod BS, Nair HH, Vijayakurup V, Shabna A, Shah S, Krishna A, et al. Resveratrol chemosensitizes HER-2-overexpressing breast cancer cells to docetaxel chemoresistance by inhibiting docetaxel-mediated activation of HER-2-Akt axis. Cell Death Dis 2015;1(1):1–9. https://doi.org/10.1038/cddiscovery.2015.61.

[76] Wu H, Xin Y, Xu C, Xiao Y. Capecitabine combined with (−)-epigallocatechin-3-gallate inhibits angiogenesis and tumor growth in nude mice with gastric cancer xenografts, Exp Ther Med 2012;3(4):650–4. Available from: http://www.ncbi.nlm.nih.gov/pubmed/22969946.

[77] Stearns ME, Wang M. Synergistic effects of the green tea extract epigallocatechin-3-gallate and taxane in eradication of malignant human prostate tumors, Transl Oncol 2011;4(3):147–56. Available from: https://www.sciencedirect.com/science/article/pii/S1936523311800161.

[78] Toden S, Tran HM, Tovar-Camargo OA, Okugawa Y, Goel A. Epigallocatechin-3-gallate targets cancer stem-like cells and enhances 5-fluorouracil chemosensitivity in colorectal cancer. Oncotarget 2016;7(13):16158–71.

[79] Somers-Edgar TJ, Scandlyn MJ, Stuart EC, Le Nedelec MJ, Valentine SP, Rosengren RJ. The combination of epigallocatechin gallate and curcumin suppresses ERα-breast cancer cell growth in vitro and in vivo. Int J Cancer 2008;122(9):1966–71.

[80] Chang L, Huo B, Lv Y, Wang Y, Liu W. Ginsenoside Rg3 enhances the inhibitory effects of chemotherapy on esophageal squamous cell carcinoma in mice, Mol Clin Oncol 2014;2(6):1043–6. Available from: http://www.ncbi.nlm.nih.gov/pubmed/25279195.

[81] Yang LQ, Wang B, Gan H, Fu ST, Zhu XX, Wu ZN, et al. Enhanced oral bioavailability and anti-tumour effect of paclitaxel by 20(s)-ginsenoside Rg3 *in vivo*, Biopharm Drug Dispos 2012;33(8):425–36. Available from: http://wewww.ncbi.nlm.nih.gov/pubmed/22898996.

[82] Liu F, He Y, Liang Y, Wen L, Zhu Y, Wu Y, et al. PI3-kinase inhibition synergistically promoted the anti-tumor effect of lupeol in hepatocellular carcinoma, Cancer Cell Int 2013;13(1):108. Available from: http://www.ncbi.nlm.nih.gov/pubmed/24176221.

[83] Liu Y, Bi T, Dai W, Wang G, Qian L, Shen G, et al. Lupeol enhances inhibitory effect of 5-fluorouracil on human gastric carcinoma cells, Naunyn Schmiedeberg's Arch Pharmacol 2016;389(5):477–84. Available from: http://www.ncbi.nlm.nih.gov/pubmed/26892272.

[84] Liu YZ, Yang CM, Chen JY, Liao JW, Hu ML. Alpha-carotene inhibits metastasis in Lewis lung carcinoma in vitro, and suppresses lung metastasis and tumor growth in combination with taxol in tumor xenografted C57BL/6 mice, J Nutr Biochem 2015;26(6):607–15. Available from: http://www.ncbi.nlm.nih.gov/pubmed/25736483.

[85] Zhang Y, Zhu X, Huang T, Chen L, Liu Y, Li Q, et al. β-Carotene synergistically enhances the anti-tumor effect of 5-fluorouracil on esophageal squamous cell carcinoma in vivo and in vitro, Toxicol Lett 2016;261:49–58. Available from: http://www.ncbi.nlm.nih.gov/pubmed/27586268.

[86] Qi Q, Liu X, Li S, Joshi HC, Ye K. Synergistic suppression of noscapine and conventional chemotherapeutics on human glioblastoma cell growth, Acta Pharmacol Sin 2013;34(7):930–8. Available from: http://www.ncbi.nlm.nih.gov/pubmed/23708557.

[87] Chougule MB, Patel A, Sachdeva P, Jackson T, Singh M. Enhanced anticancer activity of gemcitabine in combination with noscapine via antiangiogenic and apoptotic pathway against non-small cell lung cancer, PLoS One 2011;6(11)e27394 Available from: http://www.ncbi.nlm.nih.gov/pubmed/22102891.

[88] Surya Sandeep M, Sridhar V, Puneeth Y, Ravindra Babu P, Naveen Babu K. Enhanced oral bioavailability of felodipine by naringenin in Wistar rats and inhibition of P-glycoprotein in everted rat gut sacs *in vitro*, Drug Dev Ind Pharm 2014;40(10):1371–7. Available from: http://www.ncbi.nlm.nih.gov/pubmed/23883365.

[89] Zhang FY, Du GJ, Zhang L, Zhang CL, Lu WL, Liang W. Naringenin enhances the anti-tumor effect of doxorubicin through selectively inhibiting the activity of multidrug resistance-associated proteins but not P-glycoprotein, Pharm Res 2009;26(4):914–25. Available from: http://www.ncbi.nlm.nih.gov/pubmed/19067124.

[90] Wang ZD, Wang RZ, Xia YZ, Kong LY, Yang L. Reversal of multidrug resistance by icaritin in doxorubicin-resistant human osteosarcoma cells, Chin J Nat Med 2018;16(1):20–8. Available from: http://www.ncbi.nlm.nih.gov/pubmed/29425587.

[91] Sun L, Chen W, Qu L, Wu J, Si J. Icaritin reverses multidrug resistance of HepG2/ADR human hepatoma cells via downregulation of MDR1 and P-glycoprotein expression, Mol Med Rep 2013;8(6):1883–7. Available from: http://www.ncbi.nlm.nih.gov/pubmed/24145579.

[92] Samykutty A, Shetty AV, Dakshinamoorthy G, Bartik MM, Johnson GL, Webb B, et al. Piperine, a bioactive component of pepper spice exerts therapeutic effects on androgen dependent and androgen independent prostate cancer cells, PLoS One 2013;8(6)e65889 Available from: http://www.ncbi.nlm.nih.gov/pubmed/23824300.

[93] Qiang F, Kang K-W, Han H-K. Repeated dosing of piperine induced gene expression of P-glycoprotein via stimulated pregnane-X-receptor activity and altered pharmacokinetics of diltiazem in rats, Biopharm Drug Dispos 2012;33(8):446–54. Available from: http://www.ncbi.nlm.nih.gov/pubmed/22927137.

[94] Esatbeyoglu T, Huebbe P, Ernst IMA, Chin D, Wagner AE, Rimbach G. Curcumin-from molecule to biological function. Angew Chem Int Ed 2012;51(22):5308–32.

[95] Jurenka JS. Anti-inflammatory properties of curcumin, a major constituent of Curcuma longa: a review of preclinical and clinical research. Altern Med Rev 2009;14(2):141–53.

[96] Gibellini L, Bianchini E, De Biasi S, Nasi M, Cossarizza A, Pinti M. Natural compounds modulating mitochondrial functions. Evid Based Complement Alternat Med 2015;2015:.

[97] He W-T, Zhu Y-H, Zhang T, Abulimiti P, Zeng FY, Zhang L-P, Luo L-J, Xie XM, Zhang HL. Curcumin reverses 5-fluorouracil resistance by promoting human colon cancer HCT-8/5-FU cell apoptosis and down-regulating heat shock protein 27 and P-glycoprotein. Chin J Integr Med 2019;25:416–29.

[98] Vinod BS, Antony J, Nair HH, Puliyappadamba VT, Saikia M, Shyam Narayanan S, et al. Mechanistic evaluation of the signaling events regulating curcumin-mediated chemosensitization of breast cancer cells to 5-fluorouracil. Cell Death Dis 2013;4(2):e505–13.

[99] Yin J, Wang L, Wang Y, Shen H, Wang X, Wu L. Curcumin reverses oxaliplatin resistance in human colorectal cancer via regulation of TGF-β/Smad2/3 signaling pathway. Onco Targets Ther 2019;12:3893–903.

[100] Zhou QM, Wang XF, Liu XJ, Zhang H, Lu YY, Huang S, et al. Curcumin improves MMC-based chemotherapy by simultaneously sensitising cancer cells to MMC and reducing MMC-associated side-effects. Eur J Cancer 2011;47(14):2240–7.

[101] Guo S, Lv L, Shen Y, Hu Z, He Q, Chen X. A nanoparticulate pre-chemosensitizer for efficacious chemotherapy of multidrug resistant breast cancer. Sci Rep 2016;6:1–11.

[102] Howells LM, Iwuji COO, Irving GRB, Barber S, Walter H, Sidat Z, et al. Curcumin combined with FOLFOX chemotherapy is Safe and tolerable in patients with metastatic colorectal cancer in a randomized phase IIa trial, J Nutr 2019;149(7):1133–9. Available from: http://www.ncbi.nlm.nih.gov/pubmed/31132111.

[103] Ruiz De Porras V, Bystrup S, Martínez-Cardús A, Pluvinet R, Sumoy L, Howells L, et al. Curcumin mediates oxaliplatin-acquired resistance reversion in colorectal cancer cell lines through modulation of CXC-chemokine/NF-κB signalling pathway. Sci Rep 2016;6:1–17.

[104] Noratto GD, Jutooru I, Safe S, Angel-Morales G. The drug resistance suppression induced by curcuminoids in colon cancer SW-480 cells is mediated by reactive oxygen species-induced disruption of the microRNA-27a-ZBTB10-Sp axis. Mol Nutr Food Res 2013;57(9):1638–48.

[105] Bayet-Robert M, Kwiatkowski F, Leheurteur M, Gachon F, Planchat E, Abrial C, et al. Phase I dose escalation trial of docetaxel plus curcumin in patients with advanced and metastatic breast cancer. Cancer Biol Ther 2010;9(1).

[106] Epelbaum R, Schaffer M, Vizel B, Badmaev V, Bar-Sela G. Curcumin and gemcitabine in patients with advanced pancreatic cancer. Nutr Cancer 2010;62(8):1137–41.

[107] Kanai M, Yoshimura K, Asada M, Imaizumi A, Suzuki C, Matsumoto S, et al. A phase I/II study of gemcitabine-based chemotherapy plus curcumin for patients with gemcitabine-resistant pancreatic cancer. Cancer Chemother Pharmacol 2011;68(1):157–64.

[108] Mahammedi H, Planchat E, Pouget M, Durando X, Curé H, Guy L, et al. The new combination docetaxel, prednisone and curcumin in patients with castration-resistant prostate cancer: a pilot phase II study. Oncologia 2016;90(2):69–78.

[109] Popat R, Plesner T, Davies F, Cook G, Cook M, Elliott P, et al. A phase 2 study of SRT501 (resveratrol) with bortezomib for patients with relapsed and or refractory multiple myeloma. Br J Haematol 2013;160(5):714–7.

[110] Berman AY, Motechin RA, Wiesenfeld MY, Holz MK. The therapeutic potential of resveratrol: a review of clinical trials. Precis Oncol 2017;1(1).

[111] Nawaz W, Zhou Z, Deng S, Ma X, Ma X, Li C, et al. Therapeutic versatility of resveratrol derivatives, Nutrients 2017;9: Available from: http://www.ncbi.nlm.nih.gov/pubmed/29109374.

[112] Paller CJ, Rudek MA, Zhou XC, Wagner WD, Hudson TS, Anders N, et al. A phase i study of muscadine grape skin extract in men with biochemically recurrent prostate cancer: safety, tolerability, and dose determination. Prostate 2015;75(14):1518–25.

[113] Howells LM, Berry DP, Elliott PJ, Jacobson EW, Hoffmann E, Hegarty B, et al. Phase I randomized, double-blind pilot study of micronized resveratrol (SRT501) in patients with hepatic metastases - safety, pharmacokinetics, and pharmacodynamics. Cancer Prev Res 2011;4(9):1419–25.

[114] Patel KR, Brown VA, Jones DJL, Britton RG, Hemingway D, Miller AS, et al. Clinical pharmacology of resveratrol and its metabolites in colorectal cancer patients. Cancer Res 2010;70(19):7392–9.

[115] Zhu W, Qin W, Zhang K, Rottinghaus GE, Chen YC, Kliethermes B, et al. Trans -resveratrol alters mammary promoter hypermethylation in women at increased risk for breast cancer. Nutr Cancer 2012;64(3):393–400.

[116] Suzuki Y, Miyoshi N, Isemura M. Health-promoting effects of green tea. Proc Jpn Acad Ser B Phys Biol Sci 2012;88:88–101.

[117] Fujiki H, Sueoka E, Watanabe T, Suganuma M. Primary cancer prevention by green tea, and tertiary cancer prevention by the combination of green tea catechins and anticancer compounds. J Cancer Prev 2015;20(1):1–4.

[118] Chen L, Ye HL, Zhang G, Yao WM, Chen XZ, Zhang FC, et al. Autophagy inhibition contributes to the synergistic interaction between EGCG and doxorubicin to kill the hepatoma Hep3B cells. PLoS One 2014;9(1).

[119] Liang G, Tang A, Lin X, Li L, Zhang S, Huang Z, et al. Green tea catechins augment the antitumor activity of doxorubicin in an in vivo mouse model for chemoresistant liver cancer. Int J Oncol 2010;37(1):111–23.
[120] Kumazaki M, Noguchi S, Yasui Y, Iwasaki J, Shinohara H, Yamada N, et al. Anti-cancer effects of naturally occurring compounds through modulation of signal transduction and miRNA expression in human colon cancer cells, J Nutr Biochem 2013;24(11):1849–58. Available from: http://www.ncbi.nlm.nih.gov/pubmed/23954321.
[121] Sun S, Dai Y, Lu Z, Li M, Zhai Z, Ren X, et al. Epigallocatechin gallate enhances 5-fluorouracil antitumor activity in MCF7 cells by regulating the expression of Bcl-xL. Int J Clin Exp Pathol 2016;9(4):4251–9.
[122] Mazumder MEH, Beale P, Chan C, Yu JQ, Huq F. Epigallocatechin gallate acts synergistically in combination with cisplatin and designed trans-palladiums in ovarian cancer cells. Anticancer Res 2012;32(11):4851–60.
[123] Anand David AV, Arulmoli R, Parasuraman S. Overviews of biological importance of quercetin: a bioactive flavonoid. Pharmacogn Rev 2016;10:84–9.
[124] Li Y, Yao J, Han C, Yang J, Chaudhry MT, Wang S, et al. Quercetin, inflammation and immunity. Nutrients 2016;8:.
[125] Li S, Zhao Q, Wang B, Yuan S, Wang X, Li K. Quercetin reversed MDR in breast cancer cells through down-regulating P-gp expression and eliminating cancer stem cells mediated by YB-1 nuclear translocation. Phyther Res 2018;32(8):1530–6.
[126] Yuan Z, Wang H, Hu Z, Huang Y, Yao F, Sun S, et al. Quercetin inhibits proliferation and drug resistance in KB/VCR oral cancer cells and enhances its sensitivity to vincristine. Nutr Cancer 2015;67(1):126–36.

PART III

Chemosensitizing agents: Computational tools and technological approaches

CHAPTER

11

Molecular docking studies and virtual drug screening of chemosensitizers

Agostino Bruno and Gabriele Costantino

Department of Food and Drug, University of Parma, Parma, Italy

Abstract

During cancer progression the rapid growth of solid tumor generates hypoxic environment and cancer cells overcome such disadvantageous conditions by a series of metabolic switches supporting the energetic demand of malignant cells. The intimate relation between cancer and unbalanced cellular metabolism is a well-established concept, and it also represents a relevant phenomenon in Multidrug Resistance (MDR) mechanisms. Mounting experimental evidences support the notion that specific metabolic switches and in particular those linked to altered pH gradient sustain most MDR mechanisms. Indeed, the search of chemosensitizers able to modulate intra- and extracellular pH is nowadays emerging as a new efficacious therapeutic regime. In this scenario, molecular modeling approaches can actively aid the identification of new chemical classes as pH-regulatory chemosensitizers.

Abbreviations

CAIs carbonic anhydrase inhibitors
C-ter C-terminal
MDR multidrug resistance
N-ter N-terminal
ROS reactive oxygen species
vHTS virtual high-throughput screening
VS virtual screening

Conflict of interest

No potential conflicts of interest were disclosed.

https://doi.org/10.1016/B978-0-12-820701-7.00013-0

Introduction

Normal cells transformation into cancer cells relies on irreversible genetic alterations triggering oncogenic signaling pathways, which in turn sustain: (i) rapid growth, (ii) proliferation, and (iii) ability to escape normal apoptotic pathways [1, 2]. In tumorigenesis, the aforementioned needs are met by a series of metabolic switches supporting the energetic demand of malignant cells [3–6]. Such phenomenon was earlier known as Warburg paradox that describes, in cancer cells, the switch from oxidative phosphorylation to glycolytic metabolism, which in turn elevates the amount of lactic acid and carbonic acids [7, 8]. Indeed, such intimate relation between cancer and unbalanced cellular metabolism is a well-established concept. However, the relevance of such metabolic switches in the ability of cancer cells to become resistant to different chemotherapeutic drugs (Multidrug Resistance, **MDR**) is less investigated, and only in the last decades become clearer that such metabolic switches, together with further genetic alterations, play a crucial role in MDR [3, 9–13]. Nowadays, MDR represents a worldwide challenge hampering the achievement of successful clinical treatment in cancer patients [14]. It is not surprising that enormous efforts have been made (and still are ongoing) in order to better comprehend the cellular mechanism underlying MDR phenomenon. The main cellular mechanisms through which cancer cells exert MDR can be summarized as follows: (i) drug inactivation, (ii) mutation of the drug target(s), (iii) altered system of drug absorption or extrusion, (iv) epigenetic alterations, (v) altered expression level of prooncogenic and/or antioncogenic systems, (vi) altered pH regulation [9, 10, 14, 15].

Increasing amount of evidence supports the notion that specific metabolic switches and in particular those linked to altered pH gradient sustains most of the aforementioned MDR mechanisms [16, 17]. Indeed, in tissues harboring solid tumor the extracellular pH is lower than in normal tissues. Such acidic pH can have, at least, two relevant consequences: (i) the acidic pH of the extracellular matrix in presence of reactive oxygen species (ROS) can lead to a reactive species burst that in a vicious circle sustains the oxidative damage of cellular macrobiomolecules (Proteins, Lipid, and Nucleic Acids); (ii) the acidic pH of the tumor microenvironment may hamper the intake of particular chemotherapeutic drugs (e.g., weakly basic drugs). In the former case, the occurrence of new genetic variants together with the selective pressure exerted by the chemotherapeutic treatment can lead to the selection of cancer cells characterized by (i) mutated drug targets and/or (ii) epigenetic modifications. In the latter case, the bioavailability of drug to its target(s) would be significantly altered, reducing the efficacy of the clinical treatment. Recently, the use of chemosensitizer has emerged as a potential alternative therapeutic approach to deal with MDR. In the NCI Dictionary of Cancer Terms, chemosensitizer is defined as "a drug that makes tumor cells more sensitive to the effects of chemotherapy" (https://www.cancer.gov/publications/dictionaries/cancer-terms/def/chemosensitizer). With the aim to identify new efficacious combinatorial therapeutic regime several strategies have been pursued, including the search of chemosensitizers able to modulate intra- and extracellular pH [3, 5, 11, 14–17]. Several cellular targets contribute to the control of pH homeostasis (Fig. 1), among which relevant proteins are: (i) hypoxia-inducible factor (**HIF**), (ii) Na^+/H^+-exchanges (**NHE1**), (iii) carbonic anhydrases (**CAIX**, **CAXII**, **CAII**), (iv) bicarbonate transporters (**NBC**), (v) proton-glucose

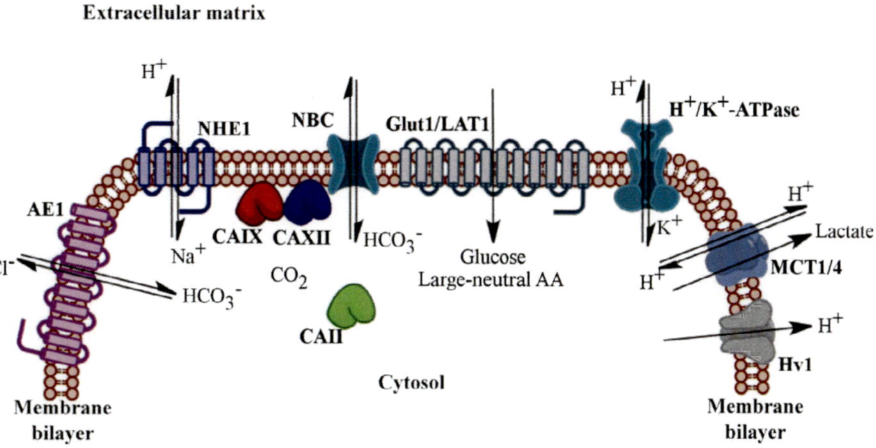

FIG. 1 Main cellular targets of pH-regulatory chemosensitizers.

and aminoacid transporters (**Glut1**, **LAT1**), (vi) anion exchanger proteins (**AE1**), (vii) monocarboxylate transporters (**MCT1**, **MCT4**), (viii) voltage-gated ion channels (**Hv1**), and (ix) hydrogen/potassium adenosine triphosphate transporters (**H^+/K^+-ATPase**) [3, 5, 11, 14–17]. The aforementioned protein systems represent already established pharmaceutical targets for which literature data are already available and, in most cases, the structural details, small-molecules modulators, the drug inhibition mechanisms are already described in detail. Taking advantages of the huge amount of data available for such cellular targets molecular modeling approaches can actively aid the identification of small molecules interfering with the protein functioning, and in turn, leading to the identification of new chemical class as pH-regulatory chemosensitizers [18–20]. Molecular modeling approaches comprehend an ensemble of in silico tools supporting, among other activities, the (i) generation of hypothesis, (ii) identification of a pattern of interaction sites within the cellular targets, (iii) identification of small molecules able to adapt into the target's binding cavity, and (iv) small-molecules optimization [19]. Docking simulations represent powerful modeling techniques useful to hypothesize/predict as possible as a reliable binding conformation of ligands into protein-binding clefts. The pillar principle on which docking simulations are based is the use of molecular mechanic background to describe the structural features underlying the interaction between ligands and protein targets [19, 21–23]. By means of scoring functions (mathematical algorithm) the interaction modes are scored and a rank of virtual hit is produced. The analysis of the ranking list led to the identification of the most promising virtual hits that are then selected for biological evaluations. In order to increase the chance to identify potential hits of interest approaches such as **vHTS** (virtual HTS, **VS**) are the method of choice. In this case a virtual library of 3-D chemical structures (from 10^3 up to 10^8 small molecules) is properly prepared and evaluated/scored by using one of more docking algorithms. For interested readers on the different docking and VS approaches we recommend the following review [18, 19, 21–24].

In the next sections of this chapter we are going to discuss docking and virtual screening studies performed on pH-regulatory targets in order to identify small-molecule chemosensitizers, we will also report successful examples of pH chemosensitizers identify by means of the aforementioned molecular modeling techniques.

Carbonic anhydrase (CA)

Carbonic anhydrases are metalloenzyme ubiquitously expressed in most, if not all, living organisms. They are encoded by seven unrelated gene families: (i) α-CAs, (ii) β-CAs, (iii) γ-CAs, (iv) δ-CAs, (v) ζ-CAs, (vi) η-CAs, and (vii) θ-CAs. All of them required metal cations to catalyze the conversion of CO_2 into HCO_3^- (Eq. 1) [25–28].

$$H_2O + CO_2 \underset{Zn^{2+} \text{ or } Fe^{2+} \text{ or } Cd^{2+}}{\overset{\text{CAs}}{\rightleftharpoons}} HCO_3^- + H^+ \quad (1)$$

Human encode for the α-CAs family only and 15 isoforms have been identified so far, 12 of them are catalytically active (CA I-IV, VA, VB, VI-VII, IX, XII, XIV) showing different degree of activity, whereas three isoforms (VII, X, and XI) are noncatalytic active, being classified as CA-related proteins (CARPs) [26, 27]. The α-CAs active site is defined by a large conical cleft, divided into two halves showing different chemical–physical properties: (i) the first one is mainly composed of hydrophobic residue, while the second one is characterized by the presence of hydrophilic residues (Fig. 2A). Such peculiar structural feature is thought to be instrumental to the catalytic activity of this class of enzyme, where the hydrophobic portion is responsible for CO_2 sequestration, whereas the hydrophilic one is necessary to create a solvent-exposed hydrogen bonds network, which facilitate proton transfer from the Zn^{2+}-H_2O/OH^- complex to the bulk [25–28]. The catalytic Zn^{2+} is located at the bottom of

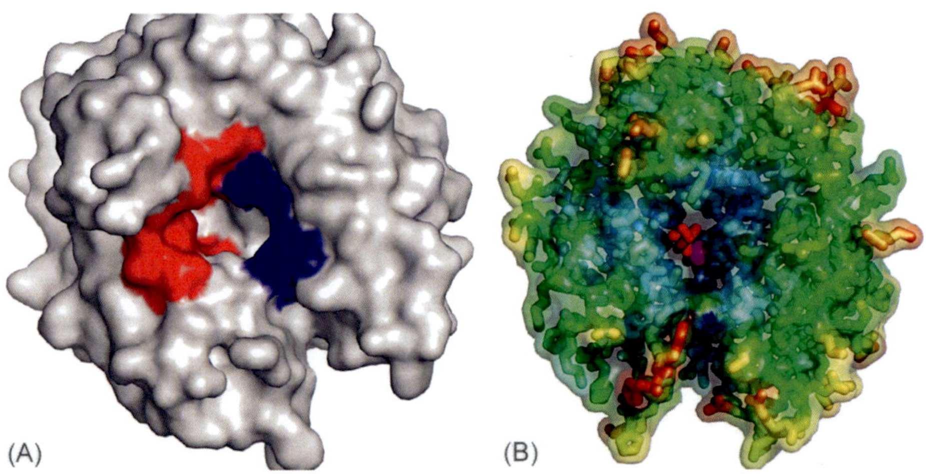

FIG. 2 (A) Hydrophilic *(red surface)* and hydrophobic *(blue surface)* halves of hCAII *(gray surface)*, (B) surface and sticks representation of hCA II *(blue-green-red scale color)* bound to ethoxzolamide *(red sticks)*, as a purple sphere the catalytic Zn^{2+} metal ion.

the active site cleft, being coordinates by three conserved His residues and H_2O/OH^-. Classical CA Inhibitors (CAIs) chelate the Zn^{2+} occupying the active site (Fig. 2B), thus, blocking its catalytic activity. Other class of CAIs are defined by small molecules that do not directly bind to Zn^{2+} but interact with the residues lining up the CAs active site [27].

CAs represent validated target for several human diseases, including cancer [25–31]. Among them, two isoforms CA IX and XII have been linked with tumor progression, metastasis and chemotherapeutic resistance [32–37]. In hypoxic condition HIF-1, the master regulator of cellular changes during hypoxia, mediates expression of CA IX, which is considered a tumor-associated protein [26, 38–41]. Indeed, CA IX expression levels are very low in normal tissues, while its expression increases in hypoxic cancer cells [29]. Noteworthy, the CA IX presence is tightly linked with radio- and chemotherapy resistance, since it is well known that the acidic tumor microenvironment affects drug uptake and the degree of damage induced by radiation [26, 38–41]. CA XII, instead, is present in physiological condition in different tissues, playing crucial physiological roles [29]. Its expression is induced in hypoxic conditions, and high CA XII expression levels were observed in different cancer cell lines [42–47]. On the basis of such experimental evidences it has been hypothesized that both CA IX and XII play crucial roles in sustaining cancer cell progression, by tuning intracellular and extracellular pH changes due to the altered cellular metabolism linked to hypoxia. It is not, therefore, surprising that enormous efforts have been made in order to identify selective inhibitors of these two isoforms. Chart 1 lists several CA IX and CA XII inhibitors identified and/or optimized by means of virtual screening and docking studies.

Compound **1** was identified by high-throughput virtual screening of a virtual library of $7*10^6$ small molecules. Salmas et al. [48] used the ZINC database (https://zinc.docking.org) to collect the 3-D virtual small molecules, the ligands were prepared using LigPrep tool of the Maestro suite (https://www.schrodinger.com), and the 3IAI X-ray crystal structure of CA IX was used. Ligands were screened by using a two-step approach: (i) the $7*10^6$ 3-D molecules were firstly evaluated by means of the Glide/HTVS (https://www.schrodinger.com) method in order to identify a small subset of interesting virtual hit (70 compounds), (ii) the identified potential hit was further evaluated using the Glide/XP (https://www.schrodinger.com) method and 19 of them were selected after visual inspection. Among the 19 selected compounds 3 were purchased and biochemically evaluated performing a spectrophotometric assay, leading to the identification of **1** as new CA IX inhibitor with a K_i in the low μM range ($K_i = 0.85$ μM). On the basis of the binding mode proposed by the authors it can be argued that **1** does not directly bind to the Zn^{2+} metal ion, but interacts with residues lining up the CA IX active site, impairing its catalytic activity.

Thiry et al. [49] in their work described a multiscale approach that led to the identification of selective CA IX inhibitors. The authors by means of the MOE (https://www.chemcomp.com/Products.htm) suite generated a pharmacophore model of CA IX inhibitors, where 10 structurally different selective CA IX inhibitors were used as training set. Flexible alignment of the training set was carried out in MOE (https://www.chemcomp.com/Products.htm), by imposing the alignment of sulfonamide moiety as main rule. The generated pharmacophoric model was defined by six pharmacophoric points: (i) three features for the sulfonamide group, (ii) one feature for hydrophobic group (such as aromatic rings), (iii) two features for Van Der Walls interaction and h-bond interactions [49]. The aforementioned model was used to screen a prefiltered subset of the ZINC database of commercially available small

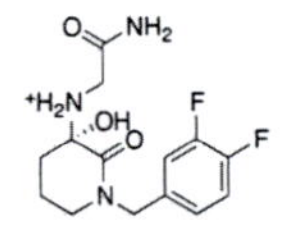

1
CA IX K_i = 0.85 μM

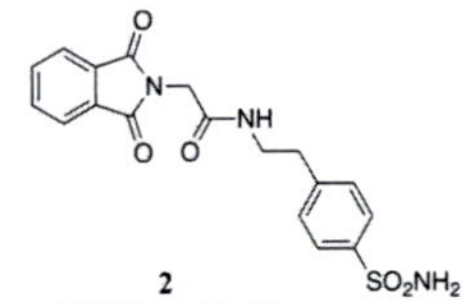

2
CA I K_i = 4.23 μM
CA II K_i = 0.66 μM
CA IX K_i = 0.29 μM

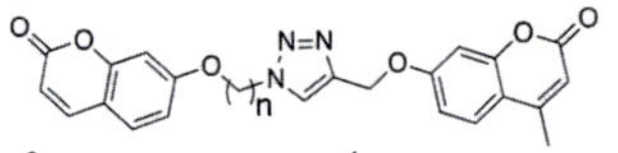

3
n = 10
CA I K_i = > 10^3 nM
CA II K_i = > 10^3 nM
CA IX K_i = 160. 9 nM
CA XII K_i = 71.5 nM

4
n = 12
CA I K_i = > 10^3 nM
CA II K_i = > 10^3 nM
CA IX K_i = 144. 6 nM
CA XII K_i = 785.2 nM

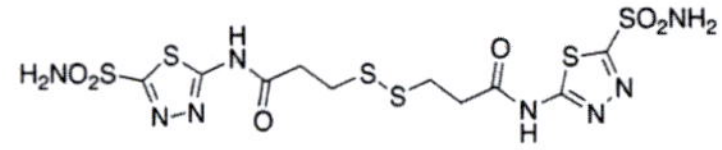

5
CA I K_i = 60 nM
CA II K_i = 61 nM
CA IX K_i = 18 nM

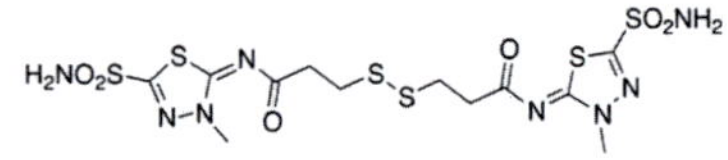

6
CA I K_i = 41 nM
CA II K_i = 42 nM
CA IX K_i = 9.3 nM

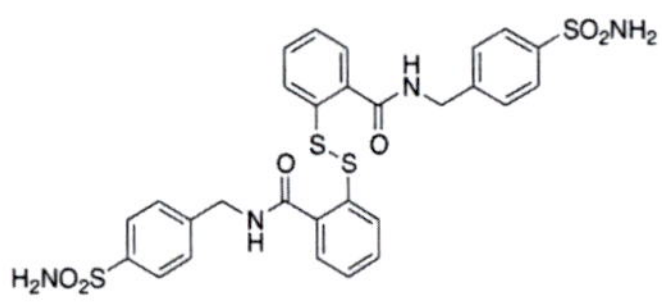

7
CA I K_i = 4350 nM
CA II K_i = 4975 nM
CA IX K_i = 653 nM

8 (Reduced form of 7, SH)
CA I K_i = 276 nM
CA II K_i = 16 nM
CA IX K_i = 9.1 nM

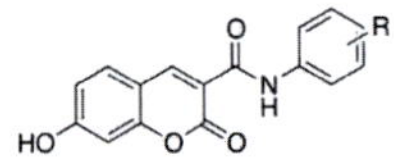

9

Compound R	Ki(nM) CA I	CA II	CAIX	CA XII
4-Br	> 1000	> 1000	0.6	7.7
4-F	> 1000	> 1000	0.7	9.0
4-Cl	> 1000	> 1000	0.2	0.2
3-F	> 1000	> 1000	0.3	7.9

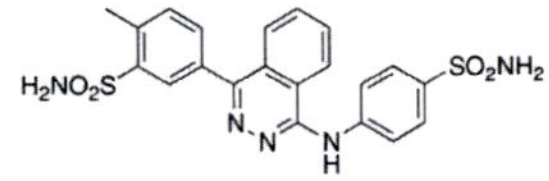

10
CA I IC_{50} = > 10^4 nM
CA II IC_{50} = 547.0 nM
CA IX IC_{50} = 0.48 nM

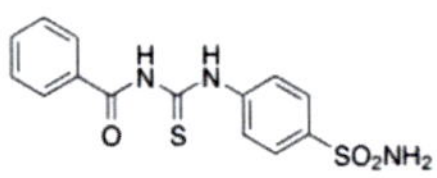

11
CA I IC_{50} = 197.64 nM
CA II IC_{50} = 60.93nM
CA IX IC_{50} = 2.86 nM

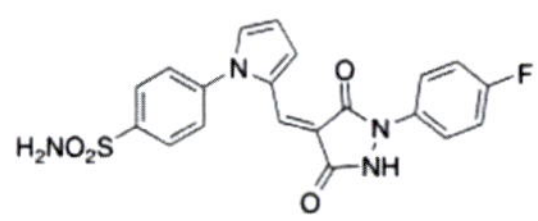

12
CA I IC_{50} = 213.69 nM
CA II IC_{50} = 41.97 nM
CA IX IC_{50} = 3.11 nM

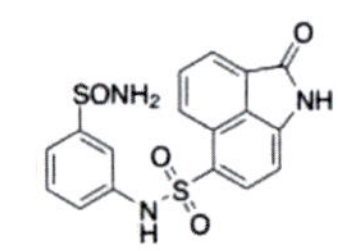

13
CA I IC_{50} = 231.41 nM
CA II IC_{50} = 3873 nM
CA IX IC_{50} = 28.35 nM

CHART 1 2-D chemical structure of CA IX and XII selective inhibitors.

molecules (https://zinc.docking.org). In this case 1176 3-D small molecules were selected and filtered by means of the pharmacophoric model, leading to the identification of 500 hits that were docked into a homology model of CA IX, using GOLD as docking software (https://www.ccdc.cam.ac.uk/solutions/csd-discovery/components/gold/). After docking studies, the compounds were selected on the basis of visual inspection and knowledge-based approaches. Six compounds were finally submitted to biochemical in vitro assays against CA I, CA II, and CA IX isoforms in order to evaluate the selectivity profile of the hits [49]. Among the tested small molecules compound **2** (Chart 1) resulted to be a promising CA IX selective inhibitor, and on the basis of the proposed binding mode structural optimization of **2** were proposed for classical med chem optimization cycles [49].

Compounds **3** and **4** represent bis-coumarins derivatives that selectively inhibit CA IX and XII [50]. At that time coumarins represented a new class of CAIs, and the authors in the efforts to identify selective CA IX and XII inhibitors designed a strategy based on a generation of a library of bis-coumarins derivatives, taking advantages of the well-known click-chemistry reaction. Sixteen derivatives were synthesized and biochemically evaluated for their ability to inhibit CA I, II, IX, and XII isoforms. Compounds **3** and **4** resulted to be the most interesting ones, for which depending on the linker length ($n=10$, 12) they achieved CA IX or CA XII selectivity. In order to better elucidate the structural features underlying the inhibition profile of the aforementioned small-molecules docking studies have been performed [50]. Since it is known that coumarins derivatives can undergo hydrolysis with the increase of incubation times, both parent compounds (**3** and **4**) and their hydrolysis products were investigated. Docking studies were performed by means of the GOLD docking software (https://www.ccdc.cam.ac.uk/solutions/csd-discovery/components/gold/), and on the basis of the obtained results the authors proposed a rational explanation accounting for the selectivity profile of such small molecules. Indeed, it is well known that to design isoform-selective CA inhibitors is a very challenging task, since the high degree of similarity of the residues lining up the CA active site [29]. Compounds **3** and **4** by binding at a binding cleft located at the entrance of the active site make interaction with regions of the CA IX and XII isoforms, which are less conserved inside the human α-CA family, thus, explaining the ability of the bis-coumarins derivatives to selectively inhibit CA IX and XII over the other ones [50].

Compounds **5–8** represent a class of prodrug that are enzymatically/chemically activated through disulfide bond reduction in hypoxic tumor tissue, exploiting aspects of tumor microenvironment [51]. Compound **7** and its reduced form **8** represent the best CA IX inhibitors of this series, and the authors through X-ray crystal structures and docking studies were able to rationalize their activity profile. Indeed, **7** in normoxic condition shows a lower affinity for CA I, II, and XII isoforms, when activated in hypoxic condition the reduction of the disulfide bond generates the formation of a thiol group in position 2 of the phenyl-amide ring. Such thiol group establishes a h-bond interaction with Gln67 residue, which is solely present in CA IX and not in CA I and CA II, thus explaining its selectivity profile [51].

The coumarin-carboxamide derivatives represented by the scaffold **9** represent a class of highly selective CA IX/XII inhibitors [52]. Such CAIs differently from **3** and **4** chelate the Zn^{2+} metal ion and do not undergo to hydrolysis of the lactone ring. The authors proposed that the presence of bulky substituents in position 3 of the coumarin ring hampers the opening of the ring, leading to a novel class of CAIs coumarin-based with a new inhibitory mechanism [52]. Docking studies were performed using CA IX and CA II X-ray crystal structures (PDB code: 3IAI and 4WW8, respectively) and Glide as docking software (https://www.schrodinger.com) [52]. On the basis of the proposed binding modes, the authors postulated that such derivatives by establishing specific interaction with nonconserved residues lining up CA IX and XII active site can account for their selectivity profiles [52].

Compound **10** represents a very highly selective CA IX inhibitor that was obtained by classical structure-based drug design optimization process (SBDD) [53]. Indeed, in previously reported works a lead compound was identified and taking advantage of the docking-derived binding mode med chem optimization cycles were undertaken. Such activities led to the identification of **10**, which represent the best in class of this series of CA IX inhibitors [53]. The authors in order to rationalize the selectivity profile of **10** aligned the docking-derived **10**-CA IX complex on the CA II X-ray crystal structures

(PDB Code: 3R16), identifying the 131 residue as responsible for the selectivity. Indeed, in CA II residue 131 is represented by a Phe, while in CA IX the same position is occupy by Val. Phe being bulkier with respect to Val hampers the accommodation of **10** into the CA II active site, making the compound less prone to inhibit such isoform. Docking studies were performed using 3IAI as CA IX X-ray crystal structure and Glide as docking software (https://www.schrodinger.com) [53].

Compounds **11–13** were identified through structure-based virtual screening techniques, where the SPECS database library (280,000 3-D small molecules) and the CA IX X-ray crystal structure (PDB Code: 3IAI) were used [54]. VS was performed using Glide as docking software (https://www.schrodinger.com) and applying a multistep approach: (i) the Glide SP method was used in order to preliminary identify a small subset of interesting virtual hits (10,000 hits), (ii) compounds were further reduced by means of the Glide XP method (500 compounds), (iii) after visual inspection 49 compounds were selected and biochemically evaluated for their ability to inhibit CA isoforms, leading to the identification of 13 small molecules as CAIs. Among them, compounds **11–13** resulted to be the most promising ones, and taking advantage of the obtained binding mode the biochemical profile of the aforementioned CAIs was rationalized [54]. The authors hypothesized that active site size and shape can account for the selectivity profile observed for **11–13** [54]. Indeed, in line with the observation reported for compounds **10** [53] the residue in position 131 is the main responsible of the observed activity profile.

Hypoxia-inducible factor (HIF)

During cancer progression the rapid growth of solid tumor generates hypoxic environment and cancer cells overcome such disadvantageous condition upregulating transcription factors responsible for the formation of tumor vasculature structure, allowing cancer cells to survive [2–4, 6, 7, 10, 12–14, 16, 17]. In this context the Hypoxia-Inducible Factor (**HIF**) is the major player in promoting the transcription of target genes regulating biological processes such as angiogenesis, energy metabolism, proliferation, apoptosis resistance, and metastasis [2, 16, 38, 55–57]. HIF belongs to the basic helix-loop-helix transcription factor family and it is composed of two subunits: (i) HIF-α for which three different isoforms exist (HIF-1α, HIF-2α, HIF-3α) and (ii) HIF-1β, which is constitutively expressed [58, 59]. The HIF-α and HIF-1β subunits present a modular structure defined by: (i) the bHLH (basic helix-loop-helix) and PAS (Per-ARNT-Sim homology) domains, which are required for heterodimerization and DNA binding to the Hypoxia Response Element (HRE) at the target gene loci; (ii) The ODD (Oxygen-dependent Degradation) domain; and (iii) the N-ter and C-ter transactivation domains (N-TAD and C-TAD), which regulate the HIF-1 target genes. HIF-α subunit expression levels are tightly regulated in normoxic conditions by a controlled rate of synthesis and degradation processes. In such condition two proline residues located at the ODD domain are hydroxylated by prolyl hydroxylase (PHD) and are recognized by specific domains of the von Hippel–Lindau (VHL) E3 ubiquitin ligase [56–59]. VHL promotes HIF-α ubiquitination and the subsequent proteasomal degradation. On the contrary during hypoxic conditions the lower O_2 levels induce PDH activity impairment, reducing the extent of HIF-α hydroxylation. In this way the stabilized forms of HIF-α translocate to the nucleus where they heterodimerize

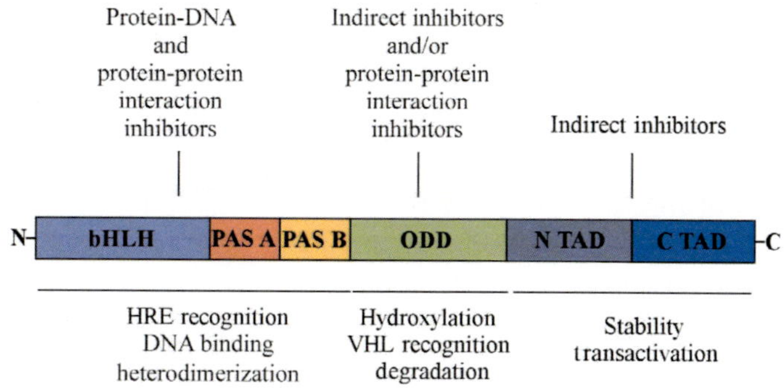

FIG. 3 Representation of the general structure of HIF-α and β. Relevant protein domain are highlighted in color-coded box.

with HIF-1β, being able to recognize the HRE at the target gene loci, and in turn, activating the transcription of the target genes [56–59]. Therefore, HIF pathway activation in hypoxic condition can trigger to the activation of mechanism (e.g., metabolic switch, angiogenetic factors synthesis, ATP-binding cassette multidrug transporters, etc.) that make cancer cells more resistant to cytotoxic chemotherapy [56–59]. Indeed, it has been reported that combination treatment with HIF inhibitors may increase the efficacy of traditional chemotherapy [3, 5, 9, 11, 14, 16, 56, 57, 60].

Several modulators of HIF activity are reported in literature and depending on their mechanism of action they are divided into direct and indirect inhibitors [56–59]. Direct inhibitors bind to HIF-1 subunits directly affecting heterodimerization, DNA binding, protein interaction (e.g., P300/CPB) (Fig. 3). Indirect inhibitors act on translation, stabilization, and degradation without directly binding to HIF-1 subunits.

Due to the complexity of the HIF proteins and their multimodular structure very few examples of validated docking and virtual screening studies have been reported in literature. Most of them focus on the retrospective analysis of small molecules interacting with the PAS B domain.

In 2016 Broekgaarden et al. [61] described the use of acriflavine as a chemosensitizers in photodynamic therapy. In their work the authors characterize the ligand mechanism of action by means of docking studies, predicting the binding of acriflavine to the PAS domains of HIF-α. Docking studies were performed by using AutoDock Vina (http://vina.scripps.edu) on different X-ray crystal structures of human HIF-PAS domains. The best docking complex was obtained with the PAS domain of HIF-2α (PDB code: 3H7W), revealing good shape complementary between the ligand and binding cleft side chains. The authors reported the presence of several hydrophobic interaction (π-π interaction) with several aromatic amino acid of the binding site. The proposed mechanism of action was validated by means of confocal microscopy, showing that acriflavine translocates from the cytosol to the nucleus under hypoxic conditions [61].

In the next future there will be enough space for computational chemists to search and identify new HIF-1 inhibitors as chemosensitizers in chemotherapy. Indeed, in the last years breakthrough on structural biology and computational techniques/algorithm paved the way for the identification of new HIF-1 modulators.

Glucose transporter 1 (GLUT1)

The sugar porter family of the Major Facilitator Superfamily (MSF) includes, among others, the Glucose transporter member 1 (**GLUT1**) [62, 63]. GLUT1 maintains the cellular basal level of glucose by facilitating the transport of monosaccharides across cell membranes along the concentration gradient [62, 63]. Experimental evidences have linked upregulated GLUT1 levels to different cancer types, [64, 65] being involved in supporting the metabolic switch from oxidative respiration to anaerobic glycolysis (Warburg effect) [7, 8]. GLUT1 is also considered a diagnostic target being able to transport biomarkers inside cancer cells (e.g., ^{18}F-DG in Positron Emission Tomography, PET) [66]. Recent technological breakthroughs in structural biology enabled the disclosure of the X-ray crystal structure of GLUT1 in complex with different substrates, [67, 68] revealing that it is a membrane-spanning protein organized in 12 transmembrane α-helices divided into two domains, namely, the N- and C-halves. Both halves shape the central cavity containing the substrate-binding site (Fig. 4).

Member of the GLUTs protein family plays crucial roles in different human diseases and the discovery of selective inhibitors is of great demand [69]. To date few virtual screening studies have been described in literature for different GLUT members, including one for GLUT1 [70–72].

Ung et al. [72] reported the generation of hGLUT1 homology model in the partially occluded outward open conformation. The 3-D structure of hGLUT1 was built using the X-ray crystal structure of the *E. coli* xylose transporter [73] as template and Modeler as software (https://salilab.org/modeller/). The 3-D model, thus obtained, was used to perform virtual screening studies using ZINC libraries (https://zinc.docking.org) composed of commercially available low-molecular-weight compounds (Fragment Now anc NCI-2007). Ligands were prepared by means of the OpenEye OMEGA software and docked using OpenEye FRED (http://www.eyesopen.com). Top-ranked molecules were evaluated for their similarity to known hGLUT1 ligands by means of the Tanimoto coefficient (Tc), using ECFP-4 fingerprints as descriptors [74, 75]. Finally, the small molecules were submitted to pain assay interference (PAINS) filters in order to exclude compound that can potentially interfere with biochemical assay [76, 77]. Starting from 548.312 3-D molecules 250-top ranked compounds were selected as described above, and after visual inspection of the binding

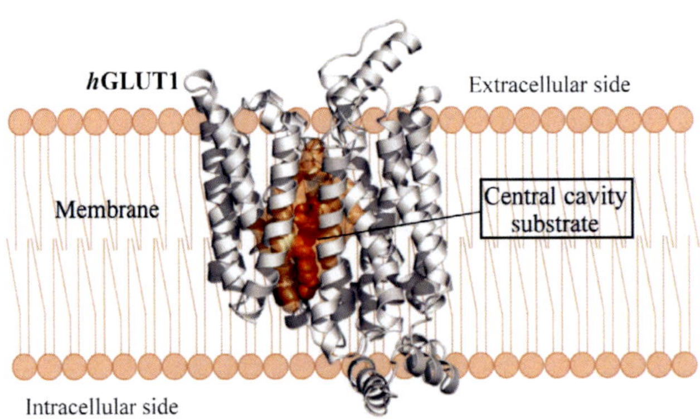

FIG. 4 Representation of the GLUT1 structure *(white cartoon)* embedded into a membrane bilayer *(orange spheres and sticks)*, highlighted as *orange surface* the central cavity hosting the substrate *(orange spheres)* binding site.

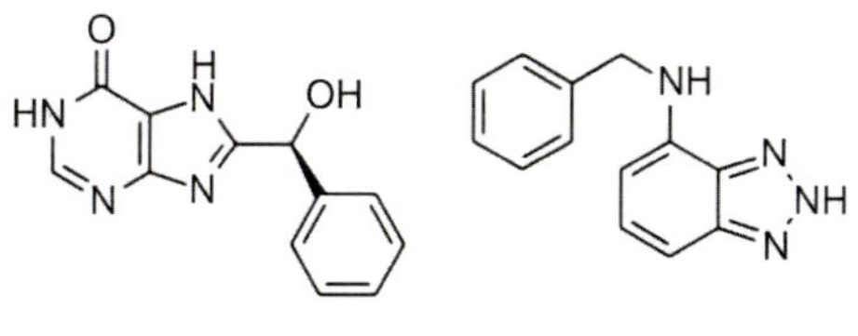

1
*h*GLUT1 IC_{50} = 0.45 μM

2
*h*GLUT1 IC_{50} = 30.00 μM

CHART 2 2-D chemical structure of hGLUT1 inhibitors identified in the work of Ung et al.

modes 19 molecules were purchased for biological testing. Among them **1** and **2** (Chart 2) resulted to be the most promising ones [72].

1 and **2** represent a new chemical class of hGLUT1 inhibitors for which the authors were able to propose binding mode suitable to further improve the activity profile of the newly identified compounds and paving the way for the development of new chemical series of hGLUT1 inhibitors as putative chemical tools useful to further investigate the pathophysiological role of this membrane protein.

Conclusions

Mounting evidence supports the notion that chemosensitizers represent an alternative option to deal with resistant cancer cells. In the previous sections of this chapter we presented and discussed different examples where docking and virtual screening studies aided the identification of modulators, as small-molecule chemosensitizers, for relevant pH-regulatory targets. Despite in most cases promising results were obtained, such pH-regulatory targets still remain difficult to treat from a molecular modeling point of view, due to the complexity of their structure and mechanism of actions. Indeed, only recently the 3-D structures of some of these protein targets have been disclosed, and for some of them no reliable chemical tools are described in literature.

Nevertheless, it is plausible to think that breakthroughs in both structural biology and computational fields will enable (i) the possibility to identify new chemosensitizers with complex mechanism of actions, by simultaneously modulating more than one therapeutic target, and (ii) the possibility to identify small molecules able to interact with targets not yet explored/drugged. In the former case, several classes of chemosensitizers are under evaluation for alternative therapeutic options, including dual-inhibitor approaches [32, 34–36, 78]. In the latter case, it is well known that molecular modeling techniques can actively sustain the identification of small molecules targeting not yet drugged targets, expanding the available chemical and biological space.

References

[1] Niederhuber JE. Abeloff's clinical oncology. 6th ed. Philadelphia, PA: Elsevier; 2019.

[2] Dixon K, Kopras E. Genetic alterations and DNA repair in human carcinogenesis. Semin Cancer Biol 2004;14:441–8.

[3] Jang M, Kim SS, Lee J. Cancer cell metabolism: implications for therapeutic targets. Exp Mol Med 2013;45:e45.

[4] Cairns RA, Harris IS, Mak TW. Regulation of cancer cell metabolism. Nat Rev Cancer 2011;11:85–95.
[5] Zhao Y, Butler EB, Tan M. Targeting cellular metabolism to improve cancer therapeutics. Cell Death Dis 2013; 4:e532.
[6] DeBerardinis RJ, Lum JJ, Hatzivassiliou G, Thompson CB. The biology of cancer: metabolic reprogramming fuels cell growth and proliferation. Cell Metab 2008;7:11–20.
[7] Warburg O, Wind F, Negelein E. The metabolism of tumors in the body. J Gen Physiol 1927;8:519–30.
[8] Vander Heiden MG, Cantley LC, Thompson CB. Understanding the Warburg effect: the metabolic requirements of cell proliferation. Science 2009;324:1029–33.
[9] Singh MS, Tammam SN, Shetab Boushehri MA, Lamprecht A. MDR in cancer: addressing the underlying cellular alterations with the use of nanocarriers. Pharmacol Res 2017;126:2–30.
[10] Gillet J-P, Gottesman MM. Mechanisms of multidrug resistance in cancer. In: Zhou J, editor. Multi-drug resist. Cancer. vol. 596. Totowa, NJ: Humana Press; 2010. p. 47–76.
[11] Wang J, Seebacher N, Shi H, Kan Q, Duan Z. Novel strategies to prevent the development of multidrug resistance (MDR) in cancer. Oncotarget 2017;8.
[12] Stäubert C, Bhuiyan H, Lindahl A, Broom OJ, Zhu Y, Islam S, et al. Rewired metabolism in drug-resistant leukemia cells: a metabolic switch hallmarked by reduced dependence on exogenous glutamine. J Biol Chem 2015;290:8348–59.
[13] Alfarouk KO. Tumor metabolism, cancer cell transporters, and microenvironmental resistance. J Enzyme Inhib Med Chem 2016;31:859–66.
[14] Saraswathy M, Gong S. Different strategies to overcome multidrug resistance in cancer. Biotechnol Adv 2013;31:1397–407. https://doi.org/10.1016/j.biotechadv.2013.06.004.
[15] Matuszcak C, Lindner K, Haier J, Hummel R. Proton pump inhibitors as chemosensitizer: new indication for a well-known medication. Cancer Cell Microenviron 2015;2:e667. https://doi.org/10.14800/ccm.66.
[16] Izumi H, Torigoe T, Ishiguchi H, Uramoto H, Yoshida Y, Tanabe M, et al. Cellular pH regulators: potentially promising molecular targets for cancer chemotherapy. Cancer Treat Rev 2003;29:541–9.
[17] Taylor S, Spugnini EP, Assaraf YG, Azzarito T, Rauch C, Fais S. Microenvironment acidity as a major determinant of tumor chemoresistance: proton pump inhibitors (PPIs) as a novel therapeutic approach. Drug Resist Updat 2015;23:69–78.
[18] Garofalo M, Grazioso G, Cavalli A, Sgrignani J. How computational chemistry and drug delivery techniques can support the development of new anticancer drugs. Molecules 2020;25:1756.
[19] Bruno A, Costantino G, Sartori L, Radi M. The in silico drug discovery toolbox: applications in Lead discovery and optimization. Curr Med Chem 2019;26:3838–73.
[20] Bruno A, Costantino G. Molecular dynamics simulations of G protein-coupled receptors. Mol Inform 2012;31:222–30.
[21] Kolb P, Ferreira RS, Irwin JJ, Shoichet BK. Docking and chemoinformatic screens for new ligands and targets. Curr Opin Biotechnol 2009;20:429–36.
[22] Irwin JJ, Shoichet BK. Docking screens for novel ligands conferring new biology: miniperspective. J Med Chem 2016;59:4103–20.
[23] Leach AR, Shoichet BK, Peishoff CE. Prediction of protein – ligand interactions. Docking and scoring: successes and gaps. J Med Chem 2006;49:5851–5.
[24] Truchon J-F, Bayly CI. Evaluating virtual screening methods: good and bad metrics for the "early recognition" problem. J Chem Inf Model 2007;47:488–508.
[25] Krishnamurthy VM, Kaufman GK, Urbach AR, Gitlin I, Gudiksen KL, Weibel DB, et al. Carbonic anhydrase as a model for biophysical and physical-organic studies of proteins and protein – ligand binding. Chem Rev 2008;108:946–1051.
[26] Supuran CT. Carbonic anhydrases: novel therapeutic applications for inhibitors and activators. Nat Rev Drug Discov 2008;7:168–81.
[27] Alterio V, Di Fiore A, D'Ambrosio K, Supuran CT, De Simone G. Multiple binding modes of inhibitors to carbonic anhydrases: how to design specific drugs targeting 15 different isoforms? Chem Rev 2012;112:4421–68.
[28] Xu Y, Feng L, Jeffrey PD, Shi Y, Morel FMM. Structure and metal exchange in the cadmium carbonic anhydrase of marine diatoms. Nature 2008;452:56–61.
[29] Nocentini A, Supuran CT. Carbonic anhydrase inhibitors as antitumor/antimetastatic agents: a patent review (2008–2018). Expert Opin Ther Pat 2018;28:729–40.

[30] Annunziato G, Giovati L, Angeli A, Pavone M, Del Prete S, Pieroni M, et al. Discovering a new class of antifungal agents that selectively inhibits microbial carbonic anhydrases. J Enzyme Inhib Med Chem 2018;33:1537–44.
[31] Annunziato G, Angeli A, D'Alba F, Bruno A, Pieroni M, Vullo D, et al. Discovery of new potential anti-infective compounds based on carbonic anhydrase inhibitors by rational target-focused repurposing approaches. ChemMedChem 2016;11:1904–14.
[32] Perut F, Carta F, Bonuccelli G, Grisendi G, Di Pompo G, Avnet S, et al. Carbonic anhydrase IX inhibition is an effective strategy for osteosarcoma treatment. Expert Opin Ther Targets 2015;19:1593–605.
[33] Kopecka J, Campia I, Jacobs A, Frei AP, Ghigo D, Wollscheid B, et al. Carbonic anhydrase XII is a new therapeutic target to overcome chemoresistance in cancer cells. Oncotarget 2015;6.
[34] Teodori E, Braconi L, Bua S, Lapucci A, Bartolucci G, Manetti D, et al. Dual P-glycoprotein and CA XII inhibitors: a new strategy to reverse the P-gp mediated multidrug resistance (MDR) in Cancer cells. Molecules 2020;25:1748.
[35] Federici C, Lugini L, Marino ML, Carta F, Iessi E, Azzarito T, et al. Lansoprazole and carbonic anhydrase IX inhibitors sinergize against human melanoma cells. J Enzyme Inhib Med Chem 2016;31:119–25.
[36] Bryant JL, Gieling RG, Meredith SL, Allen T-J, Walker L, Telfer BA, et al. Novel carbonic anhydrase IX-targeted therapy enhances the anti-tumour effects of cisplatin in small cell lung cancer: targeting CA IX as a novel treatment for SCLC. Int J Cancer 2018;142:191–201.
[37] Andreucci E, Ruzzolini J, Peppicelli S, Bianchini F, Laurenzana A, Carta F, et al. The carbonic anhydrase IX inhibitor SLC-0111 sensitises cancer cells to conventional chemotherapy. J Enzyme Inhib Med Chem 2019;34:117–23.
[38] Pastorek J, Pastorekova S. Hypoxia-induced carbonic anhydrase IX as a target for cancer therapy: from biology to clinical use. Semin Cancer Biol 2015;31:52–64.
[39] Supuran CT, Alterio V, Di Fiore A, D' Ambrosio K, Carta F, Monti SM, et al. Inhibition of carbonic anhydrase IX targets primary tumors, metastases, and cancer stem cells: three for the price of one. Med Res Rev 2018;38:1799–836.
[40] Monti SM, Supuran CT, De Simone G. Anticancer carbonic anhydrase inhibitors: a patent review (2008–2013). Expert Opin Ther Pat 2013;23:737–49.
[41] Wingo T, Tu C, Laipis PJ, Silverman DN. The catalytic properties of human carbonic anhydrase IX. Biochem Biophys Res Commun 2001;288:666–9.
[42] Haapasalo J, Hilvo M, Nordfors K, Haapasalo H, Parkkila S, Hyrskyluoto A, et al. Identification of an alternatively spliced isoform of carbonic anhydrase XII in diffusely infiltrating astrocytic gliomas. Neuro-Oncol 2008;10:131–8.
[43] Hsieh M-J, Chen K-S, Chiou H-L, Hsieh Y-S. Carbonic anhydrase XII promotes invasion and migration ability of MDA-MB-231 breast cancer cells through the p38 MAPK signaling pathway. Eur J Cell Biol 2010;89:598–606.
[44] Kivelä A, Parkkila S, Saarnio J, Karttunen TJ, Kivelä J, Parkkila A-K, et al. Expression of a novel transmembrane carbonic anhydrase isozyme XII in normal human gut and colorectal tumors. Am J Pathol 2000;156:577–84.
[45] Kivelä AJ, Parkkila S, Saarnio J, Karttunen TJ, Kivelä J, Parkkila A-K, et al. Expression of transmembrane carbonic anhydrase isoenzymes IX and XII in normal human pancreas and pancreatic tumours. Histochem Cell Biol 2000;114:197–204.
[46] Ilie MI, Hofman V, Ortholan C, Ammadi RE, Bonnetaud C, Havet K, et al. Overexpression of carbonic anhydrase XII in tissues from resectable non-small cell lung cancers is a biomarker of good prognosis. Int J Cancer 2011;128:1614–23. https://doi.org/10.1002/ijc.25491.
[47] Kivela AJ. Expression of von Hippel-Lindau tumor suppressor and tumor-associated carbonic anhydrases IX and XII in normal and neoplastic colorectal mucosa. World J Gastroenterol 2005;11:2616.
[48] Salmas RE, Senturk M, Yurtsever M, Durdagi S. Discovering novel carbonic anhydrase type IX (CA IX) inhibitors from seven million compounds using virtual screening and *in vitro* analysis. J Enzyme Inhib Med Chem 2015;1–9.
[49] Thiry A, Ledecq M, Cecchi A, Frederick R, Dogné J-M, Supuran CT, et al. Ligand-based and structure-based virtual screening to identify carbonic anhydrase IX inhibitors. Bioorg Med Chem 2009;17:553–7.
[50] Kurt BZ, Dag A, Doğan B, Durdagi S, Angeli A, Nocentini A, et al. Synthesis, biological activity and multiscale molecular modeling studies of bis-coumarins as selective carbonic anhydrase IX and XII inhibitors with effective cytotoxicity against hepatocellular carcinoma. Bioorg Chem 2019;87:838–50.
[51] De Simone G, Vitale RM, Di Fiore A, Pedone C, Scozzafava A, Montero J-L, et al. Carbonic anhydrase inhibitors: hypoxia-activatable sulfonamides incorporating disulfide bonds that target the tumor-associated isoform IX. J Med Chem 2006;49:5544–51.

[52] Thacker PS, Alvala M, Arifuddin M, Angeli A, Supuran CT. Design, synthesis and biological evaluation of coumarin-3-carboxamides as selective carbonic anhydrase IX and XII inhibitors. Bioorg Chem 2019;86:386–92.
[53] Gao R, Liao S, Zhang C, Zhu W, Wang L, Huang J, et al. Optimization of heterocyclic substituted benzenesulfonamides as novel carbonic anhydrase IX inhibitors and their structure activity relationship. Eur J Med Chem 2013;62:597–604.
[54] Wang L, Yang C, Lu W, Liu L, Gao R, Liao S, et al. Discovery of new potent inhibitors for carbonic anhydrase IX by structure-based virtual screening. Bioorg Med Chem Lett 2013;23:3496–9.
[55] Supuran CT. Carbonic anhydrase inhibition and the management of hypoxic tumors. Metabolites 2017;7:48.
[56] Burroughs SK, Kaluz S, Wang D, Wang K, Van Meir EG, Wang B. Hypoxia inducible factor pathway inhibitors as anticancer therapeutics. Future Med Chem 2013;5:553–72.
[57] Semenza GL. Hypoxia-inducible factors: mediators of cancer progression and targets for cancer therapy. Trends Pharmacol Sci 2012;33:207–14.
[58] Wilkins SE, Abboud MI, Hancock RL, Schofield CJ. Targeting protein-protein interactions in the HIF system. ChemMedChem 2016;11:773–86.
[59] Smith TG, Robbins PA, Ratcliffe PJ. The human side of hypoxia-inducible factor: review. Br J Haematol 2008;141:325–34.
[60] Kalyanaraman B. Teaching the basics of cancer metabolism: developing antitumor strategies by exploiting the differences between normal and cancer cell metabolism. Redox Biol 2017;12:833–42.
[61] Broekgaarden M, Weijer R, Krekorian M, van den IJssel B, Kos M, Alles LK, et al. Inhibition of hypoxia-inducible factor 1 with acriflavine sensitizes hypoxic tumor cells to photodynamic therapy with zinc phthalocyanine-encapsulating cationic liposomes. Nano Res 2016;9:1639–62.
[62] Mueckler M, Caruso C, Baldwin S, Panico M, Blench I, Morris H, et al. Sequence and structure of a human glucose transporter. Science 1985;229:941–5.
[63] Olson AL, Pessin JE. Structure, function, and regulation of the mammalian facilitative glucose transporter gene family. Annu Rev Nutr 1996;16:235–56.
[64] Calvo MB, Figueroa A, Pulido EG, Campelo RG, Aparicio LA. Potential role of sugar transporters in cancer and their relationship with anticancer therapy. Int J Endocrinol 2010;2010:1–14.
[65] Krzeslak A, Wojcik-Krowiranda K, Forma E, Jozwiak P, Romanowicz H, Bienkiewicz A, et al. Expression of GLUT1 and GLUT3 glucose transporters in endometrial and breast cancers. Pathol Oncol Res 2012;18:721–8.
[66] Larson SM, Schwartz LH. 18F-FDG PET as a candidate for "qualified biomarker": functional assessment of treatment response in oncology. J Nucl Med Off Publ Soc Nucl Med 2006;47:901–3.
[67] Deng D, Xu C, Sun P, Wu J, Yan C, Hu M, et al. Crystal structure of the human glucose transporter GLUT1. Nature 2014;510:121–5.
[68] Kapoor K, Finer-Moore JS, Pedersen BP, Caboni L, Waight A, Hillig RC, et al. Mechanism of inhibition of human glucose transporter GLUT1 is conserved between cytochalasin B and phenylalanine amides. Proc Natl Acad Sci 2016;113:4711–6.
[69] Schmidl S, Iancu CV, Choe J, Oreb M. Ligand screening systems for human glucose transporters as tools in drug discovery. Front Chem 2018;6.
[70] Mishra RK, Wei C, Hresko RC, Bajpai R, Heitmeier M, Matulis SM, et al. *In silico* modeling-based identification of glucose transporter 4 (GLUT4)-selective inhibitors for cancer therapy. J Biol Chem 2015;290:14441–53.
[71] George Thompson AM, Ursu O, Babkin P, Iancu CV, Whang A, Oprea TI, et al. Discovery of a specific inhibitor of human GLUT5 by virtual screening and in vitro transport evaluation. Sci Rep 2016;6.
[72] Ung PM-U, Song W, Cheng L, Zhao X, Hu H, Chen L, et al. Inhibitor discovery for the human GLUT1 from homology modeling and virtual screening. ACS Chem Biol 2016;11:1908–16.
[73] Sun L, Zeng X, Yan C, Sun X, Gong X, Rao Y, et al. Crystal structure of a bacterial homologue of glucose transporters GLUT1–4. Nature 2012;490:361–6.
[74] Bender A, Jenkins JL, Scheiber J, Sukuru SCK, Glick M, Davies JW. How similar are similarity searching methods? A principal component analysis of molecular descriptor space. J Chem Inf Model 2009;49:108–19.
[75] Bender A. How similar are those molecules after all? Use two descriptors and you will have three different answers. Expert Opin Drug Discovery 2010;5:1141–51.
[76] Baell JB, Holloway GA. New substructure filters for removal of pan assay interference compounds (PAINS) from screening libraries and for their exclusion in bioassays. J Med Chem 2010;53:2719–40.

[77] Dahlin JL, Nissink JWM, Strasser JM, Francis S, Higgins L, Zhou H, et al. PAINS in the assay: chemical mechanisms of assay interference and promiscuous enzymatic inhibition observed during a sulfhydryl-scavenging HTS. J Med Chem 2015;58:2091–113.

[78] Hamed AR, Abdel-Azim NS, Shams KA, Hammouda FM. Targeting multidrug resistance in cancer by natural chemosensitizers. Bull Natl Res Cent 2019;43.

CHAPTER

12

Nanostructures and innovative delivery systems for overcoming cancer resistance

Jean-Yves Winum

IBMM, Univ Montpellier, CNRS, ENSCM, Montpellier, France

Abstract

Nanomedicine is raising ever more and more hopes, especially in the oncology field as it offers innovative therapeutic strategies to overcome the various limitations of conventional chemotherapy, especially drug resistance. Nanostructures, in addition to being vehicles for drug delivery, can be used as pharmaceuticals as well as diagnostics. Most of the advances in targeted drug delivery have occurred in cancer therapy. This chapter provides an overview of the recent pH-sensitive nanostructures and drug delivery systems in the oncology field. In addition, innovative nanosystems targeting one of the main components responsible of the tumor microenvironment acidity, enzymes such as hCA IX and hCA XII, are presented as emerging therapeutic option and novel areas to be explored.

Abbreviations

ABC transporter	ATP binding cassette transporter
ATRAM peptide	acidity-triggered rational membrane peptide
CA	carbonic anhydrase
CAI	carbonic anhydrase inhibitor
DOX	doxorubicin
hCA	human carbonic anhydrase
HSA	human serum albumin
MCT	monocarboxylate transporter
MDR	multidrugs resistance
NIR	near infrared
NP	nanoparticle
OPNi	organic photodynamic nano-inhibitor
PDPA	poly(2-(diisopropylamino)ethylmethacrylate)
PDT	photodynamic therapy

https://doi.org/10.1016/B978-0-12-820701-7.00002-6

PDTC	ammonium pyrrolidine dithiocarbamate
PK properties	pharmacokinetic properties
PTX	paclitaxel
RGD peptide	arginylglycylaspartic acid peptide
TME	tumor microenvironment
TNBC	triple negative breast cancer

Conflict of interest

No potential conflicts of interest were disclosed.

Introduction

Cancer is a multifaceted disease and one of the leading causes of death worldwide. It is responsible for an estimated 9.6 million deaths and affecting almost 18,1 million people in 2018 [1, 2].

Conventional cytotoxic chemotherapeutic drugs (alone or in combination), radiation therapy, targeted therapy, and immunotherapy are used in the management and the treatment of cancer. Despite the considerable improvements observed the last decade with these therapies, major problems arise with inefficiency of chemotherapeutic strategies against tumors. Several types of resistance to chemotherapy have been observed in clinical practice:

(a) Intrinsic resistance occurs immediately upon administration of the first chemotherapy sessions, without an initial sensitivity phase. Many tumors are thus resistant from the outset: this is the case for melanoma, many sarcomas, and brain tumors in particular.
(b) Acquired chemoresistance: after an initial phase of high chemosensitivity, there is a secondary progression of the disease showing acquired resistance. This is the case of many solid tumors such as breast and ovarian cancers.

The molecular mechanisms explaining these resistance phenomena have been highlighted over the last 20 years. There are many different mechanisms that may explain this resistance to therapy, the most important being the following: (a) reducing drug accumulation and increasing drug export. The overexpression of membrane proteins, such as P-glycoprotein which belongs to ABC transporter family, is a pump causing a flow of the cytotoxic from the intracellular medium to the extracellular medium. The intracellular concentration of the drug is thus decreased, reducing its efficacy. The MDR gene coding for P-glycoprotein is amplified in the cancer cell that has acquired resistance, leading to overexpression of this membrane protein.; (b) altering drug targets and signaling transduction molecules. Each of the targets of anticancer drugs can be modified, undergoing qualitative or quantitative changes. This target can indeed mutate, it is then no longer recognized by the cytotoxic (qualitative modification). The target can also be produced in large quantities in the cancer cell (qualitative modification) due to amplification of the gene coding for this target. In the first case the drug no longer recognizes its target, in the second case the amount of drug is insufficient to cause significant cytotoxicity; (c) increasing repair of drug-induced DNA damage. DNA repair phenomena can be deregulated, leading to a reduction in cytotoxicity by accelerated repair of DNA damage, in particular through the amplification of genes coding for repair proteins whose level is clearly amplified; (d) promoting apoptosis evasion programs.

To resist cell death, cancer cells decrease the action of death factors (p53 for example), increase the action of survival factors (Bcl-2 for example), or perform both actions simultaneously. Other proteins that trigger apoptosis are inactivated or their expression is decreased in cancer cells in order to increase their ability to survive in the face of ongoing stress and damage [3, 4].

Resistance induced by a cytotoxic agent also extends crossresistance to a wide variety of drugs with different chemical structures and cellular targets. Once resistance emerges, high-dose systemic therapy becomes ineffective and more toxic, and resistance is enhanced further [5]. Multidrug resistance (MDR) is a severe limitation of conventional chemotherapies or targeted cancer therapies contributing to limit the survival rate to 10% of patients with metastatic cancer. It is a major factor of therapeutic failure and therefore represents a significant hurdle to achieving cures in patients with cancer, so innovative safe and effective strategies are needed to overcome MDR [6, 7].

Tumor microenvironmental conditions, which are a key factors of tumor heterogeneity, play a crucial role in all aspects of tumorigenesis (initiation, progression, and metastasis) strongly contributing to aggressive phenotype, malignant progression, and resistance to chemo- and radiotherapy treatments [8, 9].

The presence of hypoxic domains, defined as areas of oxygen deprivation (typically less than 1%), alters metabolism in solid tumors triggering acidosis of tumor microenvironment (TME), which is an essential contributor of the resistance of cancer cells to therapy [10–13].

The acidic tumor microenvironment, due to the high aerobic glycolysis rate, is known as Warburg effect. Several key proteins and buffer systems are involved in the tumoral pH regulatory process, allowing solid tumors to be resistant to extracellular acidosis (pH as low as 6.5) [14]. These proteins include monocarboxylate transporters (MCTs), isoforms of anion exchanger, Na^+/HCO_3^- cotransporters, Na^+/H^+ exchangers, and carbonic anhydrases (CAs) isoforms IX and XII. These proteins involved in the maintenance of such an altered pH gradient were proposed to be targeted for novel antitumor and diagnostic agents [15, 16].

The unique tumor extracellular acidity can be utilized to design stimuli-responsive smart nanocarriers, and can provide new opportunities for tumor treatment [17]. Hence, a drug delivery nanoplatform that could target a key player of tumor microenvironment or be triggered by the TME could be considered as a promising approach to realize targeted cancer treatment. The objective is to deliver the active substances to the desired sites of therapeutic action while reducing undesirable side effects, and overcoming the development of multidrug resistance [18]. Over the past decade, great progress has been made in the development of pH-responsive nanoplatforms as intelligent drug delivery systems. These systems have shown better performances in solving existing drug delivery problems, such as improving tumor accumulation, facilitating cell absorption or controlled release of drugs in space and time [19, 20].

The main aim of this chapter is to offer to the readers an updated overview on new drug delivery strategies aiming at addressing cancer resistance mechanisms. Promising generation of nanoparticle-based platforms and drug delivery carriers triggered by acidic tumor microenvironment, reported in very recent literature, will be emphasized. A special highlight on nano-objects targeting tumor-associated carbonic anhydrase IX and XII is also provided as both of these enzymes are associated with cancer progression, metastasis, and impaired therapeutic response [21].

pH-sensitive nano-objects targeting tumor cells

Nanodrugs offer innovative therapeutic strategies to overcome the various limitations of conventional chemotherapy, especially multidrug resistance. Due to their unique properties, they have a great potential in cancer therapy by minimizing toxicity to healthy cells, by enhancing solubility of anticancer drugs, or by overcoming multidrug resistance [22–24]. Nanosystems responsive to endogenous or exogenous stimuli have been well investigated in the last decade and pH-responsive nanoparticles are considered as the most skillful strategy to achieve reversal of MDR [25–27].

pH-sensitive nanostructures include diverse platforms enabling a plethora of rationally designed therapeutic nanosystems [28]. Their main advantages are the selective targeting of cancer cells and release of the cytotoxic drug at the tumor site. In addition, they protect the drug from degradation in the body, reducing the dose needed for treatment. In recent years, considerable efforts have been made to design stimuli-responsive smart nanoplatforms, and several papers and book chapters reviewing pH-sensitive nanostructures for targeted delivery of anticancer therapeutic compounds have been recently published [29–32].

Studies extracted from the mostly recent publications and dedicated to nanoparticles responsive to acidic pH are discussed, along with approaches for improving tumor-specific therapy.

pH-responsive polymeric nanostructures

Bioresponsive polymeric nanostructures using pH as trigger for tumors targeting have been highlighted in different recent review papers [33–38]. In this section, we have selected examples of recent studies from 2019 and 2020 literature that demonstrate the usefulness of pH-responsive polymeric nanocarriers as potential vehicles for cancer therapy, able to overcome chemoresistance and enhance therapeutic effect.

Yao and coll. Reported in 2019 multifunctional pH-responsive coassembled nanoplatforms constituted in their inner core of a poly(2-(diisopropylamino)ethylmethacrylate) (PDPA) polymer, and at the outer shell of amphiphilic lipid-poly(ethylene glycol) (lipid-PEG) with the internalizing RGD (iRGD) peptide encoded on the surface [39]. Loaded with mitoxantrone anticancer drug, the resulting large size NP platform (80 nm) was shown to be entrapped in the tumor tissues and then specifically internalized by tumor cells via iRGD-mediated ligand–receptor interactions. Endosomal pH response of the PDPA polymer induced rapid intracellular release of mitoxantrone and thus efficient inhibition of tumor growth of breast cancer in vitro and in vivo with negligible toxicities [39].

In another study, Hira and coll reported in 2019 a novel ABA-type polyethylene glycol (PEG)-b-polyketal (PK)-b-PEG block copolymer, which has shown faster degradation at acidic pH [40]. Self-assembly of this block copolymer as micelle structure of average size of 6.2 nm was loaded with anticancer drug doxorubicin to lead to block copolymeric micelles with average size of 69.9 nm. These nanosystems have shown pH-dependent elevated drug release at pH 6.4, and demonstrated tumoricidal activities against DL and RAJI lymphoma cells with significant levels of growth inhibition and apoptosis but also on the doxorubicin-resistant DL and RAJI cells. Therapy with doxorubicin-loaded copolymer reduced the tumor size and augmented the life span of the tumor-bearing animals with improved histopathological parameters, compared with the untreated control [40].

Cheng and coll. Described the coencapsulation of antitumor drugs doxorubicin and the chemosensitizer ammonium pyrrolidine dithiocarbamate (PDTC) in the pH-sensitive polymeric nanoparticles, based on poly-(orthoester urethanes) copolymers, to overcome chemoresistance and enhance the therapeutic effect [41]. These spherical dual-drug-loaded nanoparticles were shown to have an average size of 220 nm with excellent physiological stability. 2D monolayer cells and 3D multicellular spheroids experiments showed that PDTC could significantly reverse MDR, improve intracellular drugs accumulation and penetration, and then enhance doxorubicin-induced cytotoxicity and apoptosis. Superior antitumor efficacy with reduced side effects was proved with in vivo mice experiments suggesting resulting in a prolonged time in bloodstream and an increase drug accumulation at tumor site [41].

The same strategy was used in 2019 by Wan and coll. Who described a coloaded system comprising paclitaxel and cisplatin inside polymeric micelle based on amphiphilic *block* copolymer poly(2-methyl-2-oxazoline-block-2-butyl-2-oxazoline-block-2-methyl-2-oxazoline) [42]. This micelle-based nanosystem showed improvement of PK properties and an increase of tumor distribution. Improvement of antitumor efficacy was demonstrated in vivo in cisplatin-resistant human ovarian carcinoma A2780/CisR xenograft tumor and multidrug-resistant breast cancer LCC-6-MDR orthotopic tumor models [42].

Zhang and coll. Described a PEGylated nano-drug system built on the acidic responsive imine linkage of the polysaccharide hyaluronic acid with doxorubicine and a methoxy polyethylene glycol amine. High drug loading, good stability, and excellent pH-specific programmed drug release were observed. In vivo experiments showed a long cycle time and tumor-targeted accumulation capacity of the nanocarrier, with an effective tumor growth inhibition on a tumor-bearing mouse model (mouse colorectal cancer CT26 cells) [43].

Using also hyaluronic acid, Pang and Coll. developed a promising delivery system based on lipid-polymer hybrid nanoparticles decorated with this polysaccharide and loaded with erlotinib and bevacizumab for targeting non-small-cell lung cancer. The nanosystem showed low toxicity, good stability, and pharmacokinetic properties, and is able to accumulate to the tumor site causing high cytotoxic effect on tumor cells in vitro and in vivo [44].

Another system was described very recently by Palanikumar and coll., whose architecture is based on a polylactic-*co*-glycolic acid (PLGA) core covalently coated with crosslinked bovine serum albumin (BSA) shell, itself functionalized with the acidity-triggered rational membrane (ATRAM) peptide [45]. These nanoparticles, loaded with the mitochondrial targeted- doxorubicine-triphenylphosphonium, showed highly efficient pH-dependent cellular uptake, high cytotoxicity and selectivity, in vitro against cancer cells in an acidic environment. In vivo treatment showed an effective reduced 4T1 tumor volume and mass and prolonged survival, while exhibiting no obvious toxicity to healthy tissue [45].

pH-sensitive silica or gold nanoplatforms

Silica nanoparticles (porous and nonporous) are excellent nanoplatforms, which have been highly used in nanomedicine research for drug delivery. They have attracted the attention of numerous researchers due to their unique properties, such as high chemical stability, large surface area, and for mesoporous silica nanoparticle, tunable pores diameters and volumes, allowing the incorporation of large amounts of drugs, protecting them from deactivation and degradation processes. The state-of-the-art of stimuli-responsive silica nanosystems for drug delivery has been deeply reviewed in 2019 and 2020 [46–50]. One very recent example concern the work reported by Thorat and coll. Who described a pH-responsive, intracellular theranostic supra-assembled nanocargos delivery system, based on porous silica nanoparticles conjugated with anticancer drug doxorubicine, cytochrome *c* enzyme CytC (involved in life-sustaining energy production and cell apoptosis) and dual-function anticancer nucleolin-aptamer AS1411 [51]. In vitro studied showed a synergistic increase in cellular and nuclear DOX levels leading to important cytotoxicity effect against chemoresistant colon cancer HCT116 cell lines. Interesting pharmacokinetic profile was noticed and very good safety index was observed in vitro and in vivo [51].

Gold nanosystems are also very interesting promising platform in nanomedicine and have been widely explored in the last decade by the scientific community to develop sensors due to their optical properties, but also make possible the development of new hyperthermia and drug delivery systems [52–54].

For example, a very recent comprehensive study recently reported by Kumar and coll. Described the use of gold nanoparticles capped with a natural pH-responsive short tripeptide (Lys-Phe-Gly) sequence for efficient intracellular delivery of doxorubicin [55]. This interesting nanosystem used the tripeptide for the stabilization of gold nanoparticles and for the electrostatic and H-bonding interactions to doxorubicin. In vitro studies showed that loaded doxorubicin was efficiently release intracellularly upon low pH in different cancer cell lines, including the drug-resistant one [55].

Another innovative nanosystem was reported by Jiang and coll. Who elaborated multimodal nanocomposite based on gold nanorod core with triple layer coating made of mesoporous silica, polyhistidine and tocopherol polyethylene glycol succinate [56]. The nanocomposite displayed good properties for both photothermal therapy and chemotherapy, achieving pH-responsive and NIR light-triggered drug-controlled release in vitro and in vivo on multidrug-resistant colorectal cancer SW620/Ad300 cells. Synergistic chemophotothermal therapy was also evidenced both in vitro and in vivo and higher therapeutic efficacy was characterized compared to mono chemo- or photothermal therapy [56].

Nanostructure targeting CA IX/CA XII

Targeted nanosystems in the form of nanoparticles have attracted attention due to their fundamental interest and practical importance. Tumor targeting can be improved through the attachment of pH-interfering agents onto the surface of nanosystems, and subsequent binding to corresponding protein overexpressed on the surface of cancer cells. Among the important protein targets contributing to extracellular acidosis in solid tumors, carbonic anhydrases IX and XII are pivotal players.

Human carbonic anhydrase (CA, EC 4.2.1.1) IX (hCA IX) and XII (hCA XII) are transmembrane isoform proteins belonging to the human carbonic anhydrase family, which catalyze very efficiently the reversible hydration of carbon dioxide in bicarbonate with a proton release ($CO_2 + H_2O \leftrightharpoons HCO_3^- + H^+$). Both of these enzymes are hypoxia-inducible components of the tumoral pH regulatory system. Their expression is upregulated in a wide variety of human tumors under the control of HIF-1α activation cascade and results in dramatically enhanced hydrolysis of CO_2 to HCO_3^- and H^+ at the extracellular surface of cancer cells, providing a chemical buffer that is actively transported into the cell to help maintaining intracellular pH_i at 7.2 and a source of H^+, which contributes to the increasing acidosis of the extracellular space $pH_e = 6.8$ [57].

Hence CA IX and XII are associated with cancer progression, metastasis, and impaired therapeutic response. Regarding the expression patterns and functional implications of these isozymes in tumor, CA IX and XII are considered as biomarkers and therapeutic targets for various cancer types The last decade was an intense and fruitful period in the field of tumor-associated CA IX and CA XII as many key achievements have been made in drug discoveries in immunotherapy with monoclonal antibodies, or with classical small molecules inhibitors targeting CAIX/CAXII enzymatic activity [58–61].

The efficacy of CAIX inhibition as an anticancer strategy has been validated in numerous preclinical studies using a variety of cancer models and clinical trials of the sulfonamide SLC-0111 (also known as WBI-5111), which completed Phase I.

To address the selectivity and potency of CA inhibitors against the membrane-associated isoforms carbonic anhydrase IX (hCA IX) and carbonic anhydrase XII (hCA XII), the use of multivalent CA-directed nanoplatforms appears to be an interesting approach [62, 63]. Multivalent nanoplatforms coated with carbonic anhydrase inhibitors open new opportunities in the drug design of innovative selective carbonic anhydrase inhibitors and revealed to be particularly effective improving inhibitory activity and in selective targeting the isoforms expressed extracellularly, such as the tumor-associated CA IX and CA XII.

Gold nanoplatforms

The use of gold nanoplatforms to target carbonic anhydrases was first reported in 2008 with the study of functionalized gold nanoparticles coated with benzene sulfonamide inhibitors (Fig. 1). It was demonstrated that multimeric presentation of inhibitors improved inhibitory activity against the different isoforms tested (e.g., off-target isoforms CA I and CA II and also CA IX) especially with longer incubation process. Low nanomolar activity ($K_I = 2.4$ nM) and selective binding were observed in favor of CA IX. Moreover, these nanoparticles showed the important property to be unable to cross the cell membrane, remaining confined in extracellular space. This feature is of high interest for selective targeting of the extracellular active site CA IX vs intracellular offtargets CA I and CA II [64].

Improvement of the synthesis (one-pot preparation) as well as correlation study of nanoparticles properties (size and functional properties) with inhibitory activity was undertaken in 2013 by Bellissima et al. Reproducible results in terms of stability of the colloidal solution and chemicophysical properties were obtained [65].

The most advanced application on gold nanoparticles functionalized with carbonic anhydrase inhibitors was reported in 2018 by the group of M. Iies. 2-amino-1,3,4-thiadiazolyl-5-sulfonamide moiety (found in acetazolamide structure) was used as CA targeting head and connected via a spacer to gold nanoparticles (Fig. 1).

In vitro evaluation on colorectal HT-29 cell lines witnesses of a significant internalization of the nanoparticle under hypoxic conditions. This process was demonstrated to be mediated by

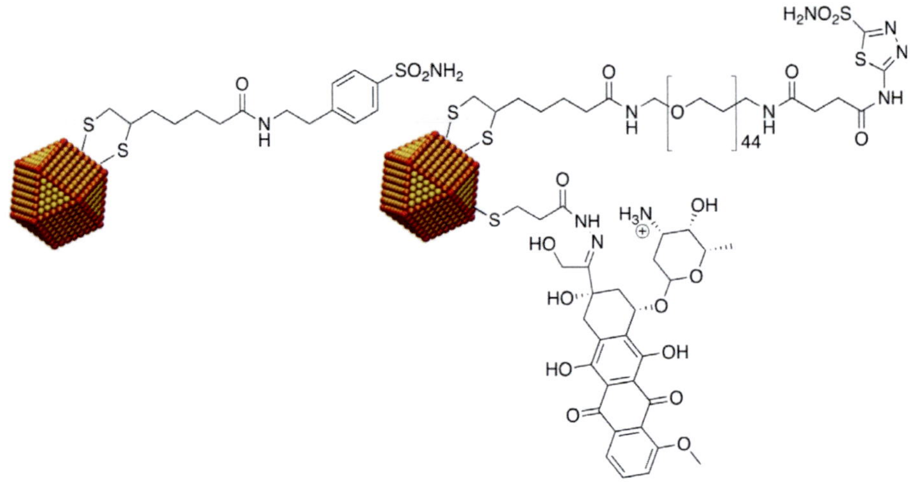

FIG. 1 Examples of gold nanoparticles targeting CA.

endocytosis uptake process triggered after binding to CA IX. Such system was then applied for doxorubicine (Dox) targeting by loading of Dox using a pH-sensitive hydrazone linker on the CAI-functionalized nanoparticles. One of the reported conjugates demonstrated interesting antiproliferative activity on HT-29 cell lines in vitro under hypoxia and on spheroid model. Enhanced intracellular delivery of Dox with this nanosystem, compared to the standard administration in solution, was brought to light using fluorescence spectroscopy [66].

In addition, derivatives of benzenesulfonamides combined with nanorods were reported in 2015. These systems showed no cytotoxicity on human colorectal carcinoma HCT116 (HCT) and human mammary adenocarcinoma MDA-MB-231 (MDR) cells after exposure to 24h. Incubation with both HCT and MDR cell lines under normoxic and hypoxic demonstrated preferential accumulation of gold nanorods on both cell lines incubated under hypoxia and treated with CAI-functionalized nanorods [67].

Silica-based nanoplatforms

Silica-based nanoplatforms are the mostly used nanomaterials, widely due to their structure (porous or nonporous), the easy functionalization of their surfaces, the range of accessible sizes as uniform particles (from 5 to 2000nm), their properties such as optically transparency, low toxicity, stability to pH, and their degradability.

In 2015, our group reported the design and the synthesis of non-porous nanometer-sized fluorescent hybrid silica (NFHS), decorated with benzenesulfonamide carbonic anhydrase inhibitors (Fig. 2). Positive multivalent effects affecting both the activity and the selectivity profile of inhibitors against cytosolic carbonic anhydrase isoforms versus the membrane bound isoforms were evidenced [68]. Other silica nano-objects such as Polyhedral Oligomeric Silsesquioxane (POSS) were also reported as potential multivalent systems with potential applications in the field of carbonic anhydrase inhibition (Fig. 2) [69].

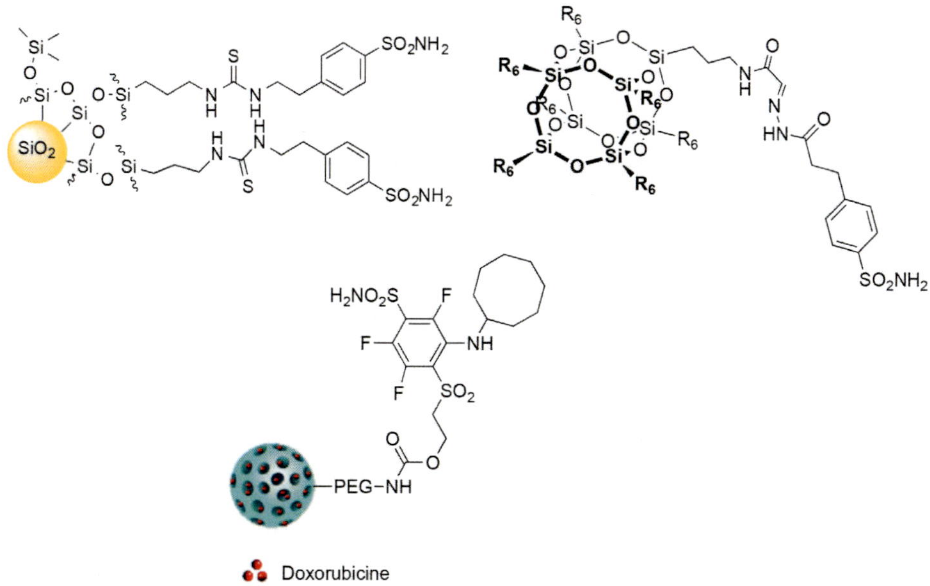

FIG. 2 Examples of silica nanoplatforms targeting CA.

Most advanced work in the field of silica nanoplatform targeting CA IX isoforms has been reported by Janoniene et al. in 2017 who investigated a porous silicon nanoparticle modified with a fluorescent carbonic anhydrase inhibitor and loaded with doxorubicin for dual hypoxia cancer imaging and therapy (Fig. 2). Binding and inhibition assays showed that the CA IX-targeted silicon nanoplatform demonstrated a high selectivity and binding affinity for its target, which translated into a strong affinity toward hypoxic human breast adenocarcinoma (MCF-7) cells and resulted in a significant reduction in cell viability as compared with its nontargeted counterpart [70].

Instead of using specific CA IX inhibitors as targeting agent, some authors reported silica nanoparticles functionalized with a monoclonal antibody IgG M75 that specifically binds to CA IX. A first study was described in 2013 by Tokárová and coll. Who reported the surface modification of fluorescent mesoporous silica nanoparticles with the monoclonal antibody (M75) known to specifically bind to the proteoglycan domain of carbonic anhydrase IX used as the antigen. Specific bioadhesion of such modified nanoparticles to surfaces containing either the PG domain of carbonic anhydrase IX (CA IX) or HT-29 cells was investigated and stronger interactions were observed, compared to the unmodified silica nanoparticles and nanoparticles coupled nonspecific antibody, BSA, or unmodified particles [71].

Lizonova and coll. reported in 2018 silica nanoplatforms functionalized with a nonimmunogenic stealth polymer poly(*N*-(2-hydroxypropyl)methacrylamide) and a monoclonal antibody IgG M75 that specifically binds to Carbonic Anhydrase IX (CA IX). Binding and inhibition assays showed that the CA IX-targeted silica nanoparticles demonstrated a high selectivity and binding affinity for its target, which translated into a strong affinity toward hypoxic human breast adenocarcinoma (MCF-7) cells and resulted in a significant reduction in cell viability as compared with its nontargeted counterpart. In vivo accumulation at the tumor site was demonstrated for targeted nanoparticles (10-fold higher concentration) compared to the nontargeted ones [72].

Iron oxide nanoparticles

Magnetic nanoparticle are also highly attractive nanomaterials for imaging and treatment of solid tumors because of their favorable biocompatibility and excellent biodegradability.

First report on iron oxide nanoparticles targeting carbonic anhydrase was reported in 2015 by Chen and coll. Who describe the preparation and characterization of magnetic nanoparticle bearing a sulfonamide function as targeting moiety. This nanosystem demonstrated no cytotoxicity in vitro and better contrast capability than nontargeting control both in vitro on HT-29 cells line and in vivo on nude mice with xenografted HT29 tumor [73].

In 2017, Khmara and coll described the preparation of magnetic nanoparticles for the detection and the targeting of tumor cells. The highly specific monoclonal antibody M75, which is able to bind to the extracellular domain of hCA IX, was coupled to magnetic nanoparticles stabilized by poly-L-lysine PLL (hydrodynamic nanoparticle diameter ranging from 100 to 200nm with different PLL/Fe_3O_4 ratios) using carbodiimide CDI procedure and utilized for targeting of the CA IX- expressing cells. Antibody-conjugated MNPs demonstrated cytostatic effect on the transfected C33 CA IX cells, control C33 neo cancer cells and a marginal reduction of proliferation on normal MRC5 human fibroblasts suggesting that their effect is cell type-specific. Indirect immunofluorescence revealed that the M75 Ab-coupled MNPs bound specifically to the CA IX antigen localized at the cell surface of the transfected C33 CA IX cells but not to the CA IX-negative C33 neo cells. Additional experiment showed intracellular accumulation of the Ab-MNPs in CA IX-positive cancer cells, Ab moiety driving the internalization of nanoparticles [74].

This work was applied by the same authors on magnetic nanoparticles coated with D, L-Lysine and conjugated with antibody (hydrodynamic diameter ranging from 26 to 29nm). In vitro experiment showed that antibody-free MNPs are not toxic for all tested cell lines and do not modify tubulin cytoskeletal structure, whereas M75-conjugated MNPs abrogate the formation of microtubules and slow down proliferation of CA IX-expressing C33 cells [75].

These studies were then extended by the same group in 2019, which reported the preparation of the same system (magnetic nanoparticles coated with poly-L-lysine and conjugated with monoclonal antibody M75) but grafted onto graphene oxide platform. Toxicological assays performed with this multifunctional platforms on B16 mouse melanoma cell line with (B16 CA IX) and without (B16-neo) CA IX, revealed no cytotoxicity effect on B16 mouse melanoma cell line without (B16-neo) CA IX, but a significant cytostatic effect on B16 mouse melanoma cell line with (B16 CA IX) after 48h. Immunofluorescence assays indirectly confirmed the bonding between graphene oxide and the magnetic nanoparticles conjugated to the monoclonal antibody toward B16 CA IX cell line providing evidence of tumor targeting [76].

Polymeric/oligomeric nanoplatforms

Alsaab et al. in 2018 developed an efficient nanoplatform constituted of an oligomer comprising of Vitamin-E- α-D-Tocopherol (TPGS) and styrene maleic anhydride (SMA) ligated with Acetazolamide. This oligomer named CA IX-SMA-TPGS was used to encapsulate CFM-4.16 (C4.16), an apoptosis inducer to obtain homogeneous nanoparticle CA IX-CFM4.16 with narrow polydispersity index that is suitable for in vivo delivery. Combination of CA IX-CFM4.16 with Sorafenib showed targeted delivery in hypoxic tumor resulting in induction of multimodal anticancer effects, including the reactivation of apoptosis, reversal of drug resistance, and reprogramming of malfunction macrophages. CA IX-CFM4.16 NP was found to be suitable for intravenous administration with superior tumor accumulation of CA IX-oligomer as compared to liver and demonstrated effective antitumor response in resistant A498 tumor. The tumor spheroid uptake study has clearly demonstrated excellent tumor core penetrating ability of CA IX-targeting oligomer, which is a critical indicator of tumor stromal disruption leading to better therapy response and immune modulation [77].

In a very recent paper, Jiang and coll reported the synthesis of organic photodynamic nano-inhibitor (OPNi) targeting hCA IX for overcome the hypoxia limitation of conventional photodynamic therapy (PDT) for synergistic cancer therapy. The nano-inhibitor (OPNi) consists of a near infra-red absorbing semiconducting polymer SP (PCVB) as the biodegradable photosensitizer and an acetazolamide (CAi)-terminated amphiphilic polymer (CAib-PEG-b-PPG-b-PEG-CAi). Selective targeting of CA-IX positive cancer cells (MDAMB-231 cells, an aggressive human breast cancer cell line with a high expression level of CA-IX) was observed with an enhanced accumulation of OPNi in tumor compared to counterpart nanoparticle (OPNc). Inhibition of the catalytic activity of CA-IX cause pH dysregulation, thereby the synergistic PDT of OPNi led to the superior therapeutic effect over monotherapy, which was indicated by the inhibited tumor growth and downregulated metastasis-related proteins [78].

Most recently, Liu and Coll. described the synthesis of a pH-cleavable amphipathic block copolymer of polycaprolactone-β-poly(ethylene glycol) that could form core-shell structure NPs (pHPP NPs) with a hydrophobic core and hydrophilic shell. When carbonic anhydrase inhibitor acetazolamide (AAZ) and the chemotherapeutic agent paclitaxel (PTX) were loaded inside these pHPP NPs (AAZ&PTX-NPs), an enhanced antitumor effect of ACE on PTX was obtained. AAZ&PTX-NPs were shown to have a long circulation time in blood, high accumulation in tumor tissue due to the EPR effect. This was explained by disintegration of the nanoparticle in the acidic tumor microenvironment and in situ release of acetazolamide as well as PTX inside the tumor. In vivo test on mice showed that treatment with AAZ and PTX coloaded NPs induce a decrease of tumor size and a higher survival rate when compared to mice treated with AAZ- or PTX-loaded NPs. This study reveals that simultaneous delivery of acetazolamide and paclitaxel to tumor tissue can upregulate the acidic pHe value, consequently enhancing the antitumor ability of chemotherapy medicine [79].

Others organic nanostructures

Several organic nanoplatforms bearing carbonic anhydrase inhibitors have been reported in the last past 4 years. Although the evaluation of these systems did not go beyond in vitro inhibitory activity against the different CA isoforms, they constitute interesting and promising systems through their potential multivalent properties to target CA IX.

Vincent's group described multimeric CA-inhibitors having different types of zinc binding functions, such as xanthate (3–6 inhibitors) or coumarin (12 inhibitors). Even if weak multivalent effects were observed for the inhibition of carbonic anhydrases (2–4 fold valence-corrected increase), improved potency (in the micromolar range) and selectivity were noticed compared to the monovalent analogues [80, 81].

In 2017, Kanfar and coll. Reported new multivalent CA inhibitors from functionalized peptide scaffolds (Fig. 3). Even if no multivalent effect was observed on tumor-associated CA IX and CA XII isoforms, enzyme inhibition assays revealed for the inhibition of hCA II a positive multivalent effect was detected since the relative potency per sulfonamide inhibitor (rp/n) increases by a factor of 3.3 comparing to the corresponding monovalent compound that contains the exact same sulfonamide-type inhibitor [82].

Multivalent inhibitory effects have also been reported the same year on the different carbonic anhydrase isoforms with a series of poly(amidoamine) (PAMAM) dendrimers grafted with carbonic anhydrase inhibitors (Fig. 3). The increase in the number of grafted inhibitors from dendrimer generation 0 to generation 3 led to inhibitory potency improvement and strong multivalent effects against hCA II, V, VII, IX, XII, and XIV [83, 84].

Others systems: Albumin or liposome base nanoplatforms

Tapiparti and coll. described in 2018 a human serum albumin (HSA)-based nanosystem labeled with carbonic anhydrase inhibitors acetazolamide (AAZ) designed to selectively deliver cancer drug paclitaxel (PTX) to the drug-resistant and invasive cell population of triple negative breast cancer TNBC tumors. In vitro evaluation against MDA-MB-231 and MDA-MB-468 cell lines, which correspond to the Triple Negative Breast Cancer (TNBC) expressing CA IX, showed a superior apoptosis-inducing capacity, with a stronger effect under hypoxic conditions (CA IX positive) compared to normoxic conditions (CA IX negative) [85].

Lin et al. developed in 2017 triptolide-loaded liposomes modified with anticarbonic anhydrase IX antibody for lung cancer therapy. The CA IX-targeted liposomes demonstrated a significant improvement in cell uptake in non-small-cell lung cancer cells (A549) under hypoxic conditions and in A549 three-dimensional tumor spheroidal models. Significant reduction in cell viability was also noticed as compared to the free drug and the nontargeted formulation. Furthermore, this CA IX-targeted liposomal platform demonstrated a significant restraint in tumor growth and increased the life span of mice with orthotopic lung tumors [86].

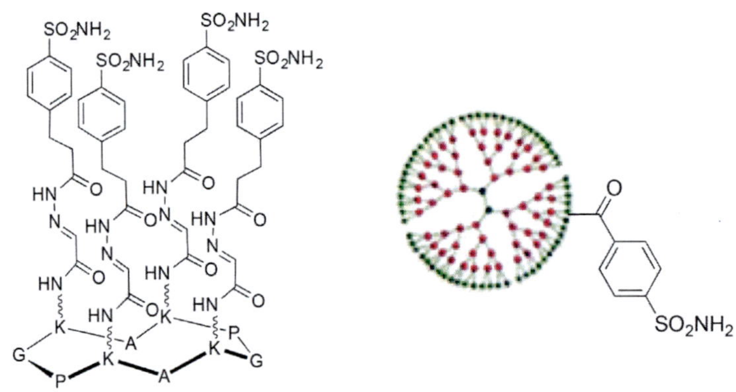

FIG. 3 Examples of cyclopeptidic and dendrimer platforms targeting CA.

Conclusion

A relevant progress has been made in developing pH-responsive nanosystems as smart drug delivery systems and the novel developments described here offer new opportunities for more efficient tumor-targeted nanocarriers and great potential for achieving better therapeutic effects in cancer treatment. We are still at an early stage with regard to the translation of these studies into clinical practice and we foresee huge obstacles on this way due to the highly heterogeneous state of tumor cells and their drug-resistance mechanisms. pH-responsive nanomedicines offer a challenging opportunity to induce on-demand drug release and have several advantages over conventional nanomedicines. All the studies described in this chapter open a new window and more intense research on the derivatization of novel platforms, or on the development of new innovative nanosystems with improved (or diverse) properties is to be expected in the future, for enhancing the antitumor ability of traditional chemotherapy in clinic. Self-assembled and bioinspired nanostructures started to give promising results and can constitute a new direction for innovative drug research and development [87]. Moreover, strategic directions to improve nanomedicine performance, translation, and exploitation have been proposed by van der Meel and coll. That could help to enhance the clinical impact and patient performance of nanomedicinal antitumoral drugs [88].

References

[1] https://www.who.int/health-topics/cancer#tab=tab_1.
[2] http://gco.iarc.fr/today/fact-sheets-cancers.
[3] Szakács G, Paterson JK, Ludwig JA, Booth-Genthe C, Gottesman MM. Targeting multidrug resistance in cancer. Nat Rev Drug Discov 2006;5(3):219–34.
[4] Ramos P, Bentires-Alj M. Mechanism-based cancer therapy: resistance to therapy, therapy for resistance. Oncogene 2015;34(28):3617–26.
[5] Wu Q, Yang Z, Nie Y, Shi Y, Fan D. Multi-drug resistance in cancer chemotherapeutics: mechanisms and lab approaches. Cancer Lett 2014;347(2):159–66.
[6] Wu G, Wilson G, George J, Liddle C, Hebbard L, Qiao L. Overcoming treatment resistance in cancer: current understanding and tactics. Cancer Lett 2017;387:69–76.
[7] Vasan N, Baselga J, Hyman DM. A view on drug resistance in cancer. Nature 2019;575(7782):299–309.
[8] Pottier C, Wheatherspoon A, Roncarati P, Longuespée R, Herfs M, Duray A, Delvenne P, Quatresooz P. The importance of the tumor microenvironment in the therapeutic management of cancer. Expert Rev Anticancer Ther 2015;15(8):943–54.
[9] Shen M, Kang Y. Complex interplay between tumor microenvironment and cancer therapy. Front Med 2018; 12(4):426–39.
[10] Junttila MR, de Sauvage FJ. Influence of tumour micro-environment heterogeneity on therapeutic response. Nature 2013;501(7467):346–54.
[11] Mumenthaler SM, Foo J, Choi NC, Heise N, Leder K, Agus DB, et al. The impact of microenvironmental heterogeneity on the evolution of drug resistance in cancer cells. Cancer Informat 2015;14(Suppl 4):19–31.
[12] Wu T, Dai Y. Tumor microenvironment and therapeutic response. Cancer Lett 2017;387:61–8.
[13] Petrova V, Annicchiarico-Petruzzelli M, Melino G, Amelio I. The hypoxic tumour microenvironment. Oncogenesis 2018;7(1):10.
[14] Ji K, Mayernik L, Moin K, Sloane BF. Acidosis and proteolysis in the tumor microenvironment. Cancer Metastasis Rev 2019;38(1–2):103–12.
[15] Neri D, Supuran CT. Interfering with pH regulation in tumours as a therapeutic strategy. Nat Rev Drug Discov 2011;10(10):767–77.

[16] Corbet C, Feron O. Tumour acidosis: from the passenger to the driver's seat. Nat Rev Cancer 2017;17(10):577–93.
[17] Shi J, Kantoff PW, Wooster R, Farokhzad OC. Cancer nanomedicine: progress, challenges and opportunities. Nat Rev Cancer 2017;17(1):20–37.
[18] Ganoth A, Merimi KC, Peer D. Overcoming multidrug resistance with nanomedicines. Expert Opin Drug Deliv 2015;12(2):223–38.
[19] Dai Y, Xu C, Sun X, Chen X. Nanoparticle design strategies for enhanced anticancer therapy by exploiting the tumour microenvironment. Chem Soc Rev 2017;46(12):3830–52.
[20] Sagnella SM, McCarroll JA, Kavallaris M. Drug delivery: beyond active tumour targeting. Nanomedicine 2014;10 (6):1131–7.
[21] Pastorekova S, Gillies RJ. The role of carbonic anhydrase IX in cancer development: links to hypoxia, acidosis, and beyond. Cancer Metastasis Rev 2019;38(1–2):65–77.
[22] Majidinia M, Mirza-Aghazadeh-Attari M, Rahimi M, Mihanfar A, Karimian A, Safa A, Yousefi B. Overcoming multidrug resistance in cancer: recent progress in nanotechnology and new horizons. IUBMB Life 2020;72(5): 855–71.
[23] Ali I, Mukhtar SD, Ali HS, Scotti MT, Scotti L. Advances in nanoparticles as anticancer drug delivery vector: need of this century. Curr Pharm Des 2020;26(15):1637–49.
[24] Navya PN, Kaphle A, Srinivas SP, Bhargava SK, Rotello VM, Daima HK. Current trends and challenges in cancer management and therapy using designer nanomaterials. Nano Converg 2019;6(1):23.
[25] Zhou L, Wang H, Li Y. Stimuli-responsive nanomedicines for overcoming cancer multidrug resistance. Theranostics 2018;8(4):1059–74.
[26] Bar-Zeev M, Livney YD, Assaraf YG. Targeted nanomedicine for cancer therapeutics: towards precision medicine overcoming drug resistance. Drug Resist Updat 2017;31:15–30.
[27] Mi P. Stimuli-responsive nanocarriers for drug delivery, tumor imaging, therapy and theranostics. Theranostics 2020;10(10):4557–88.
[28] Kalaydina RV, Bajwa K, Qorri B, Decarlo A, Szewczuk MR. Recent advances in "smart" delivery systems for extended drug release in cancer therapy. Int J Nanomedicine 2018;13:4727–45.
[29] Liu J, Huang Y, Kumar A, Tan A, Jin S, Mozhi A, Liang XJ. pH-sensitive nano-systems for drug delivery in cancer therapy. Biotechnol Adv 2014;32(4):693–710.
[30] Karimi M, Eslami M, Sahandi-Zangabad P, Mirab F, Farajisafiloo N, Shafaei Z, Ghosh D, Bozorgomid M, Dashkhaneh F, Hamblin MR. pH-sensitive stimulus-responsive nanocarriers for targeted delivery of therapeutic agents. Wiley Interdiscip Rev Nanomed Nanobiotechnol 2016;8(5):696–716.
[31] Wang Z, Zhang X, Huang G, Gao J. pH-responsive drug delivery systems. In: Singh A, Amiji MM, editors. Stimuli-responsive drug delivery systems. London: Royal Society of Chemistry; 2018. p. 51–82 [chapter 3].
[32] El-Sawy HS, Al-Abd AM, Ahmed TA, El-Say KM, Torchilin VP. Stimuli-responsive nano-architecture drug-delivery systems to solid tumor micromilieu: past, present, and future perspectives. ACS Nano 2018; 12(11):10636–64.
[33] Kaur S, Prasad C, Balakrishnan B, Banerjee R. Trigger responsive polymeric nanocarriers for cancer therapy. Biomater Sci 2015;3(7):955–87.
[34] Lu Y, Aimetti AA, Langer R, Gu Z. Bioresponsive materials. Nat Rev Mater 2016;2:16075.
[35] El-Say KM, El-Sawy HS. Polymeric nanoparticles: promising platform for drug delivery. Int J Pharm 2017; 528(1–2):675–91.
[36] Lim EK, Chung BH, Chung SJ. Recent advances in pH-sensitive polymeric nanoparticles for smart drug delivery in cancer therapy. Curr Drug Targets 2018;19(4):300–17.
[37] Avramović N, Mandić B, Savić-Radojević A, Simić T. Polymeric nanocarriers of drug delivery systems in cancer therapy. Pharmaceutics 2020;12(4).
[38] Tang H, Zhao W, Yu J, Li Y, Zhao C. Recent development of pH-responsive polymers for cancer nanomedicine. Molecules 2018;24(1).
[39] Yao Y, Saw PE, Nie Y, Wong PP, Jiang L, Ye X, Chen J, Ding T, Xu L, Yao H, Hu H, Xu X. Multifunctional sharp pH-responsive nanoparticles for targeted drug delivery and effective breast cancer therapy. J Mater Chem B 2019;7:576–85.
[40] Hira SK, Mitra K, Srivastava P, Singh S, Vishwakarma S, Singh R, Ray B, Manna PP. Doxorubicin loaded pH responsive biodegradable ABA-type amphiphilic PEG-b-aliphatic Polyketal-b-PEG block copolymer for therapy against aggressive murine lymphoma. Nanomedicine 2019;24:102128.

[41] Cheng X, Li D, Sun M, He L, Zheng Y, Wang X, Tang R. Co-delivery of DOX and PDTC by pH-sensitive nanoparticles to overcome multidrug resistance in breast cancer. Colloids Surf B: Biointerfaces 2019;181:185–97.

[42] Wan X, Beaudoin JJ, Vinod N, Min Y, Makita N, Bludau H, Jordan R, Wang A, Sokolsky M, Kabanov AV. Co-delivery of paclitaxel and cisplatin in poly(2-oxazoline) polymeric micelles: implications for drug loading, release, pharmacokinetics and outcome of ovarian and breast cancer treatments. Biomaterials 2019;192:1–14.

[43] Zhang X, Zhao M, Cao N, Qin W, Zhao M, Wu J, Lin D. Construction of a tumor microenvironment pH-responsive cleavable PEGylated hyaluronic acid nano-drug delivery system for colorectal cancer treatment. Biomater Sci 2020;8(7):1885–96.

[44] Pang J, Xing H, Sun Y, Feng S, Wang S. Non-small cell lung cancer combination therapy: hyaluronic acid modified, epidermal growth factor receptor targeted, pH sensitive lipid-polymer hybrid nanoparticles for the delivery of erlotinib plus bevacizumab. Biomed Pharmacother 2020;125:109861.

[45] Palanikumar L, Al-Hosani S, Kalmouni M, Nguyen VP, Ali L, Pasricha R, Barrera FN, Magzoub M. pH-responsive high stability polymeric nanoparticles for targeted delivery of anticancer therapeutics. Commun Bio 2020;3:95.

[46] Guimarães RS, Rodrigues CF, Moreira AF, Correia IJ. Overview of stimuli-responsive mesoporous organosilica nanocarriers for drug delivery. Pharmacol Res 2020;155:104742.

[47] Gisbert-Garzarán M, Manzano M, Vallet-Regí M. Mesoporous silica nanoparticles for the treatment of complex bone diseases: bone cancer, bone infection and osteoporosis. Pharmaceutics 2020;12(1). https://doi.org/10.3390/pharmaceutics12010083.

[48] Aquib M, Farooq MA, Banerjee P, Akhtar F, Filli MS, Boakye-Yiadom KO, Kesse S, Raza F, Maviah MBJ, Mavlyanova R, Wang B. Targeted and stimuli-responsive mesoporous silica nanoparticles for drug delivery and theranostic use. J Biomed Mater Res A 2019;107(12):2643–66.

[49] Li Z, Zhang Y, Feng N. Mesoporous silica nanoparticles: synthesis, classification, drug loading, pharmacokinetics, biocompatibility, and application in drug delivery. Expert Opin Drug Deliv 2019;16(3):219–37.

[50] Li T, Shi S, Goel S, Shen X, Xie X, Chen Z, Zhang H, Li S, Qin X, Yang H, Wu C, Liu Y. Recent advancements in mesoporous silica nanoparticles towards therapeutic applications for cancer. Acta Biomater 2019;89:1–13.

[51] Thorat ND, Bauer J, Tofail SAM, Gascón Pérez V, Bohara RA, Yadav HM. Silica nano supra-assembly for the targeted delivery of therapeutic cargo to overcome chemoresistance in cancer. Colloids Surf B: Biointerfaces 2020;185:110571.

[52] Tian L, Lu L, Qiao Y, Ravi S, Salatan F, Melancon MP. Stimuli-responsive gold nanoparticles for cancer diagnosis and therapy. J Funct Biomater 2016;7(3). https://doi.org/10.3390/jfb7030019.

[53] Zhang Y, Yang L, Yang C, Liu J. Recent advances of smart acid-responsive gold nanoparticles in tumor therapy. Wiley Interdiscip Rev Nanomed Nanobiotechnol 2020;10:e1619. https://doi.org/10.1002/wnan.1619.

[54] Wang W, Wang J, Ding Y. Gold nanoparticle-conjugated nanomedicine: design, construction, and structure–efficacy relationship studies. J Mater Chem B 2020;8:4813–30.

[55] Kumar K, Moitra P, Bashir M, Kondaiah P, Bhattacharya S. Natural tripeptide capped pH-sensitive gold nanoparticles for efficacious doxorubicin delivery both in vitro and in vivo. Nanoscale 2020;12(2):1067–74.

[56] Jiang Y, Guo Z, Fang J, Wang B, Lin Z, Chen ZS, Chen Y, Zhang N, Yang X, Gao W. A multi-functionalized nanocomposite constructed by gold nanorod core with triple-layer coating to combat multidrug resistant colorectal cancer. Korean J Couns Psychother 2020;107:110224.

[57] Sedlakova O, Svastova E, Takacova M, Kopacek J, Pastorek J, Pastorekova S. Carbonic anhydrase IX, a hypoxia-induced catalytic component of the pH regulating machinery in tumors. Front Physiol 2014;4:400.

[58] Supuran CT, Alterio V, Di Fiore A, D' Ambrosio K, Carta F, Monti SM, De Simone G. Inhibition of carbonic anhydrase IX targets primary tumors, metastases, and cancer stem cells: three for the price of one. Med Res Rev 2018;38(6):1799–836.

[59] Nocentini A, Supuran CT. Carbonic anhydrase inhibitors as antitumor/antimetastatic agents: a patent review (2008-2018). Expert Opin Ther Pat 2018;28(10):729–40.

[60] Supuran CT. Carbonic anhydrase inhibitors as emerging agents for the treatment and imaging of hypoxic tumors. Expert Opin Investig Drugs 2018;27(12):963–70.

[61] Ilies M, Winum J-Y. Carbonic anhydrase inhibitors for the treatment of tumors: therapeutic, immunologic, and diagnostic tools targeting isoforms IX and XII. In: Supuran CT, Nocentini A, editors. Carbonic anhydrases. Biochemistry and pharmacology of an evergreen pharmaceutical target. Amsterdam: Academic Press; 2019. p. 331–65 [chapter 16].

[62] Carta F, Dumy P, Supuran CT, Winum JY. Multivalent carbonic anhydrases inhibitors. Int J Mol Sci 2019;20(21). https://doi.org/10.3390/ijms20215352.

[63] Kanfar N, Bartolami E, Zelli R, Marra A, Winum JY, Ulrich S, Dumy P. Emerging trends in enzyme inhibition by multivalent nanoconstructs. Org Biomol Chem 2015;13(39):9894–906.

[64] Stiti M, Cecchi A, Rami M, Abdaoui M, Barragan-Montero V, Scozzafava A, Guari Y, Winum JY, Supuran CT. Carbonic anhydrase inhibitor coated gold nanoparticles selectively inhibit the tumor-associated isoform IX over the cytosolic isozymes I and II. J Am Chem Soc 2008;130(48):16130–1.

[65] Bellissima F, Carta F, Innocenti A, Scozzafava A, Baglioni P, Supuran CT, Berti D. Structural modulation of the biological activity of gold nanoparticles functionalized with a carbonic anhydrase inhibitor. Eur Phys J E Soft Matter 2013;36(5):48.

[66] Shabana AM, Mondal UK, Alam MR, Spoon T, Ross CA, Madesh M, Supuran CT, Ilies MA. pH-sensitive multiligand gold nanoplatform targeting carbonic anhydrase IX enhances the delivery of doxorubicin to hypoxic tumor spheroids and overcomes the hypoxia-induced chemoresistance. ACS Appl Mater Interfaces 2018;10(21):17792–808.

[67] Ratto F, Witort E, Tatini F, Centi S, Lazzeri L, Carta F, Lulli M, Vullo D, Fusi F, Supuran CT, Scozzafava A, Capaccioli S, Pini R. Plasmonic particles that hit hypoxic cells. Adv Funct Mater 2015;25:316–23.

[68] Touisni N, Kanfar N, Ulrich S, Dumy P, Supuran CT, Mehdi A, Winum JY. Fluorescent silica nanoparticles with multivalent inhibitory effects towards carbonic anhydrases. Chem Eur J 2015;21(29):10306–9.

[69] Kanfar N, Mehdi A, Dumy P, Ulrich S, Winum JY. Polyhedral Oligomeric Silsesquioxane (POSS) bearing glyoxylic aldehyde as clickable platform towards multivalent conjugates. Chem Eur J 2017;23(71):17867–9.

[70] Janoniene A, Liu Z, Baranauskiene L, Mäkilä E, Ma M, Salonen J, Hirvonen J, Zhang H, Petrikaite V, Santos HA. A versatile carbonic anhydrase IX targeting ligand-functionalized porous silicon nanoplatform for dual hypoxia cancer therapy and imaging. ACS Appl Mater Interfaces 2017;9(16):13976–87.

[71] Tokárová V, Pittermannová A, Král V, Řezáčová P, Štěpánek F. Feasibility and constraints of particle targeting using the antigen-antibody interaction. Nanoscale 2013;5(23):11490–8.

[72] Lizoňová D, Majerská M, Král V, Pechar M, Pola R, Kovářd M, Štěpánek F. Antibody-pHPMA functionalised fluorescent silica nanoparticles for colorectal carcinoma targeting. RSC Adv 2018;8:21679–89.

[73] Chen W, Han Y, Qian Y, Tang J, Hu H, Shen Y. Preparation and imaging study of tumor-targeting MRI contrast agent based on Fe_3O_4 nanoparticles. Chem Lett 2015;44:1771–3.

[74] Khmara I, Koneracka M, Kubovcikova M, Zavisova V, Antal I, Csach K, Kopcansky P, Vidlickova I, Csaderova L, Pastorekova S, Zatovicova M. Preparation of poly-L-lysine functionalized magnetic nanoparticles and their influence on viability of cancer cells. J Magn Magn Mater 2017;427:114–21.

[75] Antal I, Koneracka M, Kubovcikova M, Zavisova V, Khmara I, Lucanska D, Jelenska L, Vidlickova I, Zatovicova M, Pastorekova S, Bugarova N, Micusik M, Omastova M, Kopcansky P. D,L-lysine functionalized Fe_3O_4 nanoparticles for detection of cancer cells. Colloids Surf B: Biointerfaces 2018;163:236–45.

[76] Bugárová N, Špitálsky Z, Mičušík M, Bodík M, Šiffalovič P, Koneracká M, Závišová V, Kubovčíková M, Kajanová I, Zaťovičová M, Pastoreková S, Šlouf M, Majková E, Omastová M. A multifunctional graphene oxide platform for targeting cancer. Cancers (Basel) 2019;11(6). https://doi.org/10.3390/cancers11060753.

[77] Alsaab HO, Sau S, Alzhrani RM, Cheriyan VT, Polin LA, Vaishampayan U, Rishi AK, Iyer AK. Tumor hypoxia directed multimodal nanotherapy for overcoming drug resistance in renal cell carcinoma and reprogramming macrophages. Biomaterials 2018;183:280–94.

[78] Jiang Y, Li J, Zeng Z, Xie C, Lyu Y, Pu K. Organic photodynamic nanoinhibitor for synergistic cancer therapy. Angew Chem Int Ed Eng 2019;58(24):8161–5.

[79] Liu S, Luo X, Liu S, Xu P, Wang J, Hu Y. Acetazolamide-loaded pH-responsive nanoparticles alleviating tumor acidosis to enhance chemotherapy effects. Macromol Biosci 2019;19(2):e1800366.

[80] Abellán-Flos M, Tanç M, Supuran CT, Vincent SP. Exploring carbonic anhydrase inhibition with multimeric coumarins displayed on a fullerene scaffold. Org Biomol Chem 2015;13(27):7445–51.

[81] Abellán-Flos M, Tanç M, Supuran CT, Vincent SP. Multimeric xanthates as carbonic anhydrase inhibitors. J Enzyme Inhib Med Chem 2016;31(6):946–52.

[82] Kanfar N, Tanc M, Dumy P, Supuran CT, Ulrich S, Winum JY. Effective access to multivalent inhibitors of carbonic anhydrases promoted by peptide bioconjugation. Chem Eur J 2017;23(28):6788–94.

[83] Carta F, Osman SM, Vullo D, AlOthman Z, Supuran CT. Dendrimers incorporating benzenesulfonamide moieties strongly inhibit carbonic anhydrase isoforms I-XIV. Org Biomol Chem 2015;13(23):6453–7.

[84] Carta F, Osman SM, Vullo D, Gullotto A, Winum JY, AlOthman Z, Masini E, Supuran CT. Poly(amidoamine) Dendrimers with carbonic anhydrase inhibitory activity and Antiglaucoma action. J Med Chem 2015; 58(9):4039–45.

[85] Tatiparti K, Sau S, Gawde KA, Iyer AK. Copper-free 'click' chemistry-based synthesis and characterization of carbonic anhydrase-IX anchored albumin-paclitaxel nanoparticles for targeting tumor hypoxia. Int J Mol Sci 2018;19(3). https://doi.org/10.3390/ijms19030838.

[86] Lin C, Wong BCK, Chen H, Bian Z, Zhang G, Zhang X, Kashif Riaz M, Tyagi D, Lin G, Zhang Y, Wang J, Lu A, Yang Z. Pulmonary delivery of triptolide-loaded liposomes decorated with anti-carbonic anhydrase IX antibody for lung cancer therapy. Sci Rep 2017;7(1)1097.

[87] Li J, Shi K, Farhadi Sabet Z, Fu W, Zhou H, Xu S, Liu T, You M, Cao M, Xu M, Cui X, Hu B, Liu Y, Chen C. New power of self-assembling carbonic anhydrase inhibitor: short peptide–constructed nanofibers inspire hypoxic cancer therapy. Sci Adv 2019;5:eaax0937.

[88] van der Meel R, Sulheim E, Shi Y, Kiessling F, Mulder WJM, Lammers T. Smart cancer nanomedicine. Nat Nanotechnol 2019;14:1007–17.

Index

Note: Page numbers followed by *f* indicate figures, *t* indicate tables and *s* indicate schemes.

A

ABC. *See* ATP-binding cassette (ABC)
ABCC1 (MRP1) inhibitors, 108, 112
ABCG2 (BCRP/MXR) inhibitors, 108, 112
Acidosis, 10, 14–15
Acquired chemoresistance, 107–108, 186
Albumin base nanoplatforms, 196
Angiogenesis, 4, 14–15
Angiogenic switch, 4
Angiogenic vascular cells, 8–9
Anion exchanger (AE), 56–57, 56*f*
Anoikis resistance, 8–10
Apoptotic pathways
  B-cell lymphoma-2 (Bcl-2), 139–140
  BNC105P, 142
  cancer aggressiveness, 4
  dinaciclib, 142
  Inhibitor of Apoptosis (IAP) family proteins, 139–140
  MCL-1, 140–141
  p53 tumor suppressor gene, 139–140
  pyrazolopyridine, 140–141, 141*f*
  venetoclax, 141–142, 141*f*
ATP-binding cassette (ABC)
  ABCC1 (MRP1) inhibitors, 108, 112
  ABCG2 (BCRP/MXR) inhibitors, 108, 112
  nucleotide binding domains (NBDs), 106
  P-gp/MDR1 inhibitors, 107–112, 109*t*, 110*f*
  role of, 106–107
  topological model, 106, 107*f*
  transmembrane domains (TMDs), 106

B

Benzenesulfonamides
  acetazolamide, 87–88, 89*f*
  doxorubicin, 84
  imino linker, 82–84, 84*f*
  molecular modeling, 85–86
  nanomolar inhibitors, 82, 83*f*
  oxadiazole linker, 84, 85*f*
  phthalimide-based inhibitors, 87, 88*f*
  Psammaplin C, 84, 86*f*
  ring contraction, 86–87, 88*f*
  ring expansion, 86, 87*f*
  selective inhibitors, 85–86
  triazole linker, 82

C

Cancer aggressiveness
  apoptosis, 4
  chronic inflammation, 5
  growth suppressors, 4
  immune-surveillance, 5
  limitless replication, 4
  metabolic deregulation, 5
  self-sufficiency, growth signals, 4
  sustained angiogenesis, 4
  tissue invasion and metastasis, 4
Cancer-associated fibroblasts (CAFs), 7–8
Capecitabine, 54
Carbonic anhydrase (CA)
  α-CAs active site, 172–173
  acetazolamide (AZA), 36–37
  CA IX and XII, 173–175, 174*f*
  carbonic anhydrase inhibitors (CAIs), 36
  coumarins, 174–175
  gene families, 172
  indisulam, 36
  Glide docking software, 176
  GOLD docking software, 174–175
  hydrophilic and hydrophobic residues, 172–173, 172*f*
  MOE, 173–174
  pharmacophore model, 173–174
  SLC-0111, 37
  structure-based drug design optimization process (SBDD), 175–176
  temozolomide (TMZ), 37
  virtual screening, 173
Carbonic anhydrase II (CAII), 20
Carbonic anhydrase IX (CAIX)
  HIF1α, 22–23
  hypoxia, 17
  interaction of, 20
  proteoglycan domain, 20–21
  role of, 21–22
  SLC-0111, 22–23
  sulfonamide-based inhibitors, 22

Carbonic anhydrase-related proteins (CARPs), 16
Carbonic anhydrase XII (CAXII), 17
Chemoresistance
chemotherapy
acquired, 149, 186
autophagy, 150
epithelial-to-mesenchymal transformation (EMT), 150
intrinsic, 149
multidrug resistance (MDR), 149–150
tumor acidity and, 48–50
Chemosensitization, 130–131
Chemosensitizers
advantage, 130–131
cancer cell resistance, 130, 131*f*
carbonic anhydrase (CA)
α-CAs active site, 172–173
CA IX and XII, 173–175, 174*f*
coumarins, 174–175
gene families, 172
Glide docking software, 176
GOLD docking software, 174–175
hydrophilic and hydrophobic residues, 172–173, 172*f*
MOE, 173–174
pharmacophore model, 173–174
structure-based drug design optimization process (SBDD), 175–176
virtual screening, 173
cellular targets, 170–171, 171*f*
chemoresistance
acquired, 149
autophagy, 150
epithelial-to-mesenchymal transformation (EMT), 150
intrinsic, 149
multidrug resistance (MDR), 149–150
classification, 130–131
curcumin
acquired resistance, 153
biological properties, 152
clinical trials, 154–155, 155*t*
colorectal cancer, 154
mitomycin C, 154
oxaliplatin, 154–155
definition, 170–171
DNA repair mechanisms
breast cancer type 1 susceptibility (BRCA1) protein, 137–138
dianhydrogalactitol, 139
glioma-associated oncogene (GLI1), 137–138
human polynucleotide kinase phosphatase (hPNKP), 138–139
hypoxia, 137–138
Phosphatase and Tensin Homolog (PTEN), 139
PI3Ks, 139
regulatory proteins, 137–138, 137*f*
thymidylate kinase (TMPK), 138
docking simulation, 170–171
drug efflux transporters
drug delivery systems, 135–136, 135*f*, 137*f*
inhibitors, 133–135, 133*f*
role of, 131–132, 132*t*
epigallocatechin-3-gallate (EGCG), 157
glucose transporter 1 (GLUT1), 178–179, 178–179*f*
hypoxia-inducible factor (HIF), 176–177, 177*f*
modulators, apoptotic pathways
B-cell lymphoma-2 (Bcl-2), 139–140
BNC105P, 142
dinaciclib, 142
Inhibitor of Apoptosis (IAP) family proteins, 139–140
MCL-1, 140–141
p53 tumor suppressor gene, 139–140
pyrazolopyridine, 140–141, 141*f*
venetoclax, 141–142, 141*f*
molecular modeling, 170–171
multidrug resistance (MDR)
acidic pH, 170–171
cellular mechanism, 170
natural products, 151–152, 153*t*
nutraceuticals, 152
quercetin, 157–158
resveratrol
acquired resistance, 156
biological properties, 155
clinical trials, 156
epithelial-to-mesenchymal transformation (EMT), 155–156
melphalan, 156
side effects, 150–151
Chemotherapy
acquired chemoresistance, 186
cancer cell resistance, 130, 131*f*
chemoresistance
acquired, 149
autophagy, 150
epithelial-to-mesenchymal transformation (EMT), 150
intrinsic, 149
multidrug resistance (MDR), 149–150
classification, chemotherapeutic agents, 149
efflux transporters, 130
intrinsic resistance, 186
multidrug resistance (MDR), 187
qualitative modification, 186–187

quantitative modification, 186–187
resistance mechanism, 186–187
tumor microenvironment (TME), 187
Warburg effect, 187
Cisplatin, 124–126, 125*f*
Coumarins
human carbonic anhydrase (hCA) IX, 71–72, 71*f*
human carbonic anhydrase XII (hCA XII), 88–90, 89*f*
Curcumin
acquired resistance, 153
biological properties, 152
clinical trials, 154–155, 155*t*
colorectal cancer, 154
mitomycin C, 154
oxaliplatin, 154–155

D

DNA repair mechanisms
breast cancer type 1 susceptibility (BRCA1) protein, 137–138
dianhydrogalactitol, 139
glioma-associated oncogene (GLI1), 137–138
human polynucleotide kinase phosphatase (hPNKP), 138–139
hypoxia, 137–138
Phosphatase and Tensin Homolog (PTEN), 139
PI3Ks, 139
regulatory proteins, 137–138, 137*f*
thymidylate kinase (TMPK), 138
Docking simulation, 170–171
Drug efflux transporters
drug delivery systems, 135–136, 135*f*, 137*f*
inhibitors, 133–135, 133*f*
role of, 131–132, 132*t*
Drug transporters
ATP-binding cassette (ABC)
ABCC1 (MRP1) inhibitors, 108, 112
ABCG2 (BCRP/MXR) inhibitors, 108, 112
acquired resistance, 107–108
nucleotide binding domains (NBDs), 106
P-gp/MDR1 inhibitors, 107–112, 109*t*, 110*f*
role of, 106–107
topological model, 106, 107*f*
transmembrane domains (TMDs), 106
genetic disorders, 96
influx/efflux, 96
role in, 96
solute carrier class (SLC)
monocarboxylate transporters (MCTs), 97–102, 98*t*, 100*f*
Na^+/H^+-exchangers (NHEs), 102–106, 102*f*, 104*f*

E

Epacadostat (EPA), 70–71, 70*f*
Epigallocatechin-3-gallate (EGCG), 157
Epithelial-to-mesenchymal transformation (EMT)
chemoresistance, 150
resveratrol, 155–156
Extracellular matrix (ECM), 9–10

F

Fatty acid synthase (FASN), 55

G

Glucose transporter 1 (GLUT1), 178–179, 178–179*f*
Glycolysis
extracellular pH (pH_e), 123–124
intracellular pH (pH_i), 123
normoxia *vs.* hypoxia, 122, 122*s*
Warburg effect, 122
Gold nanoplatforms, 191–192, 192*f*

H

Hallmarks of cancer, 148
Histone deacetylases (HDACs)
buffering effect, MCTs, 124, 126
cisplatin, 124–126, 125*f*
GYY4137, 125*f*, 126
hydrogen sulfide (H_2S), 126
lysosomal membrane permeabilization (LMP), 124–126
MCT1/MCT4 inhibitor, 124, 125*f*
sodium hydrogen sulfide (NaHS), 126
structure of, 123–124, 124*f*
trichostatin A (TSA), 124–126
ZYJ1122, 125*f*, 126
Homosulfocoumarins
human carbonic anhydrase (hCA) IX, 72, 72*f*
albumin/liposome base nanoplatforms, 196
gold nanoplatforms, 191–192, 192*f*
iron oxide nanoparticles, 194
organic nanostructures, 195–196, 196*f*
polymeric/oligomeric nanoplatforms, 194–195
silica-based nanoplatforms, 192–194, 193*f*
Human carbonic anhydrase (hCA) IX
^{18}F-VM4-037, 71
coumarins, 71–72, 71*f*
deprotection, C-6 primary hydroxyl groups, 69
epacadostat (EPA), 70–71, 70*f*
vs. hCA XII
acetazolamide (AAZ), 81, 82*t*
active site, 81
carbonic anhydrase (CA) inhibitors (CAIs), 81
characteristics, 80
HIF-1, 64–65, 66*f*

Human carbonic anhydrase (hCA) IX *(Continued)*
homosulfocoumarins, 72, 72*f*
monoclonal antibodies (mAbs)
^{124}I-cG250, 66–67
A3 and CC7, 68
direct iodination, 65–66, 66*f*
girentuximab (cG250), 65–66
host-linker moieties, 66–67, 67*f*
1,4,7,10-tetraazacyclododecane-1,4,7,10-tetraacetic acid (DOTA), 68
ZCAIX:1 and ZCAIX:2, 68–69
saccharin and saccharin derivatives, 72–73, 73*f*
structure, 64, 65*f*
sulfocoumarins, 72, 72*f*
sulfonamides, 69
transcription, 64–65
ureido inhibitors, 70, 70*f*
Human carbonic anhydrase XII (hCA XII)
benzenesulfonamides
acetazolamide, 87–88, 89*f*
doxorubicin, 84
imino linker, 82–84, 84*f*
molecular modeling, 85–86
nanomolar inhibitors, 82, 83*f*
oxadiazole linker, 84, 85*f*
phthalimide-based inhibitors, 87, 88*f*
Psammaplin C, 84, 86*f*
ring contraction, 86–87, 88*f*
ring expansion, 86, 87*f*
selective inhibitors, 85–86
triazole linker, 82
characteristics, 80
coumarins and their derivatives, 88–90, 89*f*
discovery, 80
drug repurposing, 90–91
expression in
breast cancer, 80–81
hepatocellular carcinoma (HCC), 81
vs. hCA IX
acetazolamide (AAZ), 81, 82*t*
active site, 81
carbonic anhydrase (CA) inhibitors (CAIs), 81
characteristics, 80
monoclonal antibodies (mAbs), 91
P-glycoprotein (Pgp), 81
polypharmacology approach
disulfiram, 90*f*, 91
epacadostat, 90–91, 90*f*
transport metabolon, 80–81
Hypoxia
angiogenic vascular cells, 8–10
carbonic anhydrase IX (CAIX), 17
global transcriptional response, 14–15
Hypoxia-inducible factor (HIF)
chemosensitizers, 176–177, 177*f*
human carbonic anhydrase (hCA) IX, 64–65, 66*f*

I

Immune infiltrating cells, 9
Intrinsic resistance, 186
Ionophores, 38–39
Ion trapping, 22, 49–50
Iron oxide nanoparticles, 194

L

Liposome base nanoplatforms, 196
Lysosomal membrane permeabilization (LMP), 124–126

M

mAbs. *See* Monoclonal antibodies (mAbs)
MCT. *See* Monocarboxylate transporters (MCTs)
Melphalan, 156
Mitochondria, 23–24
Mitomycin C, 154
Monocarboxylate 1(MCT1), 15–16
Monocarboxylate 4(MCT4), 15–16
Monocarboxylate transporters (MCTs)
acriflavine (ACF), 101
α-cyano-4-hydroxycinnamate (CHC), 99–100
AR-C155858, 100–101
AZD3965, 101
BAY-8002, 101
biodistribution, 98–99, 98*t*
inhibitors, 99–100, 100*f*
isoforms, 98–99, 98*t*
lactate, 97
MCT1, 38, 98–99
MCT2, 99
MCT3, 99
metabolon systems, 38
physiological functions, 97
proton pump inhibitors (PPI), 55–56, 56*f*
substrate, 98–99, 98*t*
transmembrane domains, 97–98
Monoclonal antibodies (mAbs)
1,4,7,10-tetraazacyclododecane-1,4,7,10-tetraacetic acid (DOTA), 68
^{124}I-cG250, 66–67
A3 and CC7, 68
direct iodination, 65–66, 66*f*
girentuximab (cG250), 65–66
host-linker moieties, 66–67, 67*f*
human carbonic anhydrase XII (hCA XII), 91
ZCAIX:1 and ZCAIX:2, 68–69
Multidrug resistance (MDR)
acidic pH, 170–171

cellular mechanism, 170
chemotherapy, 149–150, 187

N

Na^+/HCO_3^- cotransporter (NBC), 56–57, 56*f*
Na^+/H^+ exchangers (NHE)
amiloride, 39, 103
benzoylguanidines, 103, 106
cariporide, 39
dimethylamiloride (DMA), 105–106
gastric cancers, 104–105
hexamethylene amiloride (HMA), 105
N-ethylisopropylamiloride (EIPA), 105
NHE1, 102*f*, 103
proton pump inhibitors (PPI), 56–57, 56*f*
pyrazine derivatives, 103
structure of, 102
Na^+/H^+ exchanger 1(NHE1), 18–19, 102*f*, 103
Nanodrugs, 188
Nanostructures and innovative delivery systems
advantages, 188
human carbonic anhydrase
albumin/liposome base nanoplatforms, 196
gold nanoplatforms, 191–192, 192*f*
iron oxide nanoparticles, 194
organic nanostructures, 195–196, 196*f*
polymeric/oligomeric nanoplatforms, 194–195
silica-based nanoplatforms, 192–194, 193*f*
nanodrugs, 188
pH-responsive polymeric nanostructures
hyaluronic acid, 189
paclitaxel and cisplatin, 189
polylactic-co-glycolic acid (PLGA), 189
poly(2-(diisopropylamino)ethylmethacrylate) (PDPA), 188
pH-sensitive silica/gold nanoplatforms, 189–190
Nigericin, 38–39
Nutraceuticals, 152

O

Obatoclax, 39
Omeprazole, 51–53, 52*f*
Oxaliplatin, 154–155

P

Patent survey, chemosensitizers. *See* Chemosensitizers
P-glycoprotein (Pgp)
ATP-binding cassette (ABC), 108–112, 109*t*, 110*f*
human carbonic anhydrase XII (hCA XII), 81
pH and mitochondria, 23–24
pH metabolons, 20–21
pH regulators, inhibition effect
chemotherapy, 22–23
invasion, 21–22
metastasis, 21–22
migration, 21–22
pH-responsive polymeric nanostructures
hyaluronic acid, 189
paclitaxel and cisplatin, 189
polylactic-co-glycolic acid (PLGA), 189
poly(2-(diisopropylamino)ethylmethacrylate) (PDPA), 188
pH-sensitive silica/gold nanoplatforms, 189–190
Polymeric/oligomeric nanoplatforms, 194–195
Polypharmacology approach
disulfiram, 90*f*, 91
epacadostat, 90–91, 90*f*
Proton pump inhibitors (PPI)
anion exchanger (AE), 56–57, 56*f*
clinical outcomes, 54–55
hypoxic tumor cells reprogram, 48, 49*f*
monocarboxylate transporter (MCT) inhibitors, 55–56, 56*f*
Na^+/HCO_3^- cotransporter (NBC), 56–57, 56*f*
Na^+/H^+ exchanger (NHE), 56–57, 56*f*
preclinical data, 52–54
tumor acidity and chemoresistance, 48–50
vacuolar-type H^+ ATPases (V-ATPases)
cellular functions, 50–51
clinical use, 51–52, 52*f*
eukaryotic structure, 50, 51*f*
omeprazole, activation mechanism, 51–52, 52*f*
Proton pump inhibitors (PPIs)
histamine receptor-2 antagonists (H2RA), 38
pretreatment, 37
role for, 37

Q

Quercetin, 157–158

R

Rabeprazole, 54–55
Resveratrol
acquired resistance, 156
biological properties, 155
clinical trials, 156
epithelial-to-mesenchymal transformation (EMT), 155–156
melphalan, 156
Reverse Warburg effect, 16

S

Saccharin and saccharin derivatives, 72–73, 73*f*
Salinomycin, 39
Silica-based nanoplatforms, 192–194, 193*f*

Sodium bicarbonate cotransporters, 19–20
Sodium-dependent anion exchangers, 19–20
Sodium-independent anion exchangers, 19–20
Sodium-proton exchangers (NHEs). *See* Na^+/H^+-exchangers (NHEs)
Solute carrier class (SLC)
  monocarboxylate transporters (MCTs)
    acriflavine (ACF), 101
    AR-C155858, 100–101
    AZD3965, 101
    BAY-8002, 101
    biodistribution, 98–99, 98*t*
    α-cyano-4-hydroxycinnamate (CHC), 99–100
    inhibitors, 99–100, 100*f*
    isoforms, 98–99, 98*t*
    lactate, 97
    MCT1, 98–99
    MCT2, 99
    MCT3, 99
    physiological functions, 97
    substrate, 98–99, 98*t*
    transmembrane domains, 97–98
  Na^+/H^+-exchangers (NHEs)
    amiloride, 103
    benzoylguanidines, 103, 106
    dimethylamiloride (DMA), 105–106
    gastric cancers, 104–105
    hexamethylene amiloride (HMA), 105
    N-ethylisopropylamiloride (EIPA), 105
    NHE1, 102*f*, 103
    pyrazine derivatives, 103
    structure of, 102
Structure-based drug design optimization process (SBDD), 175–176
Sulfocoumarins, 72, 72*f*

T
Transport metabolon, 80–81
Tumor angiogenesis. *See* Angiogenesis
Tumor microenvironment (TME)
  activated/recruited stromal cells, 5, 6*f*
  angiogenic vascular cells, 8–9
  cancer-associated fibroblasts (CAFs), 7–8
  chemotherapy, 187
  extracellular matrix (ECM), 9–10
  hypoxia and acidity, 10
  immune infiltrating cells, 9
  neoplasia, 5–7
  reactive stroma, 7
  tumor progression, 5, 6*f*

V
Vacuolar-type H^+ ATPases (V-ATPases)
  cellular functions, 50–51
  clinical use, 51–52, 52*f*
  eukaryotic structure, 50, 51*f*
  omeprazole, activation mechanism, 51–52, 52*f*
Vascular endothelial growth factor (VEGF), 8–9

W
Warburg effect
  chemotherapy, 187
  glycolysis, 122
  metabolic deregulation, 5
  reverse effect, 16

FSC
www.fsc.org
MIX
Papier | Fördert gute Waldnutzung
FSC® C083411

Kolibri 360 GmbH
Ferdinand-Jühlke-Straße 7
99095 Erfurt, Deutschland
produktsicherheit@kolibri360.de